MW01702837

The Star of Deep Beginnings

The Star of Deep Beginnings

THE GENESIS OF AFRICAN SCIENCE AND TECHNOLOGY

Charles S. Finch III, M.D.

Copyright © 1998 by Khenti, Inc.

All rights reserved under International and Pan-American Copyright Conventions. No part of this book may be reproduced or transmitted in any form or by any means, electronic or mechanical, including photocopying, recording, or any information storage and retrieval system, without prior permission in writing from the publisher. All inquiries should be addressed to Khenti, Inc., P.O. Box 361003, Decatur, Georgia 30036-1003.

ISBN 0-9629444-3-2

Library of Congress Catalog Card Number: 97-92687

Printed in the United States of America

I wish to dedicate this book to my two mothers: Louise Antoinette Hughes Finch, who gave me both life and whatever common sense I possess; and Maam Adji Fatou Seck, who came looking for me, found me, and led me to the Spirit of Africa

CONTENTS

FOREWORD

A New Branch of African Historiography

The appearance of *The Star of Deep Beginnings: The Genesis of African Science and Technology* marks a watershed moment in Diasporan African research into the nature and meaning of "African science." As an old friend, Dr. Charles Finch, a distinguished doctor of medicine and historian, has extended me the privilege of writing the foreword to this important book, an honor which I greatly appreciate. Dr. Finch, whose reputation for promoting the study of the history of medicine in Africa since pharaonic times has been firmly established through his scholarship, has made yet another pathbreaking contribution to the exposition of the African contribution to human knowledge. Scientists, historians, and educators have awaited focus and direction with regard to the history of science in Africa. Any such forthcoming scientific historiography has to be systematic, complete, and critical.

By its diverse methods and subject areas, Dr. Finch's book gives the historian and the layman alike a livelier sense of the range of forms that the history of science has taken in Africa and of the values which might be expected to generate from African scientists. Indeed, this scientific exegesis, this uncovering and predicting of both the form and nature of scientific inquiry in Africa, not only explains the methods of science but also and chiefly explains how scientific facts and discoveries become a real part of scientific knowledge, thereby enriching the human experience which has come to be described as "civilization."

Before proceeding further into the importance of Dr. Finch's effort, however, we must pause to consider the fundamental question: what is the history of science? Simply put, the history of science is the systematic study of the development of science from its very birth. Science can be described as the process through which men and women apply various material or immaterial (i.e., intellectual) instruments to solve the challenges of human inquiry and existence in the order of things. Each successive instrument, each method of application, reveals a crystallization of human genius. Genius in science is essentially the ability to select properly one problem at a time, in succession from the simple to the complex. The scientific procedure of selection, analysis, and solution is, since prehistoric times, always analytic, synthetic, and, equally important, intuitive. Science is the timeless convergence and fruit of humanity's essential distinguishing characteristics: love, faith, reason, and imagination.

There are at least two main reasons for studying the history of science: a historical impulse to analyze the development of civilization and to thereby understand human progress, and a philosophical impulse to understand the deeper meaning of that progress, to apprehend the deeper meaning of science. It is only by considering the history of science as an entire tapestry of human inquiry that one can appreciate the distinct nature and contribution of various levels of scientific inquiry from particular periods, peoples, and/or counries. The historian of general scientific facts and ideas can probably appreciate better than anyone else the spatial and temporal progression and continuity of science. This person is thereby better equipped to estimate the progress of humanity. Dr. Finch's book is a good point of departure for the study of the history of science in Africa. And what can be more fundamental to any serious project than a good beginning?

The historian of science cannot produce quality research without being grounded first in the general history of civilization: it is absolutely necessary and indispensible. To understand, for example, the origins and development of the sciences of geography and cartography, one has to factor in the historical circumstances into and out of which each area was born and emerged. Questions of the influence of ritual practices; the links between festivals and the understanding of the cosmos; the infusion of information from explorers' missions by land and by sea—all of these factors contribute to the understanding of how humans have mapped their physical environments, both terrestial and celestial. The evolution of many medical ideas often requires a knowledge of the history of diseases and epidemics as well as the study of social circumstances which may have caused each disease to emerge, proliferate, and/or wane.

In the same vein of historical contextualization, it follows logically that the history of science is interwoven with the history of technology, i.e., that body of knowledge available to human societies for the practical application of scientific knowledge. The investigation and exploration of the inner regions of the earth, i.e., the working of mines, has always exerted a deep influence on the evolution of sciences such as geology and chemistry along with the concomitant industrial activities which they spawned.

In Africa, scientific or technical ideas and religious ideas have never been separated. The vocation of scientist or engineer was never held distinct from that of priest; the observance of religious ritual, the medicinal implications of horticulture, the construction of calendars following the observation of celestial phenomena, and, above all, the hope for and belief in an integrated and happy afterlife are representative of the innumerable links between science and religion in Africa. The history of astronomy in ancient Egypt (Kemet), for example, is incomprehensible in its totality in the absence of the study of the intimately connected activities of priests (spiritual life), scribes (intellectual life), and the religious activities of the general populace.

Similarly, the effective study of the rhythms that swayed the "plastic" arts, architecture, and music aids in the understanding of scientific ideas that are often communicated in "works of art." This is why, I might be so bold as to posit, the history of science is also the history of "taste." Beauty and harmony give aesthetic value to scientific theory, beauty being understood as the apprehension of the splendor of truth. S. Chandrasekhar, a Nobel Laureate in physics, speaks for example of the "unparalleled beauty of the general theory of relativity."[1]

Dr. Finch presents the integrated evolution of the history of science in Africa with precision and accuracy, beginning with the most critical periods of humanity's continental African history. The earliest stone implements were invented in Africa; animals and plants were domesticated in Africa without external guidance or influence. The people of Egypt (Kemet) had attained a high level of intra-African, self-sufficient cultural development, including the use of writing and the accumulation of a vast amount of mathematical, astronomical, architectural, and medical knowledge. As early as the fourth millenium before Christ, the Egyptians (Kemetew) were already acquainted with a decimal system of numbers. An inscription dating from that period found at Hieranconpolis in Upper Egypt refers to 120,000 captives, 140,000 oxen, and 1,422,000 goats, each decimal unit being represented by a special symbol.

The Egyptian calendar of 365 days was established at least by 4241 B.C., if not sooner. The astronomical knowledge of the Dogon people of Mali, West Africa, is equally remarkable in age, precision, and complexity. The Dogon accumulated planetary observations which they utilized for ritual purposes. These storehouses of early knowledge were not only abundant, they were highly systematized as well. The Ishango Bone, an example of the elaborate observation of lunar phases, invites us to study as its African inventors did millenia ago. In the case of Egypt, mathematic papyri were written more than thirteen centuries before the time of the Greek mathematician Euclid. Complicated formulas were invented which solved equally complex problems; the arithmetical ingenuity of the Egyptians was astounding. The ancient literature on the history of mathematics often depicted the Egyptians as the founders of mathematics and geometry.

Pyramids stand as prodigious witnesses to Egyptian scientific genius. The construction of these massive structures required a requisite amount of mathematical and engineering (technological) skill and a wealth of experience and discipline. It also required a faith in Maat, whose universal nature was corroborated by a tremendous aggregate of astronomical observations, the celestial rhythm of which influenced the location and position of each pyramid.

Africa has produced many golden ages of science throughout time and space. In South Africa, Congo, Gabon, Kenya, Ethiopia, Egypt (Kemet), Chad, Mali, Ghana, Burkina Faso, Ivory Coast, Guinea, and a thousand other locales, the African scientific genius has flourished throughout the millenia and in

---

[1] Chandrasekhar, S., *Truth and Beauty: Aesthetics and Motivation in Science.* Chicago: The University of Chicago Press (1987, 1990), p. 73.

spite of internally and externally-imposed hardships. From the iron-smelting furnaces of Nubia and Tanzania to the Bamun script of Cameroon; from the Ife and Nupe cultures of Nigeria to the architecture of old Zimbabwe; from the mathematics of Timbuktu to the cartography of the Swahili coast; African has given the world some of the best scientific thinking in human memory.

The spirit of Africa requires at its most essential level that all intellectual activity must be undertaken with wisdom and morality that it might serve the political goals of solidarity and group prosperity. The concept of scientific idolatry is unknown in Africa. Science is not an autonomous entity unto itself: it is but a method, not an aim. Africans have always dealt with the synthesis of knowledge in the construction of a beautiful and functional present. Nowhere is this more the case than in ancient Egypt, the classical African expression of this sentiment.

Indeed, the humanization of science is not a process which renders science less important; to the contrary, it makes science more significant, more complete, more ethically relevant. The greatest achievement of humanity must certainly be the quest for greater truth and beauty. This is why *Maat*, "truth," is represented as a beautiful woman in ancient Egypt (Kemet). Science deals, in its broadest sense, with the integrated intellectual experiences of humanity with the universe. In his text, Dr. Finch brilliantly argues that the history of African science is a thread in the larger tapestry of "universal history." He uses the heuristic capacities of the comparative method of historical inquiry to their fullest advantage to demonstrate this assertion.

For example, Greek astronomy was largely inspired by the Egyptian example and instruction. The Greek preference for expressing ordinary fractions as the sum of fractions with numerator unity and the use of a special symbol for 2/3rds were obviously adapted Egyptian relics. As a matter of fact, Heron of Alexandria, a Greek of the 1st century A.D. who designed water clocks, water turbines, and two-cylinder water pumps, wrote that the Egyptian measurement system had been copied by the Greeks.[2]

With *The Star of Deep Beginnings*, Dr. Charles Finch has sounded the clarion call for a new era of research in the field of the historiography of African science. It is my sincere hope that each reader will take away from the following pages the regenerating inspiration to uncover even more long-neglected information which will continue to document the African contribution to human history.

Théophile J. Obenga, Ph.D.
Temple University
Philadelphia, Pennsylvania
April 1996

[2] See Letronne, *Recherches, Critiques, Historiques et Géographiques sur les Fragments d'Héron d'Alexandrie ou Du Systeme Métrique égyptien.* Paris: Imprimerie Nationale (1851).

ACKNOWLEDGMENTS

No one ever completes an important task by himself; the support and encouragement of the people around him make a writer's work possible. To my wife, Ellen, the heart and soul of our common enterprise; to Dr. William Cleveland, to whom helping struggling writers and artists has become a way of life; to Dr. Virginia (Ginger) Davis Floyd, who has put herself on the line for her countless friends, time without number; to Abdou Rahman Ndiaye, who gave me two lives in one; to Dr. Erick Gbodossou, my twin in this and other lives; to editors Carolyn J. Fundock and Alex Fundock III, in whose capable hands this book has taken shape; to Professor Ivan Van Sertima, who saw something in me no one else did; to Mr. Charles Merrill, who always respected contrary opinions; to my five brothers and one sister, who followed me to Coumba Lamba USA; and, finally, to my seven children, who have never given their father an unhappy moment in his life—I thank you for helping me open the pathways of destiny.

INTRODUCTION

Strange Attractors: Myth and Science in African Thought

Imagination is more powerful than knowledge.
—Albert Einstein

What is science? This question is not merely rhetorical for it may be the central epistemological question of our time. Like religion—from which science sets itself apart—it has impacted on so many human events that it is impossible to imagine modern life without it.

The word derives from the Latin *scientia* meaning simply "to know" and, because of this meaning, appears in such English words as "con*science*" and "pre*science*." It is the direct Latin translation of the Greek *gnosis*—found in words like dia*gnosis* and pro*gnosis*—but, curiously, the semantic histories of the two words seem to have diverged. The word *scientia*, evolving into the term *science*, came to refer more or less to a systematic and technical method of collecting, measuring, and/or defining observable phenomena, particularly of the physical or natural world. Prior to Descartes, science was not totally separate from philosophy, because what we today call "science," the Greeks used to call "natural philosophy."

The Greek word *gnosis* in its turn became the root of "Gnostic," a highly metaphysical concatenation of Egyptian mythology, Greek idealistic philosophy, and Christian imagery that was suppressed by the post-Nicean Church even as it was being influenced by it. The etymological evolution of the two words, originally inter-translations of each other, caused each to assume nuances radically at variance with the other.

Another clue, however, into the nature of what we call science perhaps comes to us from the Rhind Mathematic Papyrus of old Kemit, originally dating from the Middle Kingdom 4,000 years ago, and representing the oldest mathematical text on record. The rubric of the title conveys the *raison d'etre* of this papyrus, rendering the following transliteration:

> *tp-hsb n hat m khat nbt rekh ntt nbt snkt shtat nbt.* (The correct method of investigating all things in order to know all that exists, each mystery, and every secret.)[1]

---

[1] See Obenga T, *La Philosophie Africaine de la Période Pharaonique*, Paris: L'Harmattan, 1990, p. 357.

Fig. 1. Thoth, as the Kaf Ape, and the scribe Nebmertouf. Thoth was the patron of all science and learning.

Theophile Obenga, in his monumental *La Philosophie Africaine de la Période Pharaonique*, has supplied us with a marvelously erudite commentary on the terms employed:[2]

- *tp-hsb*—"correct method," "calculation or reckoning of."
- *hat m*—literally "descent into," e.g., "study," "research," "investigation."
- *khat nbt*—"all things," i.e., "natural facts."
- *rekh*—"to know" or "to understand."
- *ntt nbt*—"all that is or exists."
- *snkt*—"dark things," i.e., "mysteries."
- *shtat*—"hidden things" or "secrets."

Four thousand years ago, as seen in this succinct evocation of the domain and purpose of mathematics, there seems to have existed a seamless unity of *scientia* and *gnosis*. The rigor, precision, and methods of science are in-

[2] Ibid., pp. 359–60.

voked; the intent to delve into the hidden domains of religion and mysticism is clearly enunciated.

"What is the nature of existence?" seems to be the issue, and the title rubric of the Rhind Papyrus infers that mathematics can illuminate the answer(s). It is an entirely "modern" perspective; indeed all the speculations of contemporary physics—of things seen and unseen—depend on mathematics for demonstration. Moreover, we find that certain aspects of quantum theory border on mysticism.[3] It is thus clear that the epistemological separation of science and religion, empiricism and myth, matter and spirit is a recent occurrence, not fully realized until the time of Descartes 350 years ago. But if modern quantum theory is any indication, the rupture does not seem to have been permanent; particle physics now acknowledges a "web of existence" in which all phenomena are related to one another.[4]

The interrelation of mythic fact and empirical reality has been brought into sharp relief by the work of Marcel Griaule and Germaine Dieterlen among the Dogon of Mali. We will have a great deal to say about the Dogon world-system before our task is finished, but it is worth taking a look now at a sample of their epistemology. We will find that in their system, as with that of Old Kemit, there is no distinction between myth and reality or science and religion. To grasp this point, it is necessary to turn to Griaule and Dieterlen's *The Pale Fox*:

> The myth, *so tanie*, "astonishing word," which the Dogon consider to be "real" history . . . constitutes here the whole of coherent themes of creation; this is why, by virtue of their coherence and their order of succession, they make up a "history of the universe," *aduno so tanie*.[5]

The authors make it clear that we cannot think of myth in the typical manner if we are to plumb the depths of Dogon knowledge:

> By no means here " . . . should the word myth be understood in its ordinary sense, as a childlike or fantastic, somewhat absurd poetic form. The myth is, for the Blacks, only a means by which to explain something; it is a consciously composed lore of master ideas. . . . It conceals clear statements and and coherent systems reserved for initiates, who alone have access to the 'deep knowledge.'"[6]

---

[3] See Capra F, *The Tao of Physics*, Berkeley: Shambhala, 1975, for a discussion of work in theoretical physics bringing it closer to Eastern mystical traditions; see also Davies P and Gribbin J, *The Matter Myth*, New York: Simon & Schuster/Touchstone, 1992.

[4] See Hawking S, *A Brief History of Time* New York: Bantam Books, 1990; see also Griaule M and Dieterlen G, *The Pale Fox*, translated by Stephen C. Infantino, Chino Valley, Arizona: Continuum Foundation, 1986, to compare the parallel concept of the "web of existence" in African cosmogony.

[5] Griaule M and Dieterlen G, op. cit., p. 60.

[6] Ibid., p. 61.

Thus, the acquisition of knowledge occurs in ascending stages—to those qualified by intellect *and* character to acquire it—through what may be called the Four Words:

> The *giri so*, "fore-word," is a first source of knowledge with simple explanations in which mythical personages are often disguised, their adventures simplified or fantasized, all this in seemingly unrelated parts. It deals with visible things and deeds, with rituals and modern materials.
>
> The *benne so*, "side-word," includes the 'words which were in the *giri so*' and the deeper explanation of certain parts of the rites and representations. Its coordination appears only within the greater divisions of knowledge, which remain partly unrevealed.
>
> The *bolo so*, "back-word," completes the preceding knowledge on the one hand, and furnishes synthesis applicable to greater parts of the whole on the other hand. It does not, however, contain the very secret parts.
>
> The *so dayi*, "clear word," concerns itself with the edifice of knowledge in its ordered complexity.[7]

My own research among the Serer traditional healers of Fatick, Senegal, reveals that most of them claim to have acquired knowledge of their "empirical" medicinal preparations in dreams. While this may not be easy to understand it is nonetheless true that the pharmacological efficacy of scores of traditional medicinal plants have been reliably confirmed in the laboratory.[8] Such venues of knowledge are not unknown in the history of Western science either. Alfred Russel Wallace claimed to have discovered a theory of evolution based on natural selection and survival of the fittest in a flash of spontaneous insight, almost in a "vision."[9] Though he had been corresponding with Charles Darwin, the latter had been too timid to announce to anyone (Wallace included) his theory of evolution—exactly the same as Wallace's—arrived at nearly 20 years earlier. When Darwin realized he was about to be "scooped," he hastily arranged to announce the theory that was to revolutionize biology. Wallace graciously (and rightly) yielded to Darwin's claim to priority.[10] It is perhaps worth noting in this discussion of the interconnection between myth and science that both Darwin and Wallace began their careers as naturalists and avid collectors of beetles; the scarab beetle of Old Kemit, *Khepera*, signifies "coming into existence," or "becoming," in a word, "evolution."

---

[7] Ibid., pp. 69–70.

[8] Sofowara A, *Medicinal Plants and Traditional Medicine in Africa*, New York: John Wiley and Sons Ltd., 1982, passim. See also Andoh A, *The Science and Romance of Selected Herbs used in Medicine and Religious Ceremony*, San Francisco: North Scale Institute, 1986, passim.

[9] See Edey MA and Johanson DC, *Blueprints*, Boston: Little, Brown and Company, 1989, pp. 79–80.

[10] Ibid.

August Kekule, the discoverer of the benzene ring (1865), formulated this important chemical model, consisting of a six-carbon ring, after a dream in which he saw a whirling snake with its tail in its mouth.[11] Two of France's greatest mathematicians, Rene Descartes and Henri Poincare, were said to have experienced their most important theories as spontaneous flashes of insight.[12] Finally, no less a person than Albert Einstein himself was said at age 16 to have experienced a "waking dream" in which he saw himself riding a beam of light. This image led him down the path on which he eventually discovered that light travels at the same constant speed for all observers, moving or not. This insight produced the Special Theory of Relativity in 1905 that changed physics forever.[13] Clearly, these examples demonstrate how critical the intuitive, noncognitive—or in Jungian terms, "acausal"—faculty is in uncovering important scientific knowledge. It is so evident one wonders why Western scientists have always insisted so strenuously that such psychic activity has no place in the cold and sober realms of science. When Hamlet tells Horatio that there are "more things in heaven and earth than are dreamt of in your philosophy . . . " he could just as well been speaking to the modern scientist.

I have come to see that not only religion but science is the offspring of myth; to fathom the depths of both these realms of psycho-mental experience, we are impelled to return to myth. The modern penchant for trivializing myth has done the world a terrible disservice. But since myth is intrinsic to human consciousness, or, more precisely, human unconsciousness, it can never be "exorcized" in the way modern man has attempted to; it merely ramifies into bizarre and abnormal forms for want of appropriate expression.[14] Myth actualizes history—indeed, human existence itself—and science is but one of the components of myth. Two historians of science who seem to have understood this truism are Hertha von Dechend and Giorgio de Santillana, co-authors of *Hamlet's Mill*, an absorbing and erudite exploration of the interaction of myth and science in time and space. Another is the biologist James Lovelock who recently formulated the Gaia Hypothesis, depicting earth as a unified organism working purposefully to keep all life, all nature, in a homeostatic balance. Gaia was the ancient Greek Earth-goddess, mother of gods and men. Thus we find a respected modern scientist couching his theory in mythic terms.

It is also of no small interest to find the noted molecular biologist Rebecca Cann naming her "mitochondrial mother" (i.e., the 200,000-year-old East African Khoisan woman whose genes survive in all six billion of her present

---

[11] Jung CG, editor, *Man and His Symbols*, Garden City: Doubleday & Company, Inc., 1964, p. 38.
[12] Ibid.
[13] Capra, op. cit., p. 167.
[14] Jung, op. cit., passim.

descendants), "Eve."[15] Even modern cosmology describes the creative "Big Bang" as issuing from a "primal egg."[16] Nor will I ever forget the first day of my undergraduate biology lab when the lab instructor, seeking to forestall humanitarian objections to the dissection of animals, encouraged the students to think of such "necessary" death as a "sacrifice." Thus it is that science deplores or belittles the ritual sacrifices that attend religious ceremonial while coolly endorsing its own.

It is impossible not to think mythically, indeed *be* mythical, for the mythic impulse is innate to humanity, suffusing all human endeavor, including science. It is why, in the coming pages, we will mine African myth for science, while, at the same time, bringing to light scientific advances in Africa standing outside the context of myth—on their own, as it were. From iron mines in Swaziland 40,000 years ago to Cesarean-sections in Uganda 113 years ago, science and complex technology were not alien to Africa, whether encased in myth or not. Today we place so much emphasis on the primacy of science in the "intellectual adventure" of modern man that it is important to show, as in so many other domains, that Africa is veritably the crucible of modern science and technology.

---

[15] Cann RL, Stoneking M, and Wilson AC, "Mitochondrial DNA and Human Evolution," *Nature* (352), January 1, 1987.

[16] " . . . the universe began its existence as a condensed droplet of matter at an extremely high density and temperature . . . [George] Gamow named this primordial egg "ylem"—the name Aristotle gave to the basic substance out of which the ancients believed all matter was derived", Jastrow R, *Red Giants and White Dwarves*, New York: New American Library, 1969, p. 68.

1

Nile Genesis: The Continuity of Culture from the Great Lakes to the Delta

> *From Gizeh and Turah for 350 miles to Naqadeh there is no difference in the protodynastic work; and the same is true of the earlier prehistoric times for more than 300 miles from Gerzeh to Naqadeh. Moreover 200 miles farther south in Nubia the styles of this age are perfectly continuous, although mixed with other types which belong to Nubia. Thus for over 500 miles the prehistoric civilization seems to have been so well organized and unified that the same tastes, ideas, patterns and materials prevailed throughout. This shows that there were not isolated and warring tribes, which prevented intercourse and trade, but rather a peaceful, if not a united, rule over all Egypt and Nubia.*
>
> —W.M. Flinders Petrie, *Prehistoric Egypt,* p. 3

There is a concept used in contemporary physics called a "paradigm shift." Fundamentally, it means that an accepted and approved model explaining the physical nature of the universe requires radical modification because a whole new data set does not fit into it. Paradigm shifts can be momentous or modest but they often compel dramatic changes in our vision of things. The concept seems applicable to any scholarly or scientific pursuit, and in recent years our understanding of ancient history has indeed been undergoing such a shift. We are being forced to radically readjust our optic on ancient history because more and more we are stumbling upon evidence that shows that Africa is the true crucible of modern human culture. Tidy little categories such as "paleolithic," "neolithic," and "prehistoric" no longer seem adequate to explain the emerging evidence for the extreme antiquity of advanced culture in Africa.

Physics seems comparatively unencumbered by "political" baggage; not so the study of ancient history. Though paradigm shifts in physics may engender intense, even acrimonious debates, the weight of accumulating evidence usually carries the day. A paradigm shift affecting our image of ancient humanity is another matter; too much racial mythology and ideology are wrapped up in it. Often, new empirical advances in the fields of ancient history, no matter how compelling, are ignored or rejected without a hearing. Still, the tendency is clear: the more we explore the archaeology of Africa, the more we find that modern human culture took root there first. Since the human species was born in Africa, the unfolding of the traits of advanced human culture there is entirely consistent.

Ishango and Katanga: Cradlelands of African Culture

Alison Brooks and John Yellen are a married team of archaeologists from George Washington University who have been working for some years in northeastern Zaire, along the shores of Lake Edward in the neighboring districts of Katanda and Ishango.[1] Their scientific credentials are impeccable and they possess a reputation for methodical, meticulous field research. Yet their findings at archaeological sites in the lake districts of Zaire have somewhat strained their natural caution. The artifacts they are uncovering and testing by the most refined radiocarbon techniques available are producing results that may force the jettisoning of the conventional chronology of paleolithic East African culture. This is not a situation that, from a scholarly point of view, is to be taken lightly.

From the prehistoric rubble of Katanda where the team was working, Yellen, in 1988, extracted a finely-wrought, fossilized bone harpoon point.[2] This was the first of several such finds:

> . . . two more whole points and fragments of five others turned up, all of them elaborately barbed and polished. A few feet away, the scientists uncovered pieces of an equally well-crafted daggerlike tool. In design and workmanship the harpoons were not unlike those at the very end of the Upper Paleolithic, some 14,000 years ago. But there was one important difference. Brooks and Yellen believe the deposits . . . were at least *five times that old.* (emphasis added)[3]

Their conviction that a relatively advanced culture existed in east-central Africa 70,000 years ago underscores earlier work by Brooks on the related Ishango site. This site, just four miles from Katanda, was first excavated by a Belgian, Jean de Heinzelin, in the late 1950s, where he found a harpoon-rich "aquatic civilization."[4] His original date for the site was 8,500 B.P., that is, 8,500 years before the reference or "present" date of 1950 A.D. But Brooks re-dated the site more than 25 years later with more sensitive radiocarbon instruments and found, instead, that the site dates back to 25,000 B.P., making it three times older than originally estimated.[5] Furthermore, the stratigraphy of the Katandan deposits is much deeper than that of the adjacent Ishango site and the Brooks-Yellen team thinks that the artifacts there are more nearly 70,000 years old. These dates are perplexing because they mean that the Katanda cultures had already reached a level of technical development not achieved outside Africa until 54,000 years later.

Storms of controversy persist around the locus of modern human origins;

---

[1] See Shreeve J, "The Dating Game," *Discover*, September 1992, pp. 76–83.
[2] Ibid., p. 78.
[3] Ibid.
[4] See De Heinzelin J, "Ishango," *Scientific American* (106), June 1962, pp. 105–16.
[5] Shreeve, op. cit., p. 82.

several researchers continue to look for sites outside Africa for evidence of modern human beginnings.[6] But the most compelling evidence fixing the geographic region where humanity first appeared continues to trickle in from the lacustrine regions of east-central Africa. Even where Africa's role as the biological womb of humanity is conceded, many still assume that complex culture evolved elsewhere.[7] However, this recent work by Brooks and Yellen more firmly situates the cultural cradle of humanity in Africa. Indeed, Brooks's redating of the Ishango site puts the famous Ishango Bone in an entirely new light.

When De Heinzelin first excavated the Ishango site with its cache of bony harpoon heads, he came across a bone tool displaying groups of peculiar horizontal markings on it. A close examination convinced him that these markings represented an arithmetic system revealing a familiarity with prime numbers, decimals, and addition by duplication.[8] Harvard archaeologist Alexander Marshack could not accept evidence of "mesolithic" mathematics in central Africa but he did become convinced that the marks on the bone represented notations marking the lunar cycle.[9] Very likely, there is truth in both explanations and, therefore, the evidence of the Ishango Bone reveals an intellectual sophistication that no one would have accorded mesolithic populations living in east-central Africa some 9,000 years ago.

Brooks' recent radiocarbon datings have fermented the brew even more: since Ishango dates back 25,000 years, these people were not mesolithic but paleolithic. De Heinzelin had noted that the Ishango artifacts seemed archaic by comparison to other mesolithic populations and Brooks' datings cleared up the mystery. But proto-mathematics and lunar calendars in Africa in the Upper Paleolithic period? With a tantalizing possibility that such scientific skills in that region might be older yet? The conventional model of cultural evolution advanced by Marshack may have to be reconsidered.

As Brooks and Yellen are not unmindful of the minefield they have to cross. Brooks herself has said, "It is precisely because no one believes us that we want to make our case airtight before we publish. We want dates confirming dates confirming dates."[10]

De Heinzelin's study of the harpoon heads and the Ishango Bone convinced him of something else: the culture revealed by these artifacts was carried virtually intact northward into the Nile Valley region of Africa. Thus, de Heinzelin tells us:

---

[6] Kittles R, *Mitochondrial DNA and Human Origins: Assessing the Debate*, Spring 1993, unpublished.

[7] Shreeve, op. cit., p. 80.

[8] De Heinzelin, op. cit., p. 111.

[9] See Marshack A, *The Roots of Civilization*, Mt. Kisco, N.Y.: Moyer Bell Limited, 1991, pp. 21–32.

[10] Shreeve, op. cit., p. 82. Since these lines were first written, two articles, one by Brooks and the other by Yellen, were published in *Science* which affirm categorically that the dates for the Katanda artifacts are now known to be 90,000 years old! Brooks and Yellen, as the articles attest, arrived at secure dates from thermoluminescence and electron spin resonance test-

> From central Africa the [harpoon] style seems to have spread northward. At Khartoum near the upper Nile there is a site that was occupied considerably later than Ishango. The harpoon points found there show a diversity of styles . . . [other harpoons] have the notches that seem to have been invented first at Ishango. Near Khartoum, at Es Shaheinab, is a Neolithic site that contains harpoon points bearing the imprint of Ishango ancestry; from here the Ishango technique moved westward along the southern border of the Sahara.[11]

De Heinzelin goes on to say:

> The technology also seems to have followed a secondary branch northward from Khartoum along the Nile Valley to Nagada in Egypt. This site has both bone and copper harpoons. Made in the Neolithic period before the Egyptian dynasties began, many of them are notched at the head. Others show the influence of the Near Eastern Natufian technique and the Fayum technique which is closely related to it.[12]

It would seem, if de Heinzelin is correct, that a well-defined culture, like humanity itself, flowed out of the lacustrine African womb into the rest of the continent (see Fig. 2). The Nile River seems to have functioned "umbilically" to link the northern tier of the continent to its East African "placenta-land." It has to be wondered if this is the reason why the ancient Egyptians called Africa south of the Nile Valley *Ta-Kenset*, literally "placenta-land."

The reference to the Natufian harpoon technique in the above quote by de Heinzelin is of interest because the Natufian culture of Syrio-Palestine can be dated back 10,000–12,000 years and the human fossil remains associated with this culture denote a clear Africoid type.[13] The similarities of Nile Valley and Natufian harpoon manufacture probably indicate a broader cultural kinship, and if the source of these harpoon styles is the Lake Edward horizon of east-central Africa, it is not implausible to assume that the Ishango/Katandan culture radiated beyond Africa directly into western Asia (see Figs. 3 and 4).

The peopling of the Nile Valley from the Africa's Great Lakes region must have occurred over and again in waves. The population wave from the Great Lakes directly ancestral to the historical Nile Valley peoples probably began to settle north of the second cataract no later than 15,000 years ago. This settlement did not preempt later migrations, of course, but most of these

---

ing. Thus, the first clear indications of human cultural behavior emerge in Africa. See Brooks AS, et al, "Dating and Context of Three Middle Stone Age Sites with Bone Points in the Upper Semliki Valley, Zaire," *Science* (268) 28 April 1995, pp. 548–52, and Yellen JE, et al, "Middle Stone Age Worked Bone Industry from Katanda, Upper Semliki Valley, Zaire," *Science* (268) 28 April 1995, pp. 553–6.

[11] De Heinzelin, op. cit., p. 109.

[12] Ibid.

[13] See Diop CA, "Processus de Sémitisation," *Parenté Génétique de l'Egyptien Pharaonique et des Langues Négro-Africaines*, Dakar: Nouvelle Éditions Africaines, 1977, pp. xxix-xxvii.

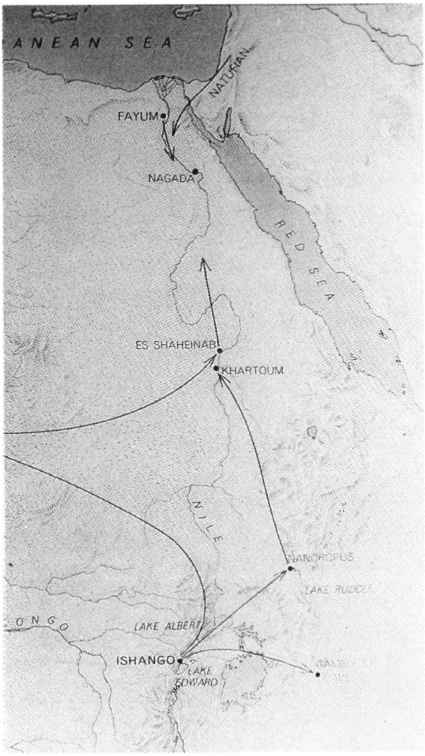

Fig. 2. Route of spread of the prehistoric harpoon-making techniques and styles northward from the Great Lakes (Ishango) along the Nile.

probably came in from the west, that is from the Sahara, beginning about 10,000 years ago.[14] Another line of migration apparently trickled in intermittently from the horn of Africa along the Blue Nile from 6,000 to 10,000 years ago. The dynastic Egyptians' never-ending fascination with *Punt*, their name for the Horn of Africa, and clear attestations from them regarding their kinship to the people of Punt, compel us to look in that direction for an ancestral lineage of the peoples who created pharaonic culture. There seems to have been a touch of religious awe associated with Punt since it was also referred to as *Ta-Neter*, i.e., "Land of the God."

One last word regarding the Ishango Bone: de Heinzelin thinks that the bone demonstrates a knowledge of addition by the method of duplication.

---

[14] See Wendorf F, Close AE, and Schild R, "Prehistoric Settlements in the Nubian Desert," *American Scientist* (73), March-April 1985, pp. 132–41, for a discussion of archaeological evidence of a pottery-making, cattle-rearing Neolithic people inhabiting the lands west of the Nile as far back as 8,000 B.C.

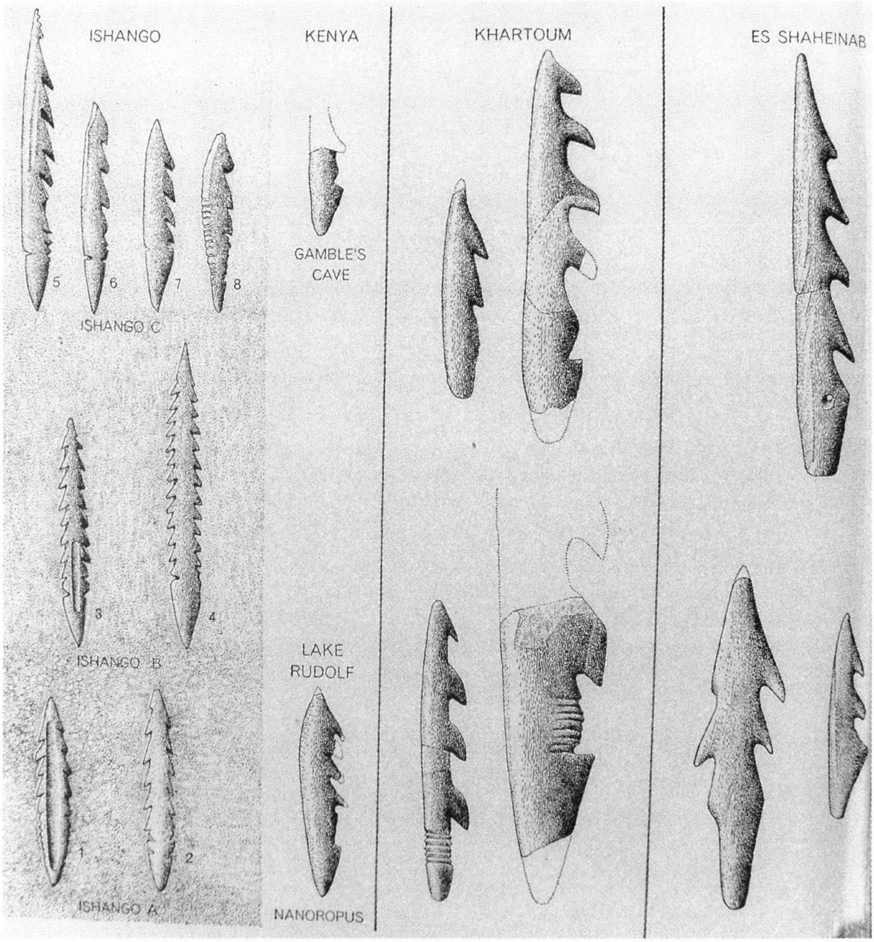

Fig. 3. Samples of harpoons in East and Northeast Africa apparently derived from Ishango prototypes in northeastern Zaire 70-90,000 years ago.

This is of surpassing interest because the very same method of addition is utilized in the Rhind Mathematical Papyrus from Egypt, dating back nearly 4,000 years ago.[15] It is the oldest mathematical treatise on record. Is it possible that this mathematical procedure reached the Nile Valley directly from the lacustrine horizons of East Africa?

---

[15] See Gillings R, *Mathematics in the Time of the Pharaohs*, New York: Dover Publications, 1982.

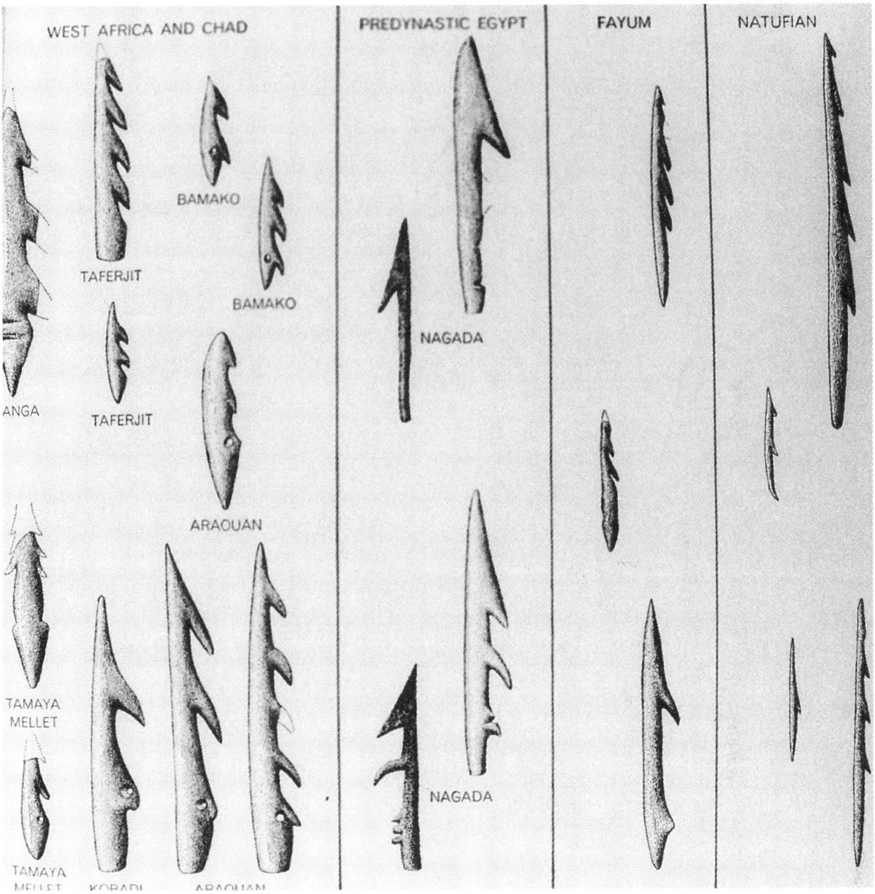

Fig. 4. Additional samples of harpoon points derived from Ishango models in West Africa, Northeast Africa, and Western Asia.

The Cultural Horizons of the Nile Valley During the Pre-Dynastic Epoch

The founder of pre-dynastic Nile Valley archaeology, William Flinders Petrie, excavated at Nagada and Ballas in Upper Egypt 100 years ago, unearthing the first artifactual evidence of the cultures of pre-existing historical pharaonic society. Petrie eventually excavated more than 1,200 pre-dynastic graves. Other pre-dynasticists, such as Quibbell, De Morgan, Brunton, and Caton-Thompson, refined and extended his research considerably during the next 40 years, but Petrie established the basic principles of pre-dynastic archaeology and worked out a dating method that is still used.[16]

---

[16] See Flinders Petrie WM, *Prehistoric Egypt*, London: Quaritch, 1920, pp. 3–6.

At Nagada and Ballas, Petrie found an almost invariant burial scheme: (1) the body contracted into the fetal position, (2) the head positioned toward the south, and (3) the face pointed westward. Moreover, the graves always contained the personal possessions of the deceased. On the strength of this artifactual evidence alone, it seems clear that many of the ideas featured in later pharaonic religion existed in prototype among these pre-dynastic people at Nagada. The fetal position evokes the idea of rebirth. The southward orientation of the body echoes the dynastic references to the south as *Khenti*, "the land of beginnings." The south is also *Ta-iakhu*, "the land of the spirits," where the souls of ancestors dwell. The significance of the face pointing westward is indicated in the place name *Amenta*, "the hidden land," for *Amenta* is the western region where the soul of the departed journeys after it quits the body. Finally, dynastic Egypt was notorious for the amount of worldly goods interred with the deceased. The elaborate funerary cult of dynastic times seems to have evolved directly from pre-dynastic practices.

Petrie found thousands of vases and pots of varying styles in these Nagada graves. After six years of continuous work, Petrie, in 1900, managed to group all the pottery into nine classes and, more impressively, assign a correct temporal sequence to each class. This method of classification became famous as Petrie's Sequence Dating, a method still in use. Sequence dating allows the determination of the temporal relationship among different groups of pre-dynastic vases—and therefore of the cultures that produced them—but *not* the absolute dates of their manufacture. Petrie created sequence dates (S.D.) from 20 to 79, with 30 representing the earliest pre-dynastic pottery styles in his sample and their associated culture(s), and 79 representing the 1st Dynasty. He very sensibly reserved S.D.'s under 30 for pottery sequences and cultures yet to be discovered. Petrie assigned a tentative date for his earliest pottery sequences, e.g., S.D. 30, of 8,000 B.C. This chronology was, within his lifetime, downdated considerably to 5,500–4,500 B.C. Petrie, to the end, never accepted such low dates.[17] The low dates, nonetheless, have held sway up to now, though recent research has been compelling a reconsideration.

The names given by Petrie to the two pre-dynastic cultural sequences delineable by his pottery classes were "Amratian" and "Gerzean." Modern pre-dynasticists often use the terms Nagada I and Nagada II to refer, respectively, to Petrie's Amratian and Gerzean. These and other pre-dynastic cultural horizons extended into the Sudan.

As more and more pre-dynastic material was recovered in Egypt and Nubia, it became clear to Petrie at least that there was a uniformity of culture that spanned Egypt and Nubia. It was Petrie's conviction that there was "a peaceful, if not a united, rule all over Egypt and Nubia" during the entire pre-dynastic period.[18] More recent research has tended to confirm that assessment.

---

[17] See Jones WD, *Venus and Sothis*, Chicago: Nelson-Hall, 1982, p. 80.

[18] Flinders Petrie, op. cit., p. 3.

In fact, there is evidence that this pre-dynastic cultural unity encompassed all of the lands hugging the Nile from the confluence of the White and Blue Niles at Khartoum to the apex of the Delta.[19] That parts of this Nile Valley, or Nilotic, culture-complex can be traced to Africa's Great Lakes region has already been touched upon.

Though pre-dynasticists such as DeMorgan and Quibell found and excavated additional graves at Nagada and Ballas, no older cultures were unearthed until the excavations of Brunton and Caton-Thompson in 1928 at Badari, 100 miles north of Nagada. The artifacts discovered were clearly pre-dynastic and eventually determined to be more archaic than those recovered from Nagada and Ballas. These cultures became known, respectively, as "Tasian" and "Badarian" and were assigned S.D.'s between 20 and 29. Though at first they seemed to be distinct horizons, with the Tasian being older and more primitive, they were eventually linked together as the same culture in different phases, e.g., the "Tasian-Badarian" sequence. Under Petrie's schema, this cultural sequence would be at least 12,000 years old (10,000 B.C.), but modern dating, though not consistent, takes it back maximally to 5,500 B.C. and minimally to 4,000 B.C.[20]

The Badarians were culturally a thriving people, most renowned for their handmade pottery. Their vases were fashioned with a durability, finish, and extraordinary thinness that were never to be surpassed. They were also notable for well-turned ivory utensils and the use of copper. Theirs is the longest pre-dynastic sequence in the Nile Valley.

The Amratians succeeded the Badarians and their pottery was characterized by distinctive red bodies with black tops and also a style with white, crossed lines. It is in the Amratian sequence that stone vases appear for the first time. In an advance over the Badarians, the Amratians not only worked extensively and more variously in copper but also in gold. Faience and blue glazing of quartz appears with them.

The Gerzeans, the last distinct cultural sequence preceding the dynasties, reveal themselves to have been a people on the brink of formal civilization. Their pottery shows unique features: buff coloration with red zigzag lines and spirals, designs of many-oared boats, and representations of flamingoes and sometimes of people. Stone vases are more varied in shape and fashioned from more diverse materials. Copper use is general, as is gold, and small quantities of silver—probably imported from Western Asia—appear. Ornamental iron beads are present and glass-making appears for the first time in the Nile Valley.

---

[19] See De Heinzelin, op. cit., p. 109, and Wendorf, Close, and Schild, op. cit., passim.

[20] See Hoffman MA, *Egypt Before the Pharaohs*, London: Ark Paperbacks, 1980, p. 142, for evidence of the higher Badarian date of 5500 B.C.; also, Hassan FA, "The Predynastic of Egypt," *Journal of World Prehistory* (2)2, June 1988, pp. 138–41, for dating results of the Badarian period at 4,000 B.C.

Given the received 1st Dynasty date of 3,100 B.C., the extremely low dating of 4,000 B.C. for the entire pre-dynastic era recently advanced creates a crowding problem.[21] If accepted, we are then required to assume that three successive, well-defined cultural horizons took root and then vanished in the Nile Valley in the space of 900 years. When, as will be noted below, evidence surfaces that pushes the dynastic period back to 3500 B.C., the crowding becomes even more severe, leaving only 500 years for the entire pre-dynastic era.

A word should be said about the pre-dynastic horizons discovered in Lower Egypt. The first was found near Merimde, just west of the Delta, and dated to 4,880 B.C.[22] The Merimdens were a mixed farming and hunting people, roughly contemporaneous with, but culturally distinct from, the Badarians to the south. The difference seems so pronounced that the Merimdens have seemed "un-Egyptian" to some. Phenotypically, they were of small stature and rounder of head than the dolicocephalic Badarians, but more significant was their very different burial practices. The Merimdens buried their dead inside their settlements instead of in separate graveyards and included no personal belongings or food offerings with the deceased. As will be discussed below, a high proportion of mother-child burials among the Merimdens seems to reflect a "matristic" or matrilineal cultural focus, but this element seems to have originated from a place very different from where the Badarians came from. Since their manner of house-building and their agriculture shares similarities with Neolithic Near Eastern societies, many have looked toward Palestine for the original home of these people. But there exists an older horizon in the Nile Valley that furnishes clues to the provenance of the Merimdens.

About 150 miles south of Merimde as the crow flies lies the Fayum depression. In antiquity, it served as a water reservoir in the dry season for dynastic Egyptians, but it is also significant for two related pre-dynastic cultures discovered there. In archaeological terms, these horizons are called "Fayum A" and "Fayum B"—confusing because Fayum B is actually older, dated to 6,300 B.C. The Fayum horizons seemed to have possessed distinct affinities with Neolithic Saharan cultures, known to have been in existence as early as 8,100 B.C.[23] The Fayum, Merimde, and Near Eastern horizons all seemed to share certain traits and may all have been extensions of the Saharan cultural complex.

As noted, the Lower Egyptian pre-dynastic horizons at Fayum and Merimde appear markedly different from the Upper Egyptian/Nubian ones. In fact, they are not strictly considered pre-dynastic at all but rather late Neolithic horizons existing at a lower cultural stage than the Badarian or Nagadan se-

[21] Hassan, ibid.
[22] Hoffman, op. cit., p. 169.
[23] Wendorf, Close, Schild, op. cit., p. 132.

quences. Lower Egyptian pottery is very crude by Badarian standards and neither Fayum nor Merimde, unlike the Badarians and Nagadans, knew of metal. The consensus is that the civilizing impulse leading to dynastic pharaonic culture emerged in the south. It might be said that classical Nile Valley civilization evolved primarily in the region between Assiut (Badari) and the 2nd cataract (Ta-Seti).

Despite clear and unequivocal indicators, the geo-ethnic provenance of the creators of Nile Valley civilization continues to be a contentious issue. For this reason, some space must be devoted to a discussion of it. As we have seen, Petrie was of the opinion that there was a kind of cultural unity linking the pre-dynastic horizons of Egypt and Nubia. Sir Alan Gardiner shared this opinion:

> However striking the change from the one stage to the other may appear, the continuity of the evolution as a whole must be affirmed with all emphasis . . . there was an affinity between Libyans, Egyptians, and Nubians which confirms our description of the earliest culture of the Nile Valley as *essentially African*. (emphasis added)[24]

Others echo this assessment. The noted pre-dynasticist Fekri A. Hassan asserts:

> The earliest Neolithic levels in the Nile Valley . . . shared a common Saharan-Nilotic base, as well as some Near Eastern affinities. . . . These regional cultural differences are not associated with biological differences. Investigations of crania from Badarian settlements indicated that they were similar to those from Nagada, Abydos, El-Amrah, and Hu.[25]

Petrie himself observed that the numerous female figurines found at Nagada—associated with the early Amratian and Gerzean sequences beginning S.D. 34—showed the steatopygy typical of the southern African Khoisan type[26] (see Figs. 5 and 6). While it is doubtful that these early Nile Valley dwellers were pure Khoisan, it seems evident that their place of origin is to be sought in Africa's southerly regions. Moreover, skull studies of the Badarians after their discovery by Brunton and Caton-Thompson have almost unanimously identified them as southern in origin.[27]

Recently, an excellent review of the data on the "skull metrics" of Nile Valley/North African populations by S.O.Y. Keita has shed additional light on these questions.[28] Keita notes that previous studies of the Badarian and

---

[24] Gardiner A, *Egypt of the Pharaohs*, New York: Oxford University Press, 1978, p. 135.
[25] Hassan, op. cit., p. 159.
[26] Flinders Petrie, op. cit., p. 8.
[27] See Gardiner, op. cit., p. 392.
[28] Keita, SOY, "Studies of Ancient Crania from Northern Africa," *American Journal of Physical Anthropology* (83), 1990, pp. 35–48.

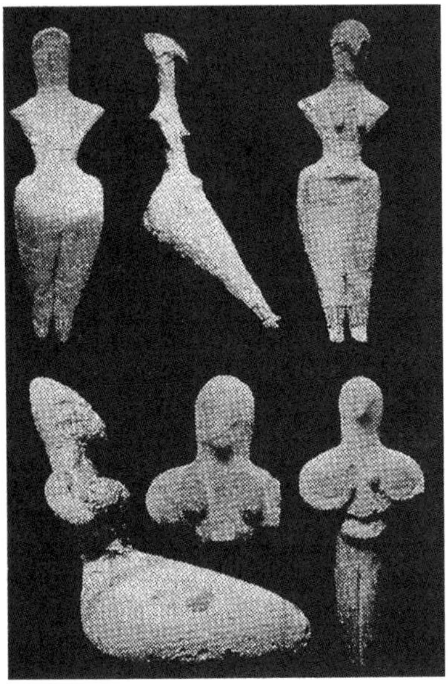

Fig. 5. Early pre-dynastic figurines of steatopygous women, circa 8,000 B.C. (Petrie)

Nagada I (Amratian) series of skulls, when compared to those disinterred from Kerma in the northern Sudan, show nearly identical phenotypes.[29] Keita conducted his own multivariate analyses of 10 skull series from North Africa (west of Egypt), the Nile Valley, East Africa, and West Africa to ascertain phenotypic clusterings on a broader continental scale. The Nile Valley series included 72 pre-dynastic skulls, 22 from Badari and 50 from Nagada. Keita duly noted the interpretive problems occasioned by the small sample size but he still found, consistent with earlier data, that both the Badarian and Nagadan skull series clustered phenotypically around those of Kerma (Nubia), East Africa (Kenya), and West-Central Africa (Gabon).[30] This phenotypical clustering extended to dynastic Abydos, the mythical home of Osiris: "The Nagada and Kerma [Nubia] series are so similar that they are barely distinguishable in the territorial maps; they subsume the first dynasty series from Abydos."[31]

In a follow-up article, Keita discussed the 1st Dynasty Abydos series more fully, finding that 57% of the skulls in that series could be assigned a southern origin.[32] But throughout his writings, Keita expressed a caveat also articulated by the late C.A. Diop, to wit: a "monotypic" classification cannot

[29] Ibid., pp. 36–8.

[30] Ibid., pp. 38–46.

[31] Ibid., p. 40.

[32] Keita SOY, "Further Analysis of Crania from Ancient Northern Africa: An Analysis of Crania from First Dynasty Egyptian Tombs, Using Multiple Discriminant Functions" *American Journal of Physical Anthropology* (87), 1992, p. 250.

Fig. 6. Front and back views of a pre-dynastic female figurine, circa 7,500 B.C. (Petrie)

be used to determine absolutely the predominant ethnicity of the dwellers of the Nile Valley.[33] In this case the monotype of the oft-caricatured "Negro," that is the type possessing an elongated cranium, jutting jaw, crispy hair, flaring nostrils, everted lips, bandy legs, and black skin, cannot be considered fully representative of native Africans. As Keita suggests, native African populations are phenotypically "polytypic," that is, there exists in tropical Africa a variety of phenotypes that deviate significantly from the caricature. A casual perusal of almost any indigenous African group can convince even an untrained observer of the truth of this assertion. Dolicocephalic skulls and subnasal prognathism, as seen in many pre-dynastic Nile series, do establish a southern provenance but a deviation from or absence of this distinct phenotype in certain Nile Valley skull series does not necessarily "de-Africanize" them.

Another feature of the pre-dynastic graves, from the Badarian epoch down through the Gerzean, that points to a southern origin is the unmistakable "matrifocal" ethos. A number of pre-dynasticists have noted the superior wealth of women's graves and the feminine gender of the greater number of pre-dynastic figurines.[34] Fekri Hassan gives us the clearest exposition of the implications of the feminine "bias" of these artifacts:

[33] See above-mentioned articles by Keita.

[34] Cf. Hoffman, op. cit., p. 328; also, Hassan, op. cit., p. 170.

> Ritual and supernatural beliefs associated with agricultural life must have emerged with the beginnings of the Neolithic. A study of figurines and mortuary goods from the Nagada region indicates that such beliefs and rituals were closely related to women. Lineage was probably matrilineal as indicated by the exclusive internment of women and children at Merimda. This may have been because women were associated with grinding grain and making bread and gruel as well as brewing beer and, perhaps, initially planting grain.[35]

Thus, even the Lower Egyptian horizons, like the one at Merimde, evince a definite matristic character. As is well-attested, this matrifocality/matrilineality, at least in social and domestic relations, continued in Egypt to the end of dynastic history. Matrilineality, in fact, was the almost invariable rule of succession governing the pharaonocracy. The custom undoubtedly derived from African antecedents.[36] Since a linkage between the Merimden and Saharan horizons is evident, it can be inferred that the custom was more or less general throughout Neolithic Africa.

Early in this essay, evidence was brought forward linking the Ishango/Katandan horizons with those in the Nile Valley, Sahara, and Syrio-Palestine. In fact, the Natufians of 10,000 years ago in Palestine shared phenotypic affinities closer to the Badarians than they did the Merimdens, whose culture is said to have been more closely related to those of the Near East. It has been customary to separate the Near East from Africa ethno-culturally though, in the light of increasing Neolithic evidence, it is perhaps more nearly correct to consider the lands between Khartoum in the south and the Tigris-Euphrates in the north as constituting one broad horizon in the period between 10,000 and 5,000 B.C. This broad horizon was composed substantially of "Nilo-Saharan" ethno-cultural elements. Regional differences and variations were certainly evident in this larger cultural complex, but ongoing techno-commercial relations linked the various groups of this horizon.

It can be said that what is known as the Near East is more properly thought of as Africa's "Northeast Extension," because geologically and geographically that is in fact what it is. It was the main corridor of human migration out of Africa into the rest of the world beginning 100,000 years ago and it makes sense to find that the earliest definable Near Eastern Neolithic populations, the Natufians, are indisputably Africoid. There seems to be no good reason not to believe that prior to 5,000 B.C., this "Northeast Extension" participated in the various cultural horizons of Africa. Thus, when we look at the Neolithic Near East, we are looking culturally at a "province" of Neolithic Africa.

---

[35] Hassan, ibid., p. 169.

[36] Cf. Finch CS, *Echoes of the Old Darkland: Themes from the African Eden*, Decatur, GA: Khenti, Inc., pp. 57–87.

Radiocarbon Dating of Nile Valley Horizons: Old Arguments, New Finds

As has already been intimated, the absolute chronology of the Nile Valley is a hotly contested subject. Despite his status as the founding father of pre-dynastic Nile Valley studies, Petrie's chronology very early underwent criticism and revision, resulting in considerable down-dating of his original estimates of pre-dynastic time. Recent work by Fekri Hassan, based on extensive radiocarbon datings of pre-dynastic sites, has pulled the Badarian epoch down to 4,400–4,000 B.C. Even these dates are not consistent with thermoluminescent tests giving a Badarian date of 5,500 B.C. Nonetheless, Hassan, a respected authority, published no fewer than six radiocarbon studies of pre-dynastic sites between 1980 and 1988. High-resolution radiocarbon instruments have brought much greater precision to dating techniques but anomalies still persist, judging from other work.

One such anomaly has arisen from recent radiocarbon datings of organic materials extracted from the Great Pyramid and numerous other pyramids, temples, and tombs in the Giza/Sakkara area by Herbert Haas and his colleagues.[37] Sixteen organic samples were extracted from the Great Pyramid and 13 of these were of charcoal. After instruments in the two testing laboratories were calibrated, results were obtained on the 16 samples that gave a range of ages the researchers found startling. The oldest radiocarbon date on the Great Pyramid came from a charcoal fragment from the 198th course, giving a date of 3,809 B.C. The youngest date came from a mortar of limestone from the second course giving a date of 2,853 B.C. Averaging all the dates on the 16 samples together, Haas and his team determined that, by radiocarbon testing, the pyramid was *minimally* 389 years older than the given "historical date," that is, it was built around 2,978 B.C. In fact, all 10 of the monuments/temples evaluated were older than the received "historical" dates by an average age of 374 years. If, in round numbers, it can be said that these Old Kingdom monuments are about 400 years older than previously believed, then the beginning of the 1st Dynasty goes back *minimally* to 3,500 B.C. As we have seen with other evidence adduced above, it is hard to reconcile these monumental radiocarbon data of Haas's with Hassan's radio-carbon readings giving a date for the beginning of the pre-dynastic era of 4,000 B.C. Even a somewhat remoter allowable date of 4,400 B.C. (Hassan) for the Badarian period hardly seems plausible. The thermo-luminescent date of 5,500 B.C. given by Hoffman for the Badarian period at least has the virtue of creating more chronological space for the pre-dynastic era.

---

[37] Haas H, Devine J, Wehnke R, Lehner M, et al, "Radiocarbon Chronology and the Historical Calendar in Egypt," *Chronologies in the Near East*, Evin J and Hours F, eds., BAR International Series, 1987, pp. 585–606.

Ta-Seti and the Beginning of Nile Valley Civilization

Complicating matters is the seldom referred to work by Bruce Williams concerning the Ta-Seti kingdom discovered at Qustul.[38] The site was excavated for the first time by Keith Seele in 1962 who, even then, thought that the size and wealth of the tombs marked them as royal.[39] Systematic analysis of the artifacts, however, did not take place until 15 years later by Bruce Williams at the University of Chicago's Oriental Institute, adding a hitherto unknown chapter to Nile Valley history. In one sense, however, the Qustul artifacts confirmed what Egyptian annals had already attested: there were whole dynasties that immediately preceded the 1st Dynasty under Menes.

Qustul lies in Nubia, more than 100 miles south of Aswan and the 1st cataract, and is now covered over by Lake Nasser. Qustul was a well-known site for what is known as A-group Nubian culture, contemporaneous with and related to the Gerzean pre-dynastic complex of Upper Egypt. The archaeological context of the tombs found in a burial spot known as "Cemetary L" showed clearly that they were older than the 1st Dynasty. When the tombs were opened, a startling variety of goods and materials were present, including expensive, finely-crafted jewelry, five different kinds of pottery, copper weapons, gold bracelets, and incense burners. All or parts of 1,000 separate ceramic vases and 100 stone vessels were recovered. As Williams said, the "range of these and other fragments from the plundered cemetary began to indicate a wealth and complexity that could only be called royal."[40]

The place and date of this find contravened all accepted models for the appearance of pharaonic civilization in the Nile Valley, and since Williams' original article in 1980, a kind of disbelieving intransigence on the part of Egyptologists has marginalized the Qustul finds:

> Tombs of this size, wealth, and date in Egypt would have been immediately recognized as royal. Their extraordinarily varied contents would have been taken as evidence of a complex culture exposed to wide outside connections. But because the discovery was made in Nubia at a time and a place when kingship was thought impossible, further proof of royalty is necessary.[41]

Most Egyptologists seem to find it difficult to divest themselves of the traditional assessment of Nubia as a "second-rate" Egypt.

Hieroglyphic signs on a seal from another related find at Siali, also located in Nubia just above Aswan, furnished a name for the kingdom to which these royal tombs belonged: Ta-Seti, "Land of the Bow." Inscribed scenes from the

---

[38] Williams B, "The Lost Pharaohs of Nubia," *Archaeology* (33)5, September/October 1980.
[39] Ibid., p. 14.
[40] Ibid.
[41] Ibid., p. 16.

seal and from an incense burner show Ta-Seti kings engaged in military campaigns in Upper Egypt and Libya. Moreover, some of the vases in the tomb were of Syrio-Palestinian manufacture. These pieces of evidence give a picture of a powerful regional kingdom conducting far-flung commercial, diplomatic, and military activities as distant as 1,000 miles away, even beyond Africa itself.

A sequence of tombs affiliated with 12 pharaohs comprise Cemetary L at Qustul. This complex of royal tombs led Bruce Williams to believe that the ruling dynasty of Ta-Seti stretched back 300 years prior to the 1st Dynasty. What Williams didn't categorically say, but what logically follows from the evidence, is that Ta-Seti represents the beginning of pharaonic civilization and is truly a founding dynasty, segueing directly into the unification dynasties inaugurated by Aha-Menes. The line of pre-dynastic pharaohs mentioned in certain annals must be that of Ta-Seti and therefore Ta-Seti belongs in the scheme of "proto-dynasties" that governed early Nile Valley history. Bruce Williams himself implies something of the sort in the original article on the subject: apparently, the demise of Qustul coincides with the campaign of Aha in Nubia, the first king of the Egyptian First Dynasty, who recorded the smiting of Ta-Seti."[42] Ta-Seti was therefore brought down by its successor dynasty, that of Aha-Menes, the one that unified Egypt and inaugurated the formal dynastic era. Ta-Seti can be said to represent the transition between the late pre-dynastic Gerzean cultural sequence and the formal dynastic period.

The temporal relationship of Ta-Seti to the 1st Dynasty is ascertained by the same contextual analysis that gave rise to Petrie's sequence dating. Thus, we have no absolute chronology for Ta-Seti; its dating is relative to that of the 1st Dynasty. Evidence has been adduced above that pushes the 1st Dynasty back to 3,500 B.C. Thus, if Ta-Seti is known to have appeared 300 years before the 1st Dynasty, its date of origin goes back to 3,800 B.C. With the appearance Ta-Seti, formal civilization ascends the stage of history in the Nile Valley. These additional chronological facts compress the pre-dynastic era too severely to entertain a Badarian date of 4,400 B.C.

Chronology and the Calendar

The other factor bearing on the complexities of early Nile Valley chronology is that of the calendar. Since the author has treated this question at length elsewhere, it is only necessary to devote but a little space here to this aspect of the problem.[43] It is enough to say that the earliest identifiable calendar date we have is 4,241 B.C. from Egypt's famed Sothic calendar, the

---

[42] Ibid., p. 21.

[43] Finch, op. cit., pp. 115-127.

one still in use today. The existence of such a sophisticated calendar, a seamless calendric interdigitation of the solar and Sirian movements, presupposes an equally sophisticated culture creating it. Though it is possible that a late, relatively advanced pre-dynastic culture such as the Gerzean was capable of such an astronomical feat, it seems more plausible that such a calendar was the product of a developed civilization. It is just another conundrum among many that chronologists and historians of the Nile Valley have wrestled with.

New Dates for the Sphinx: New Dates for Nile Valley Civilization?

In spite of the contradictory mass of facts introduced so far, more recent evidence yet threatens to blow *all* the carefully contrived schemes of Nile Valley chronology out of the water. This evidence comes from a wholly unexpected source that has sat literally in front of the very eyes of travellers, writers, and scholars for as long as there has been an Egypt; that is to say, the enigmatic and legendary Sphinx. One is reminded of the remonstrance of Sherlock Holmes to Watson: "You see but you do not observe." However, an independent Egyptologist named John Anthony West, acting on a clue from the writings of R.A. Schwaller de Lubicz, did see *and* observe. What is more, from what he saw and observed, West drew an obvious conclusion.

The late R.A. Schwaller de Lubicz, the aforementioned mentor of John Anthony West, was a decidedly heterodox metaphysician and antiquities scholar who boldy challenged some of the most cherished principles of Egyptology. Predictably, the academic Egyptologists have reacted to the ideas of Schwaller de Lubicz and his school with derision. Among other things, this school holds that Egyptian civilization is millenia older than textbook Egyptologists claim, and West found a startling confirmation of this conviction in and around the Sphinx.[44]

The evidence for a higher antiquity of Nile Valley civilization has been "hiding in plain sight," as it were. The Sphinx reposes in an enclosure or quarry just below and to the east of the plateau of Giza, the site of the three great pyramids of Egypt. Even a casual visitor to the Sphinx cannot fail to notice the deep and numerous fissures on the wall of the ditch surrounding it, and similar patterns of wear on its body. What could have caused such wear? To answer this question, West enlisted the aid of a team of geologists led by Robert M. Schoch of Boston University who had this to say about the erosive patterns present on the Sphinx and the surrounding enclosure wall:

> Geologists date landforms (including huge sculptured statues) by analyzing weathering features and correlating them with the region's climatic history. In the case

[44] The story of West's efforts to have his theory confirmed appears in the February 1993 issue of *Conde Nast Traveler*, in a cleverly written and illustrated article entitled "Civilization Rethought," pp. 100–5, 168–77.

> of the Sphinx, the sides of its body and the walls of its enclosure clearly exhibit a rolling, undulating vertical profile on the horizontal rock layers, which in turn are cut by vertical crevices. *These are textbook examples of rain-induced weathering.* (emphasis added)[45]

The question that Schoch had to pose once he got a good look at the Sphinx and its surrounding ditch, was how could the Sphinx and the walls of its enclosure sustain such deep, rain-induced weathering when Old Kingdom tombs immediately to the south of them and cut from the same limestone rock showed only wind-weathering (see Plate 1)? Nothing like the deep, vertical fissuring caused by coursing water could be found on any other monument, temple, or statue in the vicinity of the Sphinx. The only reasonable conclusion—the one West arrived at 20 years ago—is that the Sphinx was built in an entirely different epoch than those other well-known Old Kingdom structures.

Northeast Africa has not always been subjected to desert-like conditions. From 10,000 B.C. to 3,000 B.C., the Nile Valley experienced the "Nabtian Pluvial," a period of relatively heavy rainfall, before succumbing to an extremely arid climate with very sparse precipitation.[46] It is only during this pluvial period that sufficient rain could have fallen on the quarry where the Sphinx sits to have caused the deep fissuring present on the surrounding quarry walls and on its body. Since this pluvial period effectively ceased by 3,000 B.C., the Sphinx must have been built prior to that time, well before the conventional date of 2,500 B.C. usually given for its construction.

Schoch and his colleague Thomas Dobecki, a seismologist from Houston, accumulated data from their study that helped them arrive at a more definite age range for the Sphinx. They found that their seismic readings showed subsurface water penetration in the bedrock of the Sphinx to be six to eight feet deep in the front half of the Sphinx and four feet deep in the rear.[47] This differential could only be explained by presuming that the Sphinx had been carved in stages; the head and forepart first and the hindquarters last.

It is known from the work of Lehner and Gauri that the Sphinx had been renovated and repaired at least three times in its history, and West and Schoch convincingly argue that the first repair occurred during the Old Kingdom.[48] The masonry and construction technique evident on the rump of the Sphinx undoubtedly derive from Old Kingdom building methods, and an inscription on the rump, datable to the Old Kingdom, confirms this assessment.[49] There is every reason to believe that these Old Kingdom repairs and the carving of

---

[45] Schoch RM, "One Geologist's Review," *Conde Nast Traveler*, February 1993, p. 103.
[46] Cited in Schoch RM, "Redating the Great Sphinx of Giza," *KMT* (3)2, Summer 1992, p. 55.
[47] Ibid., pp. 56–7
[48] Cited by West, op. cit., pp. 170–1.
[49] Schoch, "Redating the Great Sphinx . . . ", op. cit., p. 57.

the hinderpart of the Sphinx were undertaken by Khafre (a.k.a. Cephren), which accounts for the tradition linking Khafre with the Sphinx. If this first repair, along with the sculpting of the hinderpart, occurred during Khafre's reign, conventionally given as 2,500 B.C., Schoch points out that it has taken 4,500 years—from Khafre's time until now—for water to penetrate to a depth of four feet into the bed-rock of the Sphinx's rear.[50]

Concerning the bedrock under the forepart of the Sphinx, it is known that the pace of water penetration slows the deeper it seeps into the bed-rock. Allowing for the slowing of water penetration to the profounder depths of six to eight feet, Schoch has proposed dates between 5,000 and 7,000 B.C. for the initial construction of the Sphinx.[51] He considers these estimates conservative.[52] It must be borne in mind that the bed-rock upon which the Sphinx sits was freed from overlying rock by human hands. Thus, any water penetration measured from the surface of this bedrock *must* date from the time of human construction, not before.

According to Schoch, the original Sphinx was built so that the head and forepart jutted outward from the limestone cliff. The rump did not yet exist and the effect was not dissimilar to the one created by the presidential busts at Mount Rushmore. The rump was sculpted later, and in the process, a rectangular ditch was completed around the body as the limestone was removed from the quarry.

Expectedly, conventional academicians dismiss this work out of hand. There have been numerous knee-jerk academic rejections unsupported by any counter-evidence but a few have endeavored to mount serious critiques of the work of the West-Schoch team. One line of argument contends, correctly, that there have been recurring, though infrequent, rainfalls in the Nile Valley since the end of the Nabtian pluvial period 5,000 years ago. These downpours, uncommon though they are, would be sufficient to account for the water erosion on the Sphinx. Cited as an example of the effect of these intermittent downpours in Egypt is a mud-brick wall recently washed away by rain.[53] This argument is easily met. Mud-brick is the most ubiquitous building agent in Egypt precisely because rainfall is so infrequent. If rain were more frequent in Egypt than it is, it would be impossible to use mud-brick because it so readily dissipates when exposed to water. Thus, the washing away of a mud-brick structure by one of the occasional downpours that occur in Egypt cannot be compared to the deep and persistent fissuring in the limestone from which the Sphinx was constructed, caused by frequently recurring rainstorms during a more remote period.

---

[50] Cited in by West, op. cit., p. 176.

[51] Schoch, "Redating the Great Sphinx . . . ," op. cit., p. 57.

[52] Ibid., p. 58.

[53] See "On the other hand . . . ", *Conde Nast Traveler*, February 1993, p. 105, for a synopsis of the counter-arguments put forward against the conclusions of the West-Schoch team.

Another argument brought forth to confute the West-Schoch demonstration involves discussion of the deterioration of the limestone due to condensation of moisture and precipitation of salt.[54] This line of reasoning fails to account for the deep vertical fissuring evident in the enclosure walls and on the body of the Sphinx. Nor does it explain the differential water penetration into different parts of the Sphinx's bed-rock. Furthermore, why isn't the same pattern of water erosion exhibited on any of the other Old Kingdom structures near the Sphinx that were built from the same formation of limestone?

Finally, the weathering patterns on the limestone body of the Sphinx are attributed to 50 million years of deterioration of the native limestone by the effects of the underground water table, and accounts for the need, in the time of Khafre, for the sculptor to encase the body of the Sphinx in granite blocks.[55] Zahi Hawass, the author of this theory, insists that other unweathered limestone edifices on Giza were not constructed from stone quarried from the immediate vicinity but from nearby areas at a higher elevation. This assertion flies in the face of Schoch's evidence which shows that the Sphinx and its neighboring Old Kingdom structures were constructed from blocks from the same limestone formation. Moreover, if the limestone in the quarry where the Sphinx was built had deteriorated so badly from the effects of the water table, why use it at all; why not use undamaged limestone so widely available from nearby formations? Moreover, the water table explanation still does not account for the deep external fissures on the walls of the ditch and on the Sphinx body.

Taken as a whole, the objections to this elegantly simple geological study seem facile and obtuse. They do not falsify the thesis. What is unusual about this particular debate, though, is that its issues are intelligible to the lay public:

> It's not often that a scientific breakthrough is easily understood by nonscientists, still rarer that it makes good spectator sport. The quest to unriddle the Sphinx is an exception. Most of the evidence involves no microscopes, telescopes, or particle accelerators. All that's needed is a naked eye and the ability to follow an argument. And most of the action takes place not in the lab, but in Egypt, in public view.[56]

The Sphinx is one of the architectural and artistic wonders of the world. Sculpting such a marvelous piece of statuary could only have been realized in a civilization equal to the task. There was nothing "prehistoric" or "primitive" about the Sphinx or the culture that created it. The Sphinx thus seems to represent *prima facie* evidence of the existence of a full-fledged, flourishing Nile Valley civilization no later than *7,000 years ago*! Quite possibly

---

[54] Ibid.
[55] Ibid.
[56] West, op. cit., p. 177.

it is as old as *9,000 years*. If the evidence holds up, the entire edifice of antiquarian scholarship will be faced with the prospect of revising all historical chronology and reinterpreting the emergence of civilization. Everything would have to be re-thought: the pre-dynastic sequences, the chronology of unification, the date of Ta-Seti, and other ancient facts too numerous to mention. Much of the terminology, such as "paleolithic" and "neolithic," would have to be re-defined or discarded. No wonder no self-respecting academician wants to touch it. West put it this way: "The theory, if correct, upsets the entire applecart of ancient history, and I could hardly expect enthusiasm from people who make a living selling apples."[57]

Conclusion

Where do all these developments leave us? In spite of 100 years of consistent work, pre-dynastic Egyptian archaeology remains a hodgepodge of often contradictory facts and interpretations. What we can say is that the received chronology of the Nile Valley, to which academic Egyptology tenaciously clings, is hopelessly obsolete. But then it never did make much sense.

New evidence is gradually bringing into focus a northeast African civilization existing farther back in antiquity than had ever been imagined. It may be plausibly presumed that this high culture evolved out of a confluence of migrations and influences coming down the Nile from the Great Lakes merging with those moving eastward out of the Sahara. But the Saharan horizon itself is ultimately traceable to the East African Great Lakes, exemplified by the Ishango-Katandan complex. That region must be considered the cradle-land of all African culture, especially that derivable from the Nilo-Saharan complex.

John Anthony West is convinced that there were one or more Nile Valley civilizations preceding that of dynastic Egypt by thousands of years. This is entirely consistent with Manetho's account of Egyptian history, recorded more than 2,200 years ago.[58] Fragments of evidence show that cultures of greater technical sophistication than previously assumed might have existed in paleolithic Africa between 12,000 and 50,000 years ago.[59] We also know that fewer than 10% of the known archeological sites on the continent have even been surveyed, let alone excavated.[60] Archaeologically, Africa is still largely *terra incognita*. When the mammoth task of systematically explor-

---

[57] Ibid., p. 170.

[58] See Waddell WG, translator, *Manetho*, Cambridge: Harvard University Press, 1980, for a compilation of and commentary on the fragments of Manetho's history that record king lists going back to 5500 B.C. and legendary dynasties going back to almost 37,000 B.C.

[59] See Finch CS, *Africa and the Birth of Science and Technology* (pamphlet), Decatur, GA: Khenti, Inc., 1993, pp. 2–5.

[60] At last report, there are supposed to be one million mounds in the northern Sudan that have never been touched by the spade! The Meroitic city of Naga is said to be so large that it may take 70 years to excavate.

ing the archaeology of old Africa is at last undertaken, West's suppositions concerning "neolithic" and "paleolithic" African civilizations may come to seem less and less farfetched.

Glossary

Horizon: An archaeological cultural complex.

Lacustrine: Of or pertaining to lakes.

Mesolithic: A word meaning "Middle Stone Age," not strictly applicable to Africa since it refers primarily to a European cultural stage between the Paleolithic and Neolithic. The time span of the European Mesolithic extends from about 7,000 to 2,700 B.C.

Neolithic: Means "New Stone Age" and generally refers to a cultural stage in Africa and the Near East existing between 8,000 B.C. (or earlier) and 4,500 B.C. Neolithic cultures were characterized by small farming villages and pottery, though little or no utilization of metals. Europe entered the Neolithic period later than Africa or the Near East.

Paleolithic: This term refers to the "Old Stone Age," encompassing the period between 2.5 million years ago and 10,000 B.C. The Paleolithic was divided into the three subperiods, "Lower," "Middle," and "Upper." The Upper Paleolithic was the final phase of the Paleolithic, beginning about 40,000 B.C. and ending 10,000 B.C. Paleolithic peoples were pre-agricultural hunters and gatherers.

Pluvial: Extended era of moderate to heavy rainfall.

Radiocarbon Dating: Method of determining the age of nonliving organic materials by measuring the amount of decay of carbon isotopes in the sample.

Appendix

In the world in which we live, discoveries cascade upon one another so fast that the most up-to-date findings are often rendered obsolete within months, and sometimes weeks. After this manuscript was completed, the author came across a news item dated December 16, 1995 in the "Science Watch" section (Section G) of the *Atlanta Journal and Constitution* that appears to overshadow all the recent work by Alison Brooks and John Yellen reported on in this chapter. To wit:

> . . . archaeologists from the University of Connecticut working near Lake Baringo, Kenya, say they have discovered long thin stone blades—hand-made tools that show signs of repeated sharpening—buried beneath a layer of volcanic ash *at least 240,000 years ago*.
>
> And several other African . . . sites have recently yielded evidence of stone spear points, as well as pigments and rhinestones that may have been used in symbolic rituals that date to more than 130,000 years ago—well before the first human cultural artifacts have been documented in Europe. (emphasis added)

The ramifications of these discoveries are immense. Among other things, they suggest that the range of dates put forward by Rebecca Cann in 1987 of 140,000 to 290,000 B.P. for the first appearance of *Homo sapiens sapiens* in East Africa now have to be weighted toward the upper end of the range. That is to say that, anatomically modern humans, who came into being in East Africa, are closer to 300,000 years old than the 200,000 years of age advocated by Cann. Cann selected the 200,000-year date as a compromise between the range of 290,000 and 140,000 years extrapolated from her mitochondrial DNA data. The recent findings also suggest that complex human culture begins to evolve in Africa at or very soon after the time of the appearance of *Homo sapiens sapiens* as a species.

2

The Followers of Horus: 40,000 Years of African Mining and Metal Technology

> *Tradition generally asserts that the god Horus of Behudet and his . . . followers, who are described as mesniu or mesenti, i.e., "metal workers," and who are to be identified with the Shemsu Heru or "followers of Horus," . . . came from the South and not the North.*
>
> —E.A. Wallis Budge, *A History of Egypt*, vol. 1, p. 44

> *It should not surprise us that large numbers of [smiths] become healers and sorcerors and that the entire profession is held to have a natural affinity toward manipulating nature's forces.*
>
> —Patrick R. McNaughton, *The Mande Blacksmiths*, p. 41

As early as 1930, Raymond Dart, the discoverer of the hominid genus *Australopithecus,* had, with a team of Italian archaeologists, found evidence in southern Africa of remarkably advanced mining techniques.[1] In Zambia, they had discovered a manganese mine at Chowa in-filled with thousands of ancient manganese tools—choppers, wedges, chisels, hammers, axes, and grindstones. When it came time to date this find, it was hard to credit the results: the age of the mine was 28,000 years! This was fully 20,000 years older than the oldest flint mining operations found anywhere else. Why mine manganese at all? Manganese can be fashioned into a rudimentary tool but in Africa, the blackish manganese ore seems to have been coveted most as a cosmetic dye.[2]

In 1964, an iron mining company digging for ore at Ngwenya, Swaziland, discovered a large cache of ancient mining tools in a pit that was clearly manmade.[3] Dart, once again, was summoned to study these finds and on this occasion was joined by two other archaeologists, Adrian Boshier and Peter Beaumont. Tunnels and adits were discovered that clearly indicated systematic mining of hematite ore, the source of metallic iron. Boshier took charge of the extended digs that eventually yielded 300,000 artifacts at the site

---

[1] Mutunhu T, "Africa: The Birthplace of Iron Mining."

[2] Dart RA and Beaumont P, "Amazing Antiquity of Mining in Southern Africa," *Nature* (216), October 28, 1967, pp. 407–8.

[3] Ibid.

known as Lion Peak; the mining cavern that had been bored into the slope of the cliff took the name Lion Cavern. Charcoal nodules from the middle to lower levels of the excavated shaft yielded a date of 28,000 B.P. by carbon dating. Dart was of the opinion that the mining works went back even farther in time:

> The worked floor must extend for at least another10 ft. beyond the outside edge of this second excavation. Its outer part, as yet unexcavated, may date back another few thousand years. Evidence of an earlier culture may also be located there; there is an enormous *Later Acheulian site, estimated to contain ten million artefacts barely a mile away.*[4] (emphasis added)

Adrian Boshier was able to obtain additional carbon datings from charcoal found at the site after more extensive excavations and, confirming Dart's suppositions, these gave a date of 43,200 B.P. (±1,600 years)![5] These finds demonstrated beyond all doubt that systematic mining was practiced in Swaziland more than 40,000 years ago. It seems evident that the entire region—encompassing Zambia, Zimbabwe, and Swaziland—had been the site of extensive mining operations repeatedly since the Upper Paleolithic. The archaeologists found evidence of mining at the same locations that persisted to relatively recent times. Castle Cavern, also in Swaziland, yielded numerous mining tools in addition to fragments of iron itself, all dated to before 400 A.D.[6] Thus, mining operations had been carried out in these regions for 40,000 continuous years. The reference to the "Later Acheulian" site in the above quotation with its 10 million artifacts—if they are from similar mining sites—may suggest that this archaic mining activity is older than 70,000 years.

What were these paleolithic African miners digging for? In Zambia, as we have seen, it was manganese. In Swaziland, it was a form of hematite, an iron-containing ore, called specularite. Specularite comes mostly in two distinct colors, black and red, both of which have been used as body dyes and cosmetics. Red specularite has been used in funerary rituals and also in traditional medicines. The accepted explanation is that these early African miners were mining hematite for its cosmetic properties. At Lion Cavern alone, more than 50 tons of hematite was mined over the millenia.

But is this all? What is potentially disconcerting to the "acccepted" explanation is the presence of charcoal at the most ancient sites. Apart from the occasional forest fire started by lightning, charcoal is not naturally occurring; it is produced from wood that has been partially and slowly burned. Until

---

[4] Ibid., p. 408.

[5] Ibid.

[6] Boshier A and Beaumont P, "Mining in Southern Africa and the Emergence of Modern Man," *Optima*, March 1972, p. 6.

the utilization of coke in the last century, charcoal was *the* fuel of choice the world over to separate iron from hematite.[7] In fact, iron-smelting and steel-making in Tanzania and Kenya, dating from 2,600 years ago, reveal a highly sophisticated charcoal-based process.[8] We might rhetorically ask, if charcoal from hematite mining sites dating back 28,000–40,000 years was *not* used to smelt iron from hematite, what *was* it used for?

We can imagine the enormous investments in labor required to sink a mine shaft, let alone excavate tons of ore. Clearly hematite of both the black and red variety was considered extraordinarily valuable. While cosmetic manufacture was undoubtedly one rationale for such herculean effort over tens of millenia, the extraction of ore for smelting purposes to make useful iron implements would make the whole enterprise that much more plausible. But our suggestion flies too much in the face of the accepted chronology for the birth of metallurgy; it thus demands sterner proof, or at least reasonable answers to the problems raised. Why, for example, don't we find iron or steel remnants from the Upper Paleolithic era? One answer is that iron is not stable like rock or flint; exposed to the elements, it rapidly rusts away. Moreover, iron and the implements made from it—if there were any—must have in this remote period been rather rare and valuable, compared to stone tools. This scarcity, therefore, combined with the rapidity of oxidation, might explain the absence of iron at archaeological levels of such antiquity.

Iron might well have had the same, or greater, intrinisic value for the paleolithic African as gold does for modern civilization. As we know, men will go to extraordinary lengths to mine gold even though its practical value—apart from its ornamental uses—is relatively limited. Iron is a much more useful metal and, at a time when such utility would have been at a premium, may have stimulated the ancient people of southeast Africa to mine it with as much fervor as we might mine for gold. Even if, as some suggest, archaic iron could only have been used for ornamental purposes it would still, at that time, have been considered as valuable as gold is to us.

We have admittedly speculated to the limit of plausibility and therefore can do little now but wait for further evidence. But it is already clear from these unbelievably ancient mining sites that the protohistory of technology begins in Africa.

Copper and Bronze Metallurgy from the Time of the Pharaohs

Copper was one of the oldest, if not *the* oldest, metal known and used by man. In the Nile Valley, its provenance dates back to Badarian times; that

---

[7] See "Industries, Extraction, and Processing," in the *New Encylopedia Brittanica*, vol. 9, Chicago: Encyclopedia Brittanica, Inc., 1992, p. 447.

[8] Van Noten F and Raymaekers J, "Early Iron Smelting in Central Africa," *Scientific American*, June 1988, pp. 104–111.

is, 10,000 B.C. by Petrie's reckoning or 5,500 B.C. by conventional reckoning.[9] A diverse array of tools, utensils, and weapons made of copper have been recovered from the entire pre-dynastic period. In the early pre-dynastic phase (Badarian), copper beads, borers, pins, bracelets, chisels, rings, harpoon heads, needles, tweezers, and other small implements and articles are attested (see Fig. 7). In the later Amratian-Gerzean period, copper weapons had come into use and by early dynastic times, complete mastery of copper-working had been achieved, as evidenced by the extensive manufacture and utilization of axe-heads, adzes, knives, daggers, spears, ewers, basins, ornaments of various kinds, and other sorts of household implements.

As Flinders Petrie asserted, there was a broadly unified pre-dynastic culture—evident even during its phase transitions from Badarian to Gerzean—that incorporated not only geographic Egypt but also the northern Sudan, that is to say, Nubia. The early utilization of copper in Nubia is attested in pre-dynastic and early dynastic times by the presence of copper axes found in "A-group" and "C-group" graves.[10] Thus, early copper-working in Nubia seems to have paralleled that of the cultures further north.

Extensive copper mines were located in two places in Nile Valley antiquity: Sinai and the eastern desert of Nubian Egypt. Though most of the mining activity was concentrated at Sinai, the richer ores were actually to be found in the Nubian desert. Mining operations appear to have been extremely efficient; the entire amount of ore mined within Nile Valley frontiers during the dynastic period seems to have nearly equalled the rest of the world's *entire output* up to 1800 A.D.[11] However, after about 1300 B.C., the bulk of copper used in Egypt seems to have been imported from Western Asia.[12]

The first true copper smelting furnaces found in Egypt were in the Sinai. The furnace consisted of a hole $2^1/_2$ feet deep surrounded by a stone wall with two draft holes cut into it. Additional copper smelters have been found at Buhen in northern Nubia (2nd cataract) that date from the end of the Old Kingdom. In the typical method, tuyeres (blowpipes) were inserted through these holes and, through attached bellows, air was blown in to fan the smelting fire to sufficiently high temperatures. Copper melts at 1,083° Celsius (C), though it is possible to smelt copper from its ores at about 800°C (see Fig. 8). Either of these temperatures was readily obtainable by ancient methods. Hardening of the smelted copper was achieved by either hot- or cold-hammering. However, metallic copper, particularly after hardening by hammering, can be brittle. One method of making the copper more malleable with-

[9] See Chapter 1.

[10] Adams WY, *Nubia: Corridor to Africa*, Princeton, N.J.: Princeton University Press, 1977, pp. 124; 169.

[11] Lucas A, *Ancient Egyptian Materials and Industries*, London: Histories and Mysteries of Man Ltd., 1989, pp. 208–9.

[12] Ibid., p. 209.

Fig. 7. Copper rings and bracelets from early to mid–pre-dynastic period, 7,500-6,500 B.C. (Petrie)

out sacrificing hardness is by the process of annealing. Annealing was a copper-working technique already known in the pre-dynastic era and occurs when the smelted metal is reheated to a certain temperature (400–800° C), maintained at that level of heat for a lapse of time, then allowed to cool to ambient temperature. This process preserves the hardness of the metal while imbuing it with greater ductility. A metal is most serviceable when it combines qualities of hardness and flexibility; annealing, sometimes repeated several times, could impose these properties on metallic copper.

One finds in ancient Egypt a decided preponderance of copper laced with traces of arsenic. It is not always possible to tell whether arsenic in copper is present as a natural impurity or whether it has been added deliberately in the smelting process. Either way, the presence of arsenic in copper, up to 15%, improves its hardness considerably. Naturally occurring coppers can show arsenic concentrations of up to 8% and it is assumed that it is these natural arsenical coppers that account for the abundant copper-arsenic alloys found in Egypt. Only William Flinders Petrie advanced the opinion that the ancient Egyptians deliberately added arsenic as an alloy during smelting.[13] What is clear is that the Egyptians knew that arsenical copper ores yielded greater hardness and selectively mined them for that reason.

Casting in copper was practiced in ancient Egypt in the pre-dynastic epoch and by dynastic times the technique had been refined to perfection. The earliest example of copper-casting in the Nile Valley is an axe-head that showed evidence of hot- and cold-hammering and annealing. Copper as a

[13] Petrie WMF, *Arts and Crafts of Ancient Egypt*, London: T.N. Foulis, 1910, p. 100.

Fig. 8. Egyptian crucible showing smelting technique by the forced draft method.

useful metal was also abundant at Kerma, the site representing the country the ancient Egyptians called *Kesh*, i.e., Cush (c. 2,000 B.C.). In the Old Kingdom, the Egyptians were creating beautiful copper statues of some size, undoubtedly by the *cire perdue*, or "lost wax," process. Briefly, this technique involves sculpting the desired object in wax, creating a clay mold around this wax object, then heating the wax so that it melts and drains out of the mold. The metal is then poured into the mold, allowed to harden, then liberated from the mold by careful chipping away. It is worth noting that fashioning of copper objects of any complexity is a much more difficult and delicate process than with bronze. Bronze is a much easier metallic casting agent than copper because it liquefies more readily and does not contract upon cooling the way copper does, rendering the latter porous and brittle. Thus, the casting of copper statues requires a more refined technique than does casting bronze statues.

Adding tin to copper up to 4–15% to make bronze improves the strength and hardness of the metal immeasurably. The oldest bronze objects are said to have been found in Sumer (Mesopotamia) as early as 3,200 B.C. whereas the earliest bronze objects in Egypt are said to date from the beginning of the 4th Dynasty around 2,700 B.C. But since Nile Valley chronological revisions are pushing the dates for the Old Kingdom upward by 400 years at least, it is not too much to infer that these Old Kingdom bronze finds date to before 3,100 B.C., making them more or less contemporaneous with the Sumer finds.[14]

The clear superiority of bronze to copper for tools and weapons meant that it rapidly replaced copper as the metal of choice for such articles throughout the civilized world. In fact, the use of bronze became so ubiquitous that it has given its name to an entire age spanning 3,100–1,000 B.C. It

---

[14] See Chapter 1. In a recently (1994) published book entitled *The Orion Mystery* by Robert Bauval and Adrian Gilbert, evidence is cited on page 235 that proves the existence of Egyptian bronze artifacts as early as the 2nd Dynasty. This takes the Bronze Age in the Nile Valley back to *before* 3,200 B.C and therefore to *before* the Sumerian Bronze Age. Depending on the chronology relied on, the Bronze Age in the Nile Valley begins around 3,400 B.C. or 3,800 B.C.

was also superior to iron as a tool-and-weapon metal, and iron could never replace bronze in this regard until the invention of steel around 1,200 B.C. Moreover, as we have seen above, bronze was far easier to cast than copper, another reason for supplanting the latter for ornamental as well as functional purposes.

As already indicated, bronze articles were being fashioned in Egypt from the early Old Kingdom. However, during the entire span of the Old Kingdom, fewer than a dozen bronze objects have been recovered; bronze does not become the metal of choice for tools until the Middle Kingdom, around 2,000 B.C. It is in fact of no little interest that even after the discovery of metals, Egyptians continued to rely on stone tools for a considerable period. For sacrificial purposes, stone knives were the only ones permitted down to the end of Egyptian history.

Beginning with the 12th Dynasty, the use of bronze for all functional and ornamental needs became general. After smelting, bronze was worked by either hammering or casting, this latter achieved by the *cire perdue* process already mentioned. The ancient Egyptians were a people who seemed to excel at virtually anything they undertook, and their bronze statuary remains among the world's finest. The best surviving examples date after the 18th Dynasty.

Bronze is evident in Nubia among the so-called "C-group" cultures beginning around 2,500 B.C. and very likely earlier. Many of the copper and bronze artifacts found in the graves of this culture seem to have been imported from Egypt but there is no reason to doubt that these people had developed their own copper-and-bronze metallurgy. By the time of Kerma (c. 2,000 B.C.), Nubia (Cush) was well into the Bronze Age, in tandem with its neighbor to the north; indeed the inhabitants of Kerma produced bronze implements of exceptionally fine quality.[15] This trend continued throughout Nubio-Cushitic history. The Cushitic 25th Dynasty, both in Egypt and in Napatan Cush, maintained bronze working at a high level (see Plate 2).[16] By the time of Meroe's ascendancy (300 B.C. to 300 A.D.), bronze was so ubiquitous as to be commonplace. Indeed, though Meroe is famous for its iron-works, more bronze implements have been recovered from Meroitic sites than iron or steel. Thus, the Nile Valley cultures never really relinquished their attachment to bronze, even after steel, a stronger metal, became generally available for tools and weapons.

Bronze made its presence felt in another African civilization, situated near the headwaters of the Blue Nile in Ethiopia: Axum. Axum made its appearance before 100 A.D. and continued as an independent civilization until after 800 A.D. Indeed, Axumite expansion led to the final downfall of Meroe

[15] Haynes JL, *Nubia: Ancient Kingdoms of Africa*, Boston: Museum of Fine Arts, 1992, p. 41.
[16] Ibid., p. 28.

a little after 300 A.D. What is interesting about Axum is that bronze was used, along with gold and silver, for coinage. Coins—currency in the modern sense—are entirely absent from the entire 4,300-year span of Nile Valley civilization as a medium of exchange. But the kingdom of Axum, drawn into an international network of trade that used specie more and more, struck its own coins after 270 A.D. The act of minting one's own coins was of profound political significance since only major empires and kingdoms like Rome and Persia were able to do so.[17] A national coinage was a statement of international political parity with the great states and kingdoms of the time. The age-old skill of Nile Valley civilizations in bronze-casting seems to have been continued at Axum. In addition to coinage, the Axumites were fond of casting statues in bronze, some more than 15 feet in height![18] However, the Axumites, unlike Meroe whom they conquered, seemed to have fully entered the Steel Age since virtually all weapons were of steel. The use of bronze was limited to art and ornamental objects, which, however were quite numerous and, as seen above, expertly casted.

The next great bronze-working culture-complex in Africa was situated in what is now Nigeria, consisting of the Igbo-Ukwu of southeastern Nigeria, the Ife culture of southwestern Nigeria, and finally the Benin culture to the north of Ife. Though the various objects casted by these peoples have been lumped under the term "bronzes," in reality the metals actually employed consisted of a number of different copper alloys.[19]

The first in the region to develop casting to a high degree were the people of Igbo-Ukwu whose bronze sculptures have been dated as far back as 800 A.D. or a little after. The alloy used in their metallurgy was actually a combination of copper, tin, and lead and most commentators assume that these various metals were imported, though no one knows from where. The bronze objects so far recovered are all artistic pieces, all casted by the well-known *cire perdue* process. Weapons were undoubtedly manufactured from native steel. But Igbo-Ukwu bronze art is utterly unique in Africa; its styles seem to have had no precedent on the continent nor did they survive the decline of Igbo-Ukwu itself. Numerous ingeniously shaped vessels feature elaborate surface decorations, complete with spirals, knobbed encrustations, and small insects and other animals grafted onto the object. Several of these vessels are shell-shaped and many of the objects convey artistically the impression of having lain at the bottom of the sea for eons of time collecting barnacles, corals, and other assorted marine life. This was an effect deliberately created by the caster and the fine, intricate detail embossed on all these pieces show a rare skill.[20]

---

[17] Kobishchanov YM, *Axum*, University Park: Pennsylvania State University Press, 1979, pp. 59, 184.

[18] Ibid., p. 137.

[19] Ekpo E and Willett F, *Treasure of Ancient Nigeria*, New York: Alfred A. Knopf, 1982, p. 9.

[20] Ibid., pp. 19–30.

Ife bronze art in southwestern Nigeria is that which is the most "pleasing" to the Western eye because, apart from their ancient counterparts in the Nile Valley, Ife bronze sculptors were almost alone in Africa for their realistic and naturalistic portraiture (see Plate 4). Such was the genius of the Ife metallurgical artists that nothing in the Renaissance surpassed their technique. Ife culture itself can be traced back to before the 9th century A.D., but their famous busts were wrought in the period between 1,000 and 1,400 A.D.

Ife "bronzes" are not really bronzes at all but copper alloys of zinc and lead. What is of interest is that the Ife artists worked with equal facility in pure copper. As seen above, copper is not an ideal metallic medium for casting because it oxidizes quickly and does not flow well.[21] Nonetheless, some of the finest Ife heads were casted in pure copper and Willett has described the probable method by which the Ife smiths casted highly artistic heads by sealing the mold onto the crucible so that there was no contact with air (see Plate 3).[22] The metallurgical art of Ife was truly royal because all of the recovered objects depict royal personages and themes. Apparently, the Oni (ruler) of Ife and his retinue exercized a royal monopoly over all the work done by casting with copper alloys.

This royal monopoly of metal casting was maintained by the Oba of Benin, the state that more or less succeeded Ife in political pre-eminence in western Nigeria. Benin became the dominant state in its region in the 1400s and maintain its position until the 19th century. Benin traditions clearly acknowledge learning the bronze casting technique from an Ife smith and, subsequently, Benin did more than any other kingdom to organize and control the large-scale production of bronze artistic products.[23] By far, the most numerous of the Nigerian busts that have been recovered are from Benin. What is more, the Benin heads are truly of bronze, i.e., of a copper-tin alloy. While Benin bronze-smiths achieved an exceptional level of skill of their own, their artistic canons were distinctly different from Ife. Their figures are rather more expressionistic and less naturalistic than those of Ife. Moreover, they seem to have dealt with somewhat more diverse themes since bronze animal figures are also well in evidence among the recovered objects.

The Iron and Steel Age in African Antiquity

All African peoples have extensive myths and traditions concerning smiths, that is, metal-smelters and toolmakers, particularly those who work iron.[24]

---

[21] Ibid., p. 33.

[22] Ibid.

[23] Ibid., pp. 42–3.

[24] See Griaule M and Dieterlen G, *Pale Fox*, trans. by Stephen C. Infantino, Chino, Arizona: Continuum Foundation, 1986, for a detailed explanation of the cosmo-mythic significance of the smith in an African world system. See also McNaughton PR, *The Mande Blacksmiths*, Bloomington, Indiana: Indiana University Press, 1988.

Iron is the most ubiquitous metal in Africa and assumed such a large place in the technical life of the continent that tremendous lore, taboos, myth, and magic came to surround the smith who worked it. The ability to extract the ore from hematite and turn it into steel was a secret, magical process and the smith at all times and places consequently was (and is) feared as a sorcerer. His position in the traditional caste systems varied with the African locale: in the Western Sudan, he was relatively "low-casted"; in the southern and eastern regions of the continent his status tended to be significantly higher.[25] Whatever his social position, the smith everywhere in Africa was surrounded by numerous taboos and was always situated somewhat outside the pale of normal society. In the Western Sudan, for example, he was not permitted to marry outside his caste. In sum, there was a clear "approach-avoidance" pattern in the smith's social position: he was respected but also feared, he was indispensible but kept at arm's length, he was often reviled but wielded tremendous social and political leverage.[26]

The role of the smith in Egypt followed the African pattern, certainly in the archaic period (c. 5,000–4,000 B.C.). One of the most interesting and significant of the traditions surrounding Horus is that at a remote, pre-historic time he and a group of blacksmiths known as *Heru-shemsu*, i.e., "the followers of Horus," came out of the south into Egypt to establish his earthly reign.[27] Thus, to the dynastic Egyptians, smithing, that is to say iron-working, was a technique that was known in pre-dynastic times and was of such importance that it was associated with Horus himself. Since every pharaoh was considered the human manifestation of Horus, it seems evident that the archaic pharaohs must themselves have been smiths or connected to smithing.

We do indeed find evidence of iron-working in pre-dynastic times in ancient Egypt. Iron can be extracted from two sources: (1) meteors that have landed on earth from the sky and (2) naturally occurring hematite ores, known as "stratum" or "terrestial" iron. Meteoric iron nearly always contains nickel in varying proportions, averaging 7.5% in content, whereas in stratum iron, nickel is nonexistent or present only in minute quantities. In fact, the way of differentiating meteoric iron from stratum iron is by determining

---

[25] For the relation of smithing to African royalty (and nobility), see Balandier G, *Daily Life in the Kingdom of the Kongo*, New York: Pantheon Books, 1968, pp. 107–10; for a brief discussion of the socially ambivalent position of the smith in African society, see Turnbull CM, *Tradition and Change in African Tribal Life*, New York: Avon Books, 1966, pp. 79–80; also McNaughton PR, op. cit., passim.

[26] McNaughton's book on the Mande smiths is a superb account of the power of the smith in traditional Sudanic society. He rightly interprets the opprobrium that is often heaped on the smith, even if only in a *pro forma* way, as a method of limiting and containing that power. Every one in a traditional Western Sudanese society keenly understands that the very life of their culture would be impossible without the smith.

[27] From the Edfu Text cited by Jackson JG, *Introduction to African Civilization*, New York: University Books, 1970, p. 93. This Horus-connection of the proto-historic Nile Valley blacksmiths echoes the royal connection of the Kongo smiths (see note 25).

the presence or absence of nickel. The pre-dynastic iron samples seem to have derived from meteoric iron:

> The specimens of early iron reported to have been found in Egypt may now be described. The earliest are two lots of small tubular beads (one lot of seven and one lot of two) of predynastic date found by Wainwright at Gerzeh. . . . Professor Gowland, who analyzed them, stated that originally they had been metallic iron and had been made by bending into tubular shape a thin strip of metal. These beads have since been analyzed by Professor Desch and found to contain 7.5 per cent of nickel, thus proving that the iron of which they were made was of meteoric origin. Also perhaps of pre-dynastic date was a ring found at Armant and subsequently lost in the post before any detailed examination could be made.[28]

Lucas seems to be of the opinion that because these iron artifacts were derived from meteoric iron instead of hematite, the find is of scant metallurgical significance because a less sophisticated extraction technique is required. But there is another slant on this question: these iron artifacts from a period of Egyptian history stretching between 6,000 and 12,000 years ago, show that ancient Nile Valley peoples had learned how to use iron; that is, they appreciated its metallic properties and had begun to shape and use them for their needs. That the source was meteoric iron does not detract from this accomplishment; these people had begun the process of extracting iron and turning it to practical use.

These pre-dynastic finds also indicate that the traditions concerning the proto-historic Followers of Horus who were the blacksmiths, were not altogether "mythical." We might legitimately pose the following question (without necessarily being able to answer it): What was the relation of pre-dynastic Nile Valley iron-working to late paleolithic iron-ore mining in eastern and southern Africa? This is particularly pertinent because the legendary Blacksmiths who were the Followers of Horus came into Egypt from the south.

Chronologically, the next critical find is a piece of iron lodged in the inner joints of the masonry on the south face of the Great Pyramid near the air channel found by J.R. Hill, a member of the Howard Vyse expedition in 1837.[29] This iron piece revealed no trace of nickel, making it nonmeteoric. Thus, it derives from stratum iron that had been mined, smelted, and fashioned into a usable tool. Because this iron tool has generated immense controversy it is important to note, as Cheikh Anta Diop did in his article entitled "Iron in the Ancient Egyptian Empire,"[30] that the discoverers went to a great deal of trouble *not* to remove this iron fragment until the site and its

---

[28] Lucas, op. cit., p. 237.

[29] See Diop CA, "Iron in the Ancient Empire," translated by Darryl Prevost, in *Great African Thinkers*, edited by Ivan Van Sertima, New Brunswick, N.J.: Transaction Books, 1986, p. 64.

[30] Ibid.

immediate surroundings were carefully investigated to ascertain whether it had intruded, by accident or design, into the joint at a later time. Their conclusion, after these admirable observational precautions, was that this iron tool had never been disturbed and had been in place since the construction of the Great Pyramid itself.[31]

What is the implication, then, of this find? The weight of "received" opinion on this pyramidal iron implement since its discovery is that it represents a tool that became lodged in that particular joint as the result of stone-removal activities of a later era. Some think that this intrusion occurred as late as the era of the Arabic conquest. Largely, these conclusions are based on the *a priori* assumption that significant iron-smelting only begins on a large scale with the Hittites around 1,500 B.C. In the first chapter, evidence was cited that pushes back the date for the construction of the Great Pyramid at least 400 years.[32] The radiocarbon evidence of Haas and colleagues dates the Great Pyramid to about 3,000 B.C. Thus, finding evidence of a smelted iron product extracted from and contemporaneous to the Great Pyramid completely dismantles the Hittite thesis for the beginnings of the Iron Age and places it squarely in Africa.[33]

More recent work by C.A. Diop suggests that African iron smelting at such an early date was not confined to the Nile Valley.[34] He points out that at the Nok sites in Nigeria, iron slag is present and several bits of charcoal found at the site have been dated, respectively, at 3,500 B.C., 2,000 B.C., and 900 B.C.[35] However, with understandable caution, Diop hesitated to affirm straight out that iron smelting (let alone steel-making) in West Africa could be attested back to the 4th millenium B.C., though he did point out that no other valid explanation has yet been forthcoming. Moreover, diggings near Kaolack, Senegal, reveal the presence of copper and iron tools with associated charcoal debris dated back to 2,800 B.C. by two separate laboratories.[36] Again, it is an *a priori* unwillingness to accept such early dating for the Metal Age in Africa that inhibits proper evaluation of such above-mentioned finds: "But, it is important to emphasize that from this mound a complete tool kit of iron and copper has been extracted, associated with the end of the

---

[31] Ibid., p. 65

[32] Cf. Haas H, Devine J, Wehnke R, Lehner M, et al, "Radiocarbon Chronology and the Historical Calendar in Egypt," *Chronologies in the Near East*, edited by J. Evin and F. Hours, BAR: International Series, 1987, pp. 587-806.

[33] In the book by Bauval and Gilbert, *The Orion Mystery* (p. 236), the authors refer to an analysis of the pyramidal iron plate performed by two British metallurgists in 1989. The conclusion, published in the *Journal of Historical Metallurgy*, was that the iron plate was indeed contemporaneous with the Great Pyramid, meaning that the Iron Age is 1,500 to 2,000 years older in the Nile Valley than among the Hittites, said to have begun the Iron Age.

[34] Diop CA, "L'Usage du Fer en Afrique," in *Notes Africaines*, Dakar: IFAN, No. 152, Octobre 1976, pp. 93-5.

[35] Ibid., p. 94.

[36] Ibid., p. 95.

Neolithic period of the Capsian tradition. The difficulty at the outset is that no one would dare situate this material back in such a remote period."[37] (translated from French by the author) Nonetheless, the evidence is more than suggestive and none other than Franz Boas himself—much to the chagrin of his colleagues—insisted that it is in Africa that we must look for the beginning of the Iron (and Steel) Age.[38]

K.C. Barraclough has rightly argued that what is conventionally termed the "Iron Age" should be more properly referred to as the "Steel Age."[39] It is not until the invention of carburized steel, that is, iron infused with carbon in a concentration of 0.5-1.7%, that iron becomes a metal whose virtues of strength, hardness, ductility, and durability surpass that of bronze. Prior to the invention of steel, iron was utilizable mainly as an ornament.

Archaeological fragments have helped metallurgical historians trace a somewhat sketchy outline of the development of iron-and-steel technology in the Nile Valley. Like the above-mentioned pyramidal iron tool, the date and provenance of some of these iron implements are still debated. One such example is a socketed iron spearhead from a tomb at Buhen in Nubia dating to 1,900 B.C. This find has been deemed "enigmatic" by archaeologists because of its age and because socketing was not practiced in the Nile Valley on copper or bronze tools and weapons of similar age. Typical of metallurgical historians' attitude concerning these items is that of Nikolaas J. Van Der Merwe who thinks that sporadic Old and Middle Kingdom iron implements are "accidental" by-products of copper smelting using iron oxide as a flux.[40]

After the advent of the New Kingdom around 1,580 B.C., iron working, though still uncommon compared to copper and bronze, appears more and more in evidence. An elongated, tanged iron projectile point was discovered in the middle palace of Amenhotep III (c. 1,400 B.C.).[41] A pair of iron bracelets decorated with dogs' heads, dating back to 1,350 B.C., was recovered from Akenaten's capital. A veritable cache of iron objects was recovered from Tutankhamen's tomb, dating back to 1,340 B.C.: an iron headrest, a dagger, and 16 miniature chisels.[42]

After 1,200 B.C., objects of steel manufacture make their appearance in Egypt. The earliest methods of steel smelting and working involved sequential techniques of carburization, reheating, cooling, quenching, and cold and

---

[37] Ibid. The original quotation: "Or, il importe de souligner que l'on a extrait de ce tumulus tout un outillage en fer et en cuivre associé à un Néolithique finissant de tradition capsienne. La difficulté vient du fait qu'au prémier abord on n'oserait pas faire remonter ce material à une époque aussi lointoin."

[38] Mutunhu, op. cit., p. 5.

[39] Barraclough KC, *Blister Steel, the Birth of an Industry*, London: The Metals Society, 1984, p. 19.

[40] Van Der Merwe N, "The Advent of Iron in Africa," in *The Coming of the Age of Iron*, edited by Theodore A. Wertime and James D. Muhly, New Haven: Yale University Press, 1980, p. 456.

[41] Lucas, op. cit., p. 239.

[42] Ibid.

hot hammering. Though steel-making is said to have been invented elsewhere, the articles found in Egypt reveal complete mastery of steel smithing. As we shall see, the earliest evidence of the quenching technique survives from Egypt, making that kingdom the most legitimate contender for its invention.

The earliest steel samples in the Nile Valley are two knife-blades dated to 1,200 B.C.[43] The iron of these blades was smelted into a bloom and then carburized to 0.7% carbon. Reheating of the blades to 900°C and then rapid air cooling hardened the final product. However, the earliest quench structure known to archaeologists is revealed in an Egyptian steel axe-head, dated to 900 B.C.[44] What is of interest here is the technical refinement achieved by the ancient metallurgist:

> . . . an Egyptian axe dated to around 900 BC which exhibits a carburized edge with about 0.9% carbon, fully hardened. As the section thickens away from the cutting edge, the carbon decreases as does the severity of the quench. . . . [the structure] . . . was edge carburized and then heated and quenched, but withdrawn from the quenching liquid before it was cold, so that the thicker portions were tempered by the residual heat while the edge remained hard. It is difficult to envisage such a practice as being a happy chance and it appears to indicate a sound appreciation of the art coupled with previous experience.[45]

This same axe-head is also the earliest example of the technique of "lamination," one in which fine layers of steel are fused, one over the other, to increase strength and flexibility.[46] An Egyptian chisel from 700 B.C. survives, revealing similar carburizing, reheating, and quenching techniques.

The ubiquitous William Flinders Petrie found a cache of 23 tools datable to the time of the 25th Dynasty (750-670 B.C.), when Egypt was ruled by pharaohs from Napatan Cush.[47] This tool kit consisted of punches, chisels, files, rasps, adzes, ferrules, scoop, cutters, center bits, and a crank piece. Five specimens were analyzed by Williams and Maxwell-Hyslop.[48] The tools were notable for their "remarkably good condition" and for their "surprisingly modern appearance."[49] Two of the five tools analyzed were composed of low carbon steels (0.1-0.2% carbon content) but even this low carbon level was sufficient to convey a hardness three times greater than iron.[50]

---

[43] Van Der Merwe, op. cit., p. 468.

[44] Tylecote RF, *A History of Metallurgy*, London, 1976, pp. 44-5.

[45] Barraclough, op. cit., pp. 20-1.

[46] Wheeler TS and Maddin R, "Metallurgy and Ancient Man," in *The Coming of the Age of Iron*, edited by T.A. Wertime and J.D. Muhly, New Haven: Yale University Press, 1980, pp. 118-20.

[47] Williams AR and Maxwell-Hyslop KR, "Ancient Steel from Egypt," *Journal of Archaeological Science* (3), 1976, pp. 276-305.

[48] Ibid.

[49] Ibid., p. 286.

[50] Ibid., p. 303.

Cush itself was the center of a burgeoning iron industry that reached its zenith during the ascendancy of Meroe beginning about 300 B.C. Iron objects in Cush datable to the period of the 25th dynasty come from noble rather than royal graves.[51] An iron bangle, tweezers, an arrow head, a hook, a blade, an axe-head, a socketed knife, and an adze have been extracted from noble gravesites dating back to 750 B.C. A royal grave at Nuri, dating to 689 B.C., yielded an unsocketed iron adze and a spearhead. However, major evidence of iron smelting on a large scale comes from slag heaps near Meroe dating as far back as 500 B.C.[52] The first archaeologists to study these remains were so impressed by the magnitude of the slag heaps, suggesting a prolific iron industry, that they dubbed Meroe the "Birmingham of Africa."[53] They immediately drew the conclusion that iron-working had diffused into the rest of Africa through Meroe.[54] It was a seductive conclusion but not strictly accurate for, as we shall see, there is compelling evidence of iron-smelting and steel-making in East Africa antecedent to the iron industry at Meroe.

Iron and Steel Metallurgy in Sub-Saharan Africa

The iron and steel technology beyond the Nile Valley was maintained at a high level in East Africa, at least from the 7th century B.C. The picture that is emerging from recent work in the region reveals a skill in steel-making that won the admiration of the wider world. In an area defined by the so-called "interlacustrine" region of East Africa, numerous smelters have been found going back beyond 2,500 years ago. In regards to this East African Iron Age technology, Van Noten and Raymaekers have this to say:

> As radiocarbon methods were more widely employed, however, a range of sites in Rwanda, Burundi, Uganda, the Sudan, Kenya, Zaire, and Tanzania were dated, and it became clear that in all these areas Early Iron Age technology had appeared by about the seventh century B.C.[55]

What is of interest ethnographically is that this extensive Iron Age technology seems to have preceded Bantu penetration into the southern third of the continent. Thus, no one really knows for certain the provenance of this technology: Was it local or derivative, possibly from the north? Remoter datings of iron technology revealed by Diop (see above) may suggest that Iron Age technology had been present throughout Africa long before the first millenium B.C. If the dating of African iron metallurgy precedes 1,000 B.C.,

[51] Van Der Merwe, op. cit., p. 472.
[52] Ibid., p. 473.
[53] Ibid., p. 471.
[54] Ibid., 473.
[55] Van Noten and Raymaekers, op. cit., p. 104.

as some evidence suggests, it would explain, at least partly, why Africa beyond the Nile Valley seems to have bypassed the Bronze Age altogether.

In East Africa, this same iron-and-steel technology has persisted to within living memory. Thus, "ethnoreconstruction" of the ancient iron smelters is still possible. In fact, the construction and use of these old-style smelters was revived during World War II when iron shipments from the colonial powers were effectively suspended.[56] This ethno-reconstruction gives us our best look into the African Iron Age technology of 2,600 years ago and allows us to observe the different stages of smelting in detail.[57]

As already mentioned, iron is derived from hematite ore where it is present as iron oxide ($Fe_2 0_3$). Surface hematite, found on certain hillsides, is plentiful in interlacustrine East Africa and seems to have been the primary source of the ore; as yet no mines similar to the paleolithic ones discovered in Swaziland and Zambia have been found. In preparation, the ore is preheated in a very hot fire for 30 minutes. Preheating allows easier separation of the metal iron from the ore and other impurities; it also renders the ore itself more brittle, making it easier to break into small pieces and thus smelt more efficiently.

Also in the preparatory phase is charcoal-making—the fuel used in the smelt—and wood from special trees is used for this purpose. A good example is the thornbush tree whose wood is solid with a high silica content, allowing it to burn slowly. Iron smelting requires large quantities of charcoal and a single smelt can consume 15 or more medium-size trees.[58] It seems likely that consistent smelting would have had a long-term environmental impact through progressive deforestation, though there is some evidence in other parts of the region that the local smiths developed fuel-saving mechanisms that retarded the deforestation process.[59] In any case, the process of making charcoal generally takes about 24 hours.

Van Noten and Raymaekers report the existence of two types of smelters from early Iron Age East Africa, one consisting simply of a large bowl set in the ground and the other of a tall shaft sitting over a pit dug into the ground.[60] Both types existed side-by-side, the tall furnace used for the initial smelt and the bowl-smelter used to refine the metal produced in the tall furnace (see Fig. 11). However, archaeologically, only the remains of the tall furnace have survived into modern times (see Fig.9).

After the pit (or bowl) had been dug, the shaft—often attaining a height of 10 feet—was constructed of a special clay containing quartz, mica, and

---

[56] Ibid., p. 106.

[57] Ibid.; see also Schmidt P and Avery DH, "Complex Iron Smelting and Prehistoric Culture in Tanzania," *Science* (201)4361, September 22, 1978, pp. 1085-9.

[58] Van Noten and Raymaekers, op. cit., p. 106.

[59] Schmidt and Avery, op. cit., p. 1089.

[60] Van Noten and Raymaekers, op. cit., pp. 105-7.

feldspar which could withstand the heat of the furnace and was fire-proof. For this purpose, the smiths constructed their shafts with the clay of anthills or termite mounds which served their requirements very well. Termite and/or anthill mud also provides a near-ideal reducing environment for the conversion of iron oxide into metallic iron because its insulating properties keep the interior of the smelter nearly oxygen-free. But these East African metallurgists had also devised construction techniques that markedly enhanced the interior reducing environment by the ingenious use of bricks and long, rope-like coils of clay. Van Noten and Raymaekers reveal how:

> The individual bricks, which were sunbaked, seem to have formed the lower part of the shaft. One reason for such an arrangement was that individual bricks would form a structure that can expand and contract more readily without cracking than one made of coils; this flexibility would be of greatest significance near the base of the shaft, where the temperatures were highest.[61]

The upper part of the furnace walls was fashioned by the "coil-building" methods. Thus, this style of construction was reserved for the tall, rather than the low, furnaces. After the erection of the furnace, the inner and outer walls were plastered to further improve insulation. Unlike the low smelters that were used but once or twice, the tall furnaces were used and reused for many years (see Fig.10). It seems likely that furnace doors were constructed to allow the removal of the finished iron without dismantling the entire furnace.

As Van Noten and Raymaekers point out, the need to provide an efficient reducing environment largely dictated the shape of the tall smelters. The cylindrically-shaped, upwardly-tapering furnace could be filled with charcoal and ore, minimizing the oxygen content within. Yet oxygen was essential for combustion, so a method of supplying it without allowing it to diffuse through the entire furnace was required. The method employed was that of the induced draft, facilitated by clay, funnel-shaped blowpipes or "tuyeres." These tuyeres were inserted into the base of the furnace and air was forced through them by a bellows mechanism. Thus, jets of air could be induced into the bottom of the furnace where the charcoal was burning without diffusing into the upper interior of the structure.

Once the initial preparations and construction of the furnace were finished, the loading began. The furnace would be filled completely from above with alternating layers of charcoal and ore, and then lighted by a glowing charcoal inserted through a tuyere into the center of the furnace. When the fire reached a sufficiently high temperature, the smelting process would begin. There were two phases to the smelt: (1) the reduction of iron oxide to metallic iron and (2) the separation of iron from associated minerals that

---

[61] Ibid., p. 109.

Fig. 9. Ten feet high iron smelter-steel maker from East Africa. Smelters of this type were used and re-used over many decades (Van Noten and Raymaekers).

would run off as slag. Oxygen, introduced as an air draft through the tuyeres, combines with the carbon in the charcoal to yield carbon monoxide (CO). This carbon monoxide extracts oxygen from the iron oxide ore, leaving carbon dioxide (CO_2)—which escapes up the shaft—and iron. This initial reduction process begins at a temperature of 800°C or 1,472° Fahrenheit (F).

The second phase, the separation of metallic iron from slag, begins at 1,150°C (2,102°F), the temperature at which the remaining impurities begin to liquefy. The temperature at this phase must be kept between 1,150°C and 1,540°C (2,804°F) because the latter temperature represents the melting point of iron. As Van Noten and Raymaekers point out, much of the technical skill of these metallurgists was exercised in maintaining the temperature inside the furnace within this range.[62] To have obtained such temperatures with such simple materials and to have kept the temperatures within a specified range was no mean feat. The resulting iron was the spongy, "bloom" variety, well-known among other Iron Age cultures outside Africa,

[62] Ibid., p. 111.

Fig. 10. Depiction of East African smelter with blow holes where drafts of pre-heated air are forced into the combustion zone (Van Noten and Raymaekers).

including Medieval Europe. The remaining impurities, slag and charcoal, were removed by hammering the hot bloom.

There was yet another technical level of iron metallurgy achieved in the same region during roughly the same period (2,000 years ago) by the Haya people of northern Tanzania. This method involved the production of carbon steel by a process significantly more advanced than the one described above. Though the techniques to be described below were probably not as widespread as those described by Van Noten and Raymaekers, it is evident that peoples other than the Haya had mastered them, most conspicuously the Yoruba of the old Oya kingdom (Nigeria).[63] The metallurgy involved reveals an iron-and-steel technology not to be surpassed in some respects until the end of the 19th century. This complex metallurgical technology was brought to light by Peter Schmidt and Donald Avery of Brown University.[64]

To reiterate, steel is a compound of iron plus carbon, minimally 0.5–0.7% and maximally 1.5–1.7%, investing steel with its characteristic strength and durability. Carbon concentrations exceeding 1.7% produce brittle cast iron and concentrations less than 0.5% create a "mild steel," softer and more malleable than high-carbon steel. As already noted, several areas of Africa, including Tanzania's Haya region, evolved a metallurgy capable of producing high-carbon steel.

[63] Van Der Merwe, op. cit., p. 491.
[64] Schmidt and Avery, op. cit.

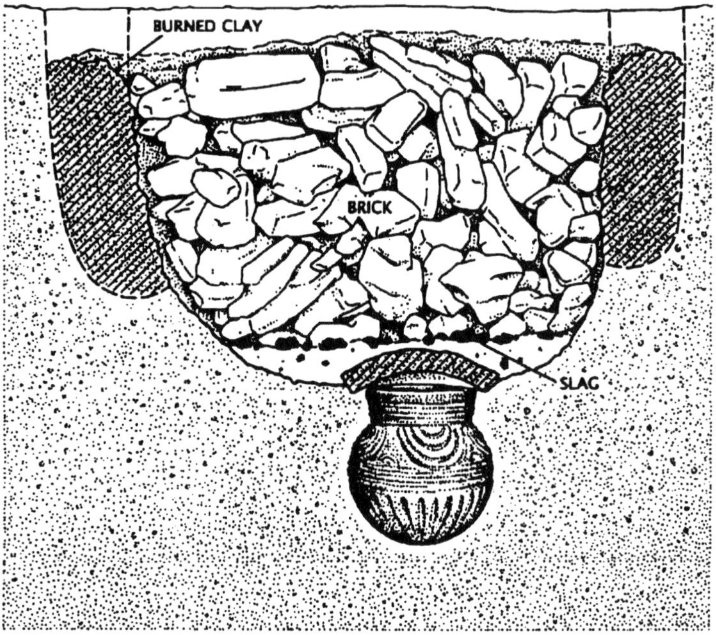

Fig. 11. Cross-section of a pit smelter (Van Noten and Raymaekers).

European steel-making processes until about 200 years ago relied on bloom iron production and steel-making by the cold-blast method. Among the Haya, metallurgists as far back as two millenia ago had developed a method of preheating air that conferred marked advantages in steel-production. Their preheating methods allowed them to achieve much better fuel economy *and* higher furnace temperatures than were possible in Europe until the mid-19th century. This level of expertise made the steel of East Africa world-famous.

The method for reclaiming this evidence was a Hayan "ethnoreconstruction" of a traditional blast furnace. Under the sponsorship of Schmidt and Avery, Haya elders erected a smelting furnace whose antecedents date back 2,500 years. Thus, the Hayas' oral traditions had preserved information concerning iron-smelting (and steel-making) for the better part of two and a half millenia. The fidelity and accuracy of this oral transmission over that length of time has been confirmed by archaeology.[65]

Schmidt and Avery, prior to the test smelt, had assumed, with other archaeologists, that the highest possible temperature obtainable in a primitive furnace was about 1,200°C.[66] And yet, preliminary test results of the slag from early Iron Age Hayan furnaces revealed a temperature obtained as high as 1,400°C.[67]

---

65 Schmidt and Avery op. cit.; Van Noten and Raymaekers, op. cit.
66 Schmidt and Avery, op. cit., p. 1086.
67 Ibid.

A tremendous boost to industrial steel manufacture in 19th-century Europe was achieved by the preheating of air through "inefficient pipe stoves" to 100°C. It became evident to Schmidt and Avery that the high-carbon steel found in certain parts of Africa indicated that the technique of preheating of air must have been utilized there as well. Following the directions afforded by oral tradition, the reconstructed Haya steel furnace exhibited eight blowholes carved into the bottom of the smelter into which tuyeres were inserted. A continuous draft of air was forced through each tuyere by bellows, eventually creating temperatures of 1,300°C in the combustion zone just beyond the tuyere and raising the temperatures of the tuyeres themselves to 1,250°C. Schmidt and Avery measured the temperatures inside the tuyere by a passing a ceramic-insulated thermocouple through it, revealing a preheating temperature of 600°C (or more). This means that the Haya had been able, more than 2,000 years earlier, to devise an air preheating method that achieved temperatures six times as high as reached in early-19th century Europe. What is more, in certain parts of the combustion zone itself (beyond the inside mouths of the tuyeres), Schmidt and Avery recorded temperatures *in excess* of 1820°C (3308°F). The highest attested temperature a blast furnace was capable of elsewhere prior to the end of the 19th century was an experimental 2nd-century Roman furnace that achieved temperatures of just 1,600°C (2,912°F).[68]

These astonishingly high temperatures explain a unique characteristic of Haya iron: contrary to bloom iron found throughout the world that forms by the sintering of fine solid particles, Haya iron forms by the precipitation of large crystals from the ore. The molten slag then undergoes a "carbon boil"—similar to the process in modern blast furnaces—resulting in the formation of two products: pure iron and steel. This is an extraordinarily efficient way of producing iron since it requires less fuel than the standard Iron Age smelting techniques.

As Nikolaas Van Der Merwe has pointed out, this methodology represents a significant technical innovation in steel manufacture that seems to have been unique to Africa.[69] There are at least four places where crystalline steel was produced: Tanzania (Haya), Nigeria (Oyo), Ethiopia (Lake Tana), and Ghana, all presumably by similar methods. Thus, the induced draft furnace with a tall shaft plus preheating of bellows-driven air are two unique African inventions. The laborious processes of carburization, quenching, tempering, and hammering so characteristic of steel-making in other parts of the world seem to have been bypassed in a technical tour de force among those African peoples mastering the "Haya technique."

It is known that well into the 20th century, the steel forged by local smiths

---

[68] Ibid., p. 1087.

[69] Van Der Merwe, op. cit., p. 500.

in Nigeria and Ghana was considered superior to the variety imported from Europe.[70] This superiority was demonstrated in spectacular, if unheralded fashion, when an attempt was made to cut through a native steel sample in Accra, Ghana:

> [A] sample was taken from a bloom kept as a curiosity in the compound of Sowle Wakyesu at Lawra. The bloom was a saucer-shaped specimen of about 4 kg. . . . One assumes that it had already been reheated and hammered together since the sponge-coral surface normally associated with a bloom is absent. In an attempt to cut a sample from this bloom in Accra, *seemingly no impression could be made on it with an alloyed rotary diamond blade.* (emphasis added)[71]

Some mention should be made of Mande iron-working and steel-making, the shape of which has been brought to light in an admirable monograph by Patrick McNaughton.[72] The Mande seem to have utilized techniques very similar to those made famous by the Haya and have achieved a comparable level of skill (see Figs. 12 and 13). The Mande, for example, seldom use the quenching method because the resulting metal is hard but brittle. But they handle iron and steel with ingenious subtlety. Certain tools and weapons are forged in such a manner that the interior is composed of softer iron while the outside is fashioned into steel. This makes the implement/weapon hard on the outside and soft on the inside, allowing it to cut very sharply through difficult substances but without breaking easily because of the internal flexibility. Even today, the Mande find that their iron-and-steel tools often to be superior to those imported from Europe though the latter are used pervasively because they are cheap and plentiful.

What can be said, however, is that steel-making in Africa, though it rose to a high technical level, seldom moved beyond subsistence production. The mass production of steel to create a surplus for mercantile trade seems not to have occurred to any significant degree in Africa with one notable exception: East Africa.

In the Medieval Period, the most famous swords in the world were the so-called Damascus swords, prized for their strength, durability, and fine, tempered quality. But, despite the name, these swords were *not* actually made in Damascus but instead in India. El Idrisi, the Arabic author and traveller of the 12th century, tells us that the steel to make these swords was obtained commercially by Indian swordsmiths *from East Africa*.[73] This was the heyday of the Zanj trading cities of the East African coast—Mombasa, Kilwa, Sofala, Malindi—who carried on a far-flung commerce on the Indian Ocean

---

[70] Ibid., pp. 491-3.
[71] Ibid., pp. 493-4.
[72] See McNaughton, op. cit., for a full discussion of the technical finesse of Mande blacksmiths.
[73] Cited in Davidson B, *Lost Cities of Africa*, Boston: Little, Brown & Co., 1970, p. 165.

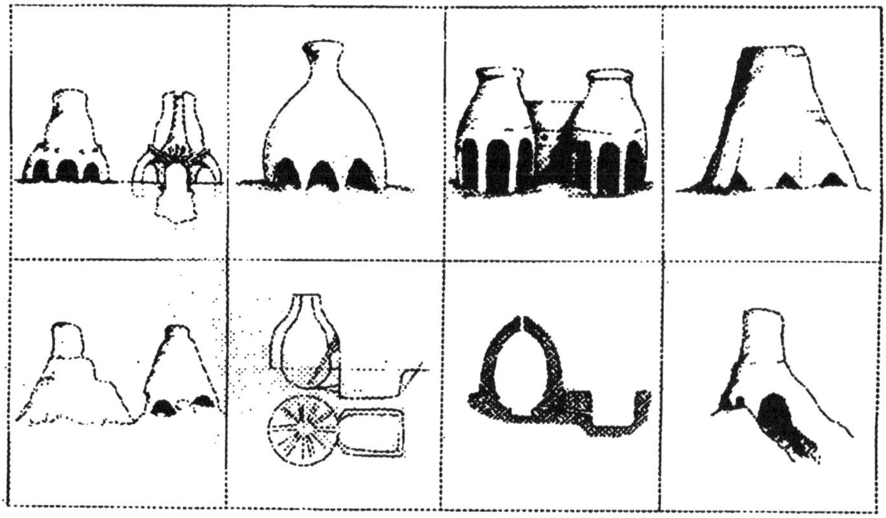

Fig. 12. Dome-style iron smelters/steel makers from the Western Sudan, shown with blow-holes at the bottom (McNaughton).

as far away as China. Thus, the period from 900 to 1,500 A.D. witnessed a thriving East African mercantile enterprise of which steel was one of the most valued commodities.

In the 19th and 20th centuries, European steel supplanted native steel in Africa, not because it was better, but because it was cheaper and more abundant. It is not without interest, however, that during World War II when imported steel became scarce in East Africa, the natives revived their traditional steel manufacture to keep themselves suitably supplied with this indispensible metal.[74]

Gold: The Royal Metal of Africa

Few metals possess the lore and romance that surround gold. No metal has excited as much naked greed or had as much blood spilt in its pursuit over the millenia as this one. Its bright, burnished yellow coloration, malleability, and relative scarcity have combined to make it the most sought-after metal of the historical era, used all over the world for jewelry, statuary, and coinage.

No continent has possessed or possesses as much gold as Africa. The richest ancient gold-bearing region in the world was the arid region to the southeast of Egypt. So fabled was this area as a source of gold that it became known as *Nubia*, a word undoubtedly derived from the ancient Egyptian word for

[74] See note 54.

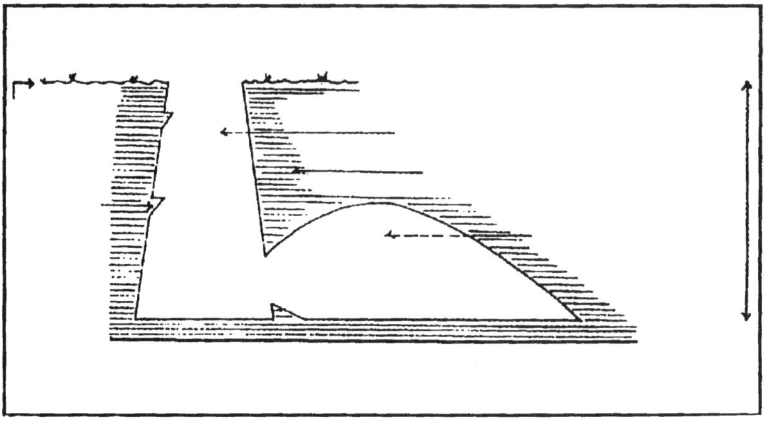

Fig. 13. Cross-section of an iron mine shaft from Mali (McNaughton).

gold, *nub*. No one doubts, furthermore, that gold sources from farther up the Nile—modern-day Ethiopia, the southern Sudan, and points further south—also supplied the Nile Valley with gold.[75]

Gold, though not as anciently known as copper, has been found in early pre-dynastic graves of the Nile Valley. This early gold was probably obtained by extraction from alluvial deposits, a simple method by which gold particles are separated from heavier gravel by sifting through running water. This method is familiar to anyone who has seen a few movie westerns. Gold was mined at a later period by cracking gold-bearing rock with fire, breaking these rocky fragments into smaller pieces with hammers and picks, crushing these pieces into pebbles, grinding them into a fine powder, then finally separating the metallic gold from the rock by washing.[76]

Petrie identified a number of gold objects from the pre-dynastic graves he excavated.[77] The earliest were beads fashioned out of gold that surrounded small cores of limestone. A gold finger ring was discovered, and in the latter pre-dynastic era, gold foil was used to cover vases and knives. Also, a gold cylinder seal has survived the depredations of the later grave robbers. Frequently, these objects consisted of gold alloyed with either copper or silver, and the practice continued through dynastic times. Silver, being harder to obtain in the Nile Valley, was more valuable than gold, and a gold-silver alloy, called *nub hedj* ("white gold") by the Egyptians and *electrum* by the later Greeks, was often used for ornamentation. An electrum spearhead is

[75] Lucas, op. cit., pp. 224-5.
[76] Ibid., p. 228.
[77] Petrie, op. cit., p. 84.

known from the artifacts of Ta-Seti collected at Qustul and analyzed by Bruce Williams.[78]

Nile Valley goldsmiths have never been excelled for the intricacy and quality of their work. Though some authorities think that actual refining (purification) of gold was mastered before 2,000 B.C., Lucas insisted that there was no evidence of gold refining in the Nile Valley before 525 B.C.[79] Whatever the case, the jewelry, masks, and statuary turned out by Nile Valley gold craftsmen are marvels of workmanship and artistry. These smiths found it relatively easy to smelt and cast gold since it melts at 1,063°C, a temperature lower than the melting point of copper. With hammering and casting, gold was put to a startling array of uses including engraving, embossing, plating, gilding, and wiring. Many objects were fashioned by soldering technique showing that from the outset of the dynastic period, the smiths had mastered the subtleties of their craft.[80]

Tiny gold granules were manufactured by the hundreds and thousands for decorative uses. In addition, the Nile Valley goldsmiths were past masters in the art of fashioning gold leaf and foil, that is, gold beaten into extremely thin sheets and used to plate a variety of articles from furniture to weapons. Not until the 18th century in Europe did artisans from any other culture succeed in producing gold foil thinner than that produced in the ancient Nile Valley. Gold leaf of 0.001 millimeter thickness has been preserved from ancient Egypt, though today, leaf of one-tenth that thickness is routinely produced.

The above-mentioned gold-silver alloy known as nub hedj or electrum was also in widespread use in the Nile Valley. The presence of an electrum spearhead in the pre-dynastic pharaonic tombs at Qustul has already been noted. Then and later, electrum was occasionally used to fashion weapon heads because it was harder than gold. Probably such electrum weapons were limited to royalty and served primarily a ceremonial function. Of course, electrum was a common constituent of jewelry. Many obelisks and also the capstone of the Great Pyramid were plated with electrum. Consequently, these structures could be seen from a far distance owing to the brilliant reflection of sunlight off the electrum.

The inventiveness, delicacy, and genius for design exhibited by these ancient gold-workers has to be seen to be believed. Though the artistic level rose and fell with the fortunes of the state, all periods are represented by masterpiece after masterpiece of gold-working. This being said, few objects arrest the attention as much as Tutankhamen's gold funerary mask. This life-size mask shows a refinement in modelling of feature and an effortless naturalism that no other gold sculpture has ever surpassed.

---

[78] Williams B, "The Lost Pharaohs of Nubia," *Archaeology* (33)5, September/October 1980, p. 20.

[79] Lucas, op. cit., p. 229.

[80] Petrie, op. cit., pp. 85-6.

The Nubian kingdoms of Napata and Meroe did not lag behind their northern neighbor in perfection of craftsmanship. Gold, if anything, was more plentiful in Nubio-Cush than in Egypt and the region remained Egypt's chief supplier. Gold was so commonplace in Nubio-Cush that it did not seem to possess the same intrinsic value as in other places and, as in Egypt, silver was much more valuable than gold. Not surprisingly, numerous examples of exquisite gold ornaments, jewelry, inlay, and sculpture have come down to us from the ancient kingdoms of Nubio-Cush.

Gold was mined, worked, and traded extensively also in the Ethiopic kingdom of Axum. Like their Nile Valley predecessors, Axumites had mastered the techniques of gold-working and were known to have casted large gold statues. But, in a significant departure from earlier kingdoms, the Axumites minted their own coinage beginning about 270 B.C. As noted above, Axumite coins were made from copper, bronze, and silver as well as gold, but gold coins did not become the principal medium of exchange until after 600 A.D. Axum had close diplomatic and trading links with Rome so their coinage reflected Roman standards. But Axum was one of but four kingdoms in that part of the world with sufficient standing to mint its own specie.

Though ancient Nubia was proverbial for its gold wealth, few places in Africa or the world could match southeast Africa—in the region of present-day Zimbabwe—for the extent and magnitude of ancient gold mining and extraction. Evidence for a thriving gold industry in Zimbabwe goes back more than 1,000 years.[81] Ancient gold mines greater than 4,000 feet in length and 150 feet deep have been discovered in Zimbabwe.[82] The presence of numerous adits driven into the sides of hills testify to the volume of extraction work done. The estimated amount of gold ore mined from the entire region by the ancients was staggering, exceeding 43 million tons.[83] This ore yielded nearly 700 tons of pure gold which today would be valued at over $7.5 billion.[84]

The method of extracting metallic gold from the ore closely resembled the method described above in the Nile Valley. Large, heavy stones of granite or diorite were used to crush the gold-bearing ore into smaller and smaller

---

[81] For a refutation of the "Phoenician Thesis" of Zimbabwe's origins, see Davidson, op. cit., pp. 299-307.

[82] Hall RN and Neal WG, *The Ancient Ruins of Rhodesia*, London: Methuen & Co., 1904, p. 55. More recently, the extent of the gold wealth of Great Zimbabwe has been questioned. However, it seems reasonable to assume that gold specie in significant quantities was available to pay for the exotic import commodities evident at Great Zimbabwe. See Garlake P, *Great Zimbabwe: Described and Explained*, Harare, Zimbabwe: Zimbabwe Publishing House, 1982, pp. 15-6, for a discussion about the limitation of Zimbabwe's gold supply.

[83] Ibid., p. 66.

[84] Ibid. In contemporary times, gold has tended to fluctuate around the figure of $350 per ounce; $300 × 21,637,500 ounces of pure gold estimated to have been been extracted by the Zimbabweans of old = $7,573,125,000.

particles. When reduced to a fine gravel, the ore was washed to separate the gold dust from the rest of the ore. The gold dust thus reclaimed was taken to the towns where gold smelting operations were conducted to put it to functional uses. Much of the gold dust found its way to the coast where undoubtedly it was used as trade currency. As seen also in the West African empires such as Ghana (see below), much of the gold was sequestered by the ruler or *Monomatapa* at the great metropolis of Zimbabwe, very likely to keep its price within a specified range.

The actual smelting of gold was conducted in the same location as that of iron and copper, often using the same furnaces. The typical African technique of inducing a preheated draft of air into the furnace through tuyeres to efficiently achieve a high temperature was employed. However, by comparison to iron-smelting furnaces known from Tanzania, these gold furnaces were relatively simple affairs, consisting primarily of round basins sunk into the floor, lined with a granite cement. Crucibles connected to these furnaces are in abundance. The presence of gold ingot molds in Zimbabwe testifies to its role in long-distance trade since ingots are the most portable form of gold.

Like earlier African civilizations, the ancient Zimbabweans manufactured gold ornaments, jewelry, and plating. As Hall and Neal asserted: "All the branches of the goldsmith's art were practised by them [Zimbabweans], including gold wire drawing, beating gold into thin sheets, plating iron and bronze with gold, and burnishing."[85] The Zimbabwean gold wiring was of many different gauges; gold thread was woven into cloth and gold chain links were fashioned.[86] Gold beads of all descriptions were produced, i.e., with flattened ends or diamond-like facets, and also punched or bent into circles. Many of these beads displayed chevrons and other designs engraved on them; some of the engraving was so fine that it is detectable only by the aid of a magnifying glass.[87] Beaten gold and gold plating were fashioned to a remarkable thinness and used to cover everything from furniture to statues to arrow heads and battle-axes. The Zimbabwean goldsmiths' tools included burnishing stones, beating hammers, and perfect stamps by which certain designs were imprinted directly onto the gold object. Most bead patterns, however, were carved with a sharp implement.

Certain aspects of Zimbabwean culture suggest ancient contacts with Nile Valley civilization and an examination of Zimbabwean gold mining, smelting, and smithing enhances that impression. While avoiding the controversies surrounding issues of diffusion versus native invention, it can be said that there seems to have been an African-style gold-working industry that had its roots in the ancient Nile Valley and was carried on in Zimbabwe as in other parts of Africa.

---

[85] Ibid., p. 93.
[86] Ibid.
[87] Ibid., p. 94.

There was only one other region of Africa, indeed the world, that rivalled, and even exceeded, southeast Africa for the sheer plentitude of gold and that was West Africa. The western quarter of sub-Saharan Africa has been the home of numerous empires, the most notable being Ghana, Mali, and Songhay. The largest, Songhay, encompassed a land area nearly double the size of Europe. The origins of the oldest of these empires, Ghana, date back to about 300 A.D., but it first came to the attention of the wider world about 800 A.D. From that time until 1076, when it was overrun by the Almoravids, Ghana was the most storied kingdom in West Africa. Most of what we know about the great West African empires comes to us from the travelogues and accounts of Arab and Moorish historians. El Bekri, writing about 1062, is the first to give an extended account of Ghana, and though he wrote from secondary sources, he, like all the writers who came after him, was flabbergasted by Ghana's gold wealth.[88]

Ghana, like its successor states, was situated in the region now called the Sahel and derived its fabled wealth from its position as middleman between the gold-rich, salt-poor forest regions immediately to its south and the salt-rich, gold-poor kingdoms of North Africa. Veritable rivers of gold flowed into Ghana and El Bekri regaled his readers with delectable tales of the king's court where horses stood beside him caparisoned in gold cloth, pages surrounded him bearing gold-mounted swords, and sons of the nobility sat in attendance with gold plaited into their hair.[89] Gold was so plentiful that the king reserved all gold nuggets for himself, leaving gold dust to be circulated for trade and craft. This expedient prevented gold from circulating so freely as to depress its value.[90] Reputedly, the king owned a nugget so large that his favorite horse could be tethered to it.[91] These attestations demonstrate the various practical and ornamental uses to which gold was put by the smiths of the realm. But it was more vital in commerce since it spawned a lucrative trade with North Africa that brought in consumer and luxury items from around the world. Indeed, most of the gold of Europe in the Medieval period found its way there from the Western Sudan through various Moorish middlemen.

The kings of Ghana do not seem to have minted their own coinage but they did tax incoming caravans from and outgoing caravans to North Africa in their own coin. Each bag of salt (or any other goods) coming in from northern Africa was assessed at one dinar; each bag of gold leaving Ghana was assessed at two. No wonder the wealth of Ghana—and all the later West African kingdoms—was proverbial in the Muslim world.

The vast holdings of gold possessed by the empires of the Western Sudan

---

[88] Davidson, op. cit., pp. 81-8.
[89] Ibid., p. 85.
[90] Ibid., p. 87.
[91] Cited in Bovill EW, *The Golden Trade of the Moors*, New York: Oxford University Press, 1970, p. 81.

were brought into international prominence most spectacularly by the hadjj of Kankan Musa, Mansa of Mali, in 1324.[92] After the Almoravid invasion, Ghana declined and disappeared, to be succeeded by the rising kingdom of Mali. Eventually, Mali took over all the lands formerly belonging to Ghana and expanded its empire until it equalled Western Europe in size. From the beginning, the Mansas of Mali were pious Muslims, and Kankan Musa, in accordance with religious duty, undertook the most talked-about hadjj in history. He was gone three years accompanied by 60,000 of his subjects. In this caravan, pack-animals carried more than 15 tons of gold and Mansa Musa and his retinue distributed this gold with a prodigal liberality in all the lands they crossed. It is said that so much gold was released into the Cairo market by the Malians that its price fell and remained depressed for 12 years afterward.[93]

West African gold miners adopted techniques similar to those found in Zimbabwe. Much alluvial gold was panned but most of the ore was obtained by sinking shafts sometimes 100 feet deep, often linked by side shafts and galleries.[94] It is estimated that the total amount of gold mined in West Africa up to 1500 was 3,500 tons, worth more than $30 billion in today's market.

Conclusion

Mining and metallurgy have an old history in Africa; iron ore mining goes back 43,000 years while manganese ore mining goes back 28,000 years. The actual use of metals for functional and ornamental purposes can be dated back to the Nile Valley pre-dynastic period 7,000–10,000 years ago. Thus, we find copper and gold implements, utensils, and ornaments recovered from pre-dynastic sites. Though "industrial" iron is customarily thought of as originating with the Hittites, important strands of evidence from northeastern and western Africa suggest that the functional uses of iron pre-date the Hittites by more than 1,500 years. Moreover, after steel was invented, Africans mastered its manufacture to such an extent that they produced some of the finest quality steel in the world up to the 20th century. Thus, the common historical image of Africa as a technological backwater is not merely distorted but definitively false.

---

[92] Ibid., pp. 85–91.
[93] Ibid.
[94] Ibid., pp. 119–31.

3

Number and Nature in African World Systems

He [Pythagoras] visited all the Egyptian priests, acquiring the wisdom each possessed. He thus passed twenty-two years in the sanctuaries of the temples, studying astronomy and geometry, and being initiated in no casual or superficial manner in all the mysteries of the Gods.
—Iamblichus

. . . if a mathematical harmony underlies the realm of tone and music. . . . Number may account for other phenomena in the cosmic order—for example, planetary motion, which was also thought of being related to the mathematical harmonia of the scale, this being the famous "Music of the Spheres."
—David R. Fideler, *The Pythagorean Source Book*, by Kenneth S. Guthrie, p. 28

. . . because Number is universal it is divine . . . because it is divine, it is universal . . .
—David R. Fideler, *The Pythagorean Source Book*, p. 21

Our word *arithmetic* derives from a Greek word meaning "number"; the word *mathematics* comes from a Greek word meaning "what is known." Indeed, the original *mathematikoi*, a term practically synonymous with "Pythagorean," were not concerned exclusively with numbers and their operations but were *philosophers* in the original Pythagorean sense of the term: "lovers of wisdom." Nonetheless, Pythagoras did say all things were number, an idea he undoubtedly inherited from the Egyptian temples where he studied for 22 years in the 6th century B.C. This notion that number is intrinsic to nature is one found in African systems of thought both in and outside of the Nile Valley.

The earliest arithmetic system of which we have any record is revealed in the Ishango bone.[1] It has been minutely studied both by its discoverer, Jean de Heinzelin, and the archaeologist Alexander Marshack. When de Heinzelin excavated the cache of ancient tools at Ishango in northeastern Zaire and found the bone, he was immediately struck by the interesting pattern of horizontal marks carved into it. He perceived at once that these marks

[1] See Chapter 1, note 15.

were not random and, in fact, represented a number system (see Figs. 14 and 15). The marks were arranged in three vertical columns on the bone—two columns on one side and a single column on the reverse side. The single column of grouped marks was arranged in the following sequence from the top downward: 3-6, 4-8, 10, 5-5, 7, showing a pairing of 3-6, 4-8, 5-5. These pairs of markings seemed, in de Heinzelin's opinion, to have been connected to the operations of adding and multiplication by doubling.

On the obverse side of the bone there were two columns, one showing a marking sequence from the top downward of 11, 21, 19, 9. According to de Heinzelin, this sequence was arrived at by the following operations: 10 + 1, 20 + 1, 20 - 1, 10 - 1. On the opposite vertical edge of the same side, the sequence from the top went 11, 13, 17, 19. This sequence appeared to represent all the prime numbers between 10 and 20. To de Heinzelin, all of these sequences taken together revealed evidence among a lacustrine East African people of knowledge of primary numbers, a base 10 (decimal) system, and multiplication by doubling.[2]

In all three columns, the sum of the sequence of numbers equals 60, suggesting that these ancient Ishango people also possessed a sexagesimal numerical system ordinarily thought to have been developed first by the Babylonians.[3] As will be seen, all the numerical patterns seen in the Ishango bone show up in later African mathematical systems.

Alexander Marshack, examining the Ishango bone in the decade after its discovery, rejected de Heinzelin's interpretation for one that explained these markings by reference to a lunar calendar system.[4] Possibly, the number 60 showing up as the sum of each column is connected to this lunar notation as a "double month." It is true that the lunar calendars are very old, much older than solar ones, and since a lunation occurs at intervals averaging 29$^1/_2$ days, lunar months vary between 29 and 30 days. But if this sum of 60 (repeated three times on the bone) is calendrical, it cannot itself be lunar because a lunar "double month" would amount to 59 days. However, the number 60 appearing three times (180) may hint at the existence of a "functional year" of 360 days in this Paleolithic Ishango culture.

De Heinzelin initially gave an age for the Ishango artifacts of 8,500 years, carrying them back to 6,500 B.C., thereby categorizing the culture as "mesolithic." As we have seen in Chapter 1, however, re-datings of the site by Alison Brooks has forced a revising of the age of Ishango to 25,000 years, pushing it back to 23,000 B.C. This puts Ishango in the Upper Paleolithic era, forcing us to radically revise our ideas of when and where humanity began to engage in formal intellectual, indeed "scientific," inquiry. Upper Paleolithic proto-mathematics and calendars in East Africa are consistent with the Upper Paleolithic

---

[2] De Heinzelin J, "Ishango," *Scientific American*, June 1962, pp. 109-11.

[3] See Neugebauer O, *Exact Sciences in Antquity*, New York: Dover Publications, pp. 17, 22, for an explanation of the Babylonian sexagesimal system.

[4] Marshack A, *The Roots of Civilization*, Mount Kisco, N.Y.: Moyer Bell, 1991, pp. 27-30.

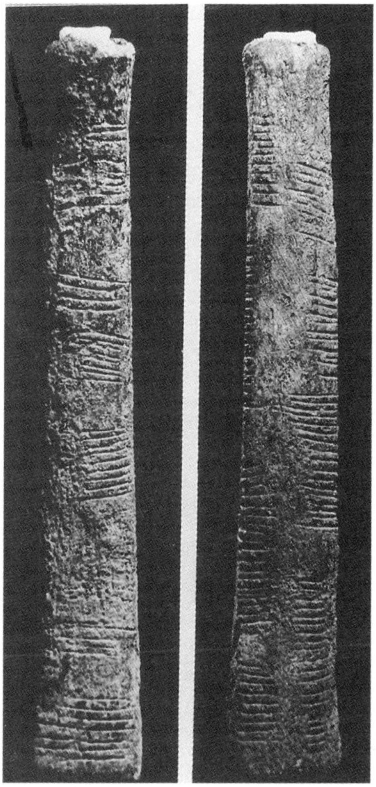

Fig. 14. Two sides of the Ishango Bone from northeastern Zaire; dates from 23,000 B.C.

evidence (41,000 to 28,000 B.C.) for systematic iron ore mining in southeast Africa (see Chapter 2), revealing the presence of a "protoscience" and a "prototechnology" in Africa older than 20,000 years. This cultural sophistication, bringing Africa to the brink of civilization tens of thousands of years earlier than anyone ever suspected, again shows Africa to be the cradleland of techniques that gradually diffused to the rest of the world.

Number in the Nile Valley: A Case Study of Mathematics in Evolution

It is in the Nile Valley that one finds the earliest examples of mathematics in the formal sense. The Greeks were unanimous in attributing the origin of mathematics to ancient Egypt, particularly geometry, which the Greeks called the "queen of sciences." There is no way of telling how many mathematical treatises were composed in the Nile Valley, but only four survive at the present day: the Rhind Mathematical Papyrus (RMP), the Moscow

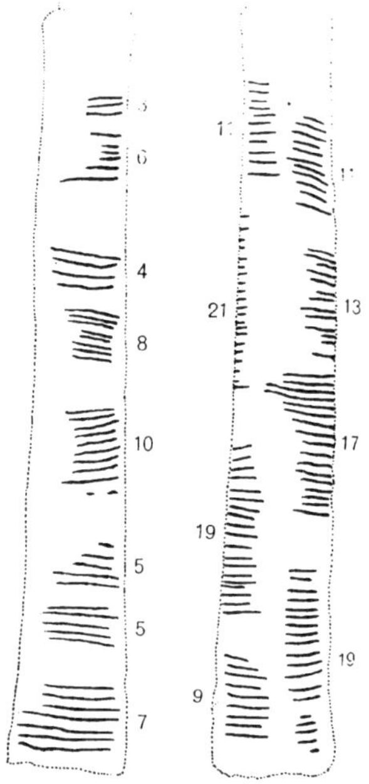

Fig. 15. Drawings of the two sides of the Ishango Bone showing the grouped markings that constitute a paleolithic African number system.

Mathematical Papyrus (MMP), the Berlin Mathematical Papyrus (BMP), and the Kahun Mathematical Papyrus (KMP). The oldest, the Rhind Papyrus, dates back to about 1,834 B.C., the time of the Middle Kingdom, but may very well be a copy of an older text. Some think it was a student's copy book and that a higher math was known by the priesthood. Aside from these four treatises, several documentary fragments containing mathematical operations have also survived.

Fragmentary evidence allows us to trace the presence of mathematics in the Nile Valley in predynastic times. Flinders Petrie, among his recovered predynastic ceramic artifacts, found that "On a basalt vase . . . there is inscribed the mouth sign and two strokes beneath it, reading in the usual hieroglyphs 'fraction one half'" (see Fig. 16).[5] Along with this, Petrie had collected numerous other predynastic weights and measures allowing the accurate calculation of distance, weight, and volume. We are therefore entitled to assume that the mathematics, particularly the geometry, familiar to us in dynastic times is already the product of a long development, stretching back into Nile Valley prehistory.

[5] Petrie WF, *Prehistoric Egypt*, London: British School of Archeaology/Bernard Quaritch, 1920, p. 27.

The Egyptian number system was fundamentally a decimal one and they possessed 15 different numerical characters. Characters 1 to 9 were simple groupings of vertical marks (see Fig. 18). After these, the Egyptians represented powers of 10 up to 1,000,000 by hieroglyphic symbols. Their expression for the concept of infinity was the twisted rope 𓍢, usually translated by the locution "millions and millions of years."

As Figure 18 shows, 10 is represented as an inverted "U" and to write "twenty," the Egyptians would write this 10-sign twice ∩∩ (Fig. 18). As Richard Gillings points out, this method seems to moderns incongruously cumbersome; to write the number "1967" would require 23 characters.[6] However, the number 20,000 could be written with just two.[7] Nonetheless, the Egyptians stuck with these numerical characters for the entire span of their history and seemed to have been able to use them with a dexterity we would hardly think possible.

For everyday accounting and computational purposes, the Egyptians used an abbreviated, semicursive form of "hieratic" numerical characters. In practical matters of everyday living, speed and concision were the overriding concerns and a hieratic method of writing seems to have been in place from the beginning of the organized state in the Nile Valley.

Of the four primary arithmetic operations, multiplication and division can be easily performed using the hieroglyphic numerical system employed by the ancient Egyptians. Adding and subtracting are a different matter, however, and though it is known that both of these operations using large numbers were carried out routinely in Ancient Egypt, it is not known exactly how. These considerations have prompted historians to assume that the scribes possessed precompiled addition and subtraction tables that allowed them to read off the answers.[8]

Though hieratic was used to record numbers and the results of numerical operations, these operations do *not* seem to have been performed in hieratic:

> Multiplication by continual doubling, performed by a simple duplication of each sign of a number, seems simple enough . . . but is feasible only in hieroglyphs, which . . . were seldom used. We are forced to the conclusion that reference tables were used for the four fundamental operations (which we indicate by +, −, ×, and ÷); and it is probable that portions of these tables were memorized much as we do today.[9]

---

[6] Gillings R, *Mathematics in the Time of the Pharaohs*, New York: Dover Publications, 1982, p. 5.

[7] Ibid.

[8] Ibid., p. 11.; see also Chace AB, *The Rhind Mathematical Papyrus*, Reston: National Council of Teachers of Mathematics, 1927, 1986, p. 40.

[9] Gillings, op. cit., p. 14.

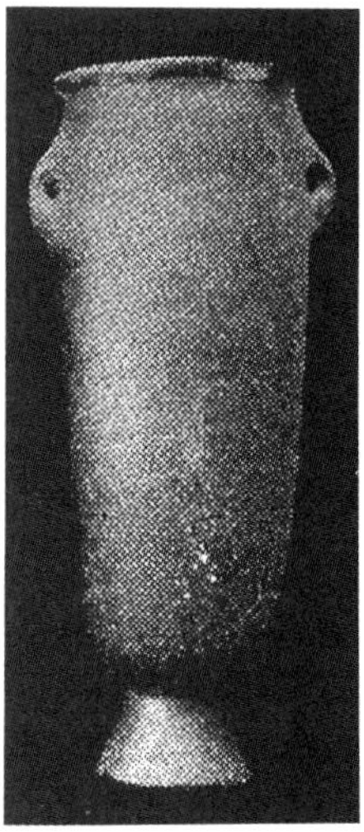

Fig. 16. Basalt vase from 7,000 B.C. with the hieroglyphic inscription "fraction one-half." The inscription shows that the pre-dynastic Nilotic culture possessed hieroglyphics, mathematical notation, and, with the jar itself, a liquid volume measure. (Photo: Petrie)

One of the Egyptians' most intriguing arithmetic operations was their technique of multiplying by doubling, mentioned above. As Gillings describes, the Egyptian, in multiplying two numbers, would make one the multiplier and the other the multiplicand. The multiplicand would be doubled repeatedly until the intermediate multipliers added up to the original; the continually doubled multiplicand then "arrived" at the correct total. The example below illustrates the method; **5** is the multiplier (left side) and **17** is the multiplicand (right side):

1/	**17/**
2	34
4/	68/
5	85

On the left hand (multiplier) side, 1 and 4 are marked because they add up to 5; the numbers opposite to them, 17 and 68, respectively, are then added to give the correct answer of 85.

To take another example we might multiply **15** (multiplier) and **9** (multiplicand):

1/	9/
2/	18/
4/	36/
8/	72/
15	135

On the left side, *all* the numbers are marked because, together, they add up to 15, the original multiplier; *all* the right hand numbers, opposite the marked numbers on the left, are added to give the correct answer of 135.

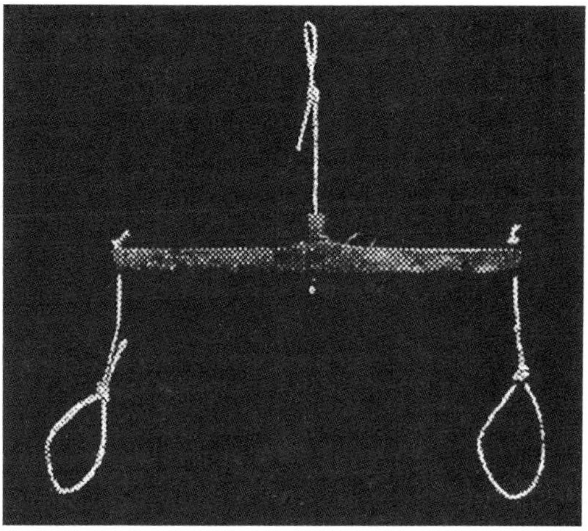

Fig. 17. Pre-dynastic balance-beam made of limestone (Petrie).

As Gillings says,

> These additions were made easier for the scribe by virtue of a special property of the series 1, 2, 4, 8, 16, 32 . . . for any integer can be uniquely expressed as the sum of some of its terms. . . . We do not know whether or not the scribes were explicitly aware of this but they certainly used it, just as do the designers of a modern electronic computer, and this is surely a somewhat sobering thought.[10]

It is worth digressing here to note that this reliance on the 2^n series[11] can be shown to have wider implications that take us to the outer limits of contemporary science. We can appreciate these implications better by examining some of the principles that have emerged from Chaos theory, a domain of science that seems to blend all disciplines.[12] The roots of Chaos theory go back to the work of Edward Lorenz, a meteorologist at the Massachusetts Institute of Technology (M.I.T.) who, in 1961, was attempting to mathematically model long-range weather forecasting in nonlinear equations. Using a computer program that allowed him to "iterate" the equation again and again for a particular weather system, he found, almost by accident, that very small differences in inputs resulted in very large differences in forecasts, once the equation was iterated beyond a certain time. For Lorenz, this was an astonishing, totally unexpected result that flew in the face of the Newtonian or-

---

[10] Ibid., p. 19.

[11] We will call the series 1,2,4,8,16,32,..., the 2^n series because *all* the numbers in the series represent powers of 2. $2^0 = 1$, $2^1 = 2$, $2^2 = 4$, $2^3 = 8$, and so on.

[12] The following discussion of Chaos will draw primarily from James Gleick's *Chaos* (New York: Penguin Books, 1987) a clear, comprehensive layman's treatment of the subject.

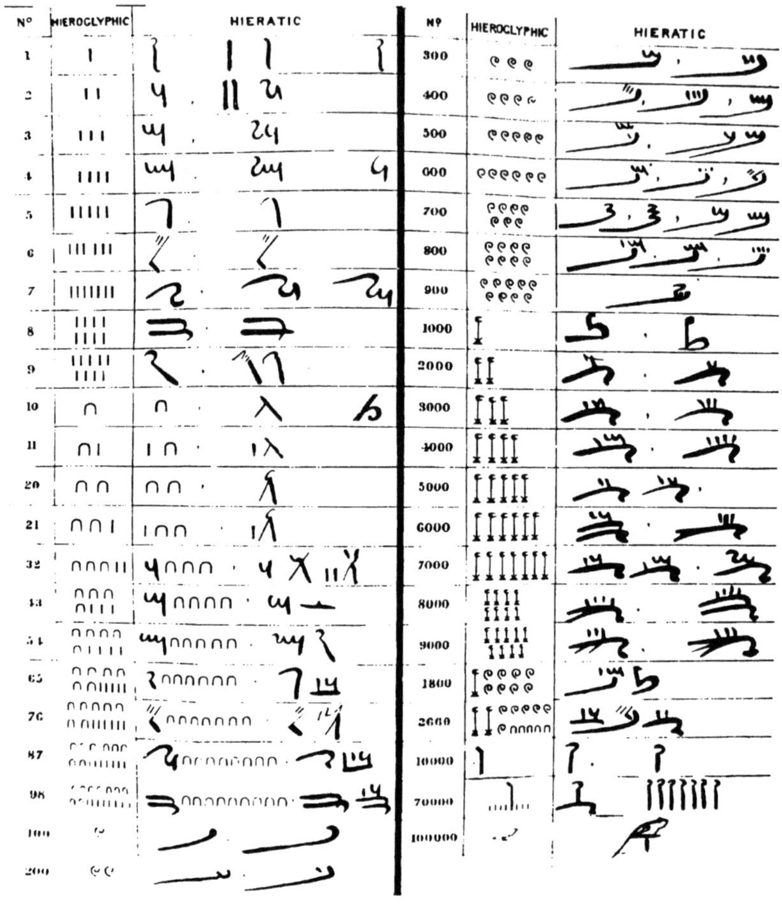

Fig. 18. Egyptian numerals and numbers in hieroglyphics and hieratic.

derliness that scientists assumed existed in the "macroverse." Relativity and quantum mechanics had demonstrated that in the subatomic world predictability was impossible, but in the world of physical objects it was still assumed that certain processes could be accurately predicted provided enough was known about the initial conditions. Lorenz's work changed all that since he showed that long-term weather patterns were dictated by "sensitive dependence on initial conditions." A weather system is so complex and contains so many dynamical elements that modify, feedback on, enhance, and distort all others that the tiniest fluctuations eventually lead to massive changes in the system, however large. As Lorenz suggested (though in slightly different terms), a butterfly flapping its wings in the Amazon could conceivably affect tornadoes in Kansas. In effect, unpredictability is intrinsic to weather systems; it is impossible to "factor out."

It is this intrinsic randomness and unpredictability that Lorenz had stumbled upon that gave rise to the notion of Chaos. In the decades following his initial discovery, Chaos was found in other systems including biological population dynamics, economic boom-bust cycles, air and fluid turbulence, and a host of other natural and technical actions. But Chaos is a process more subtle than it first appeared, for Lorenz noticed that though it was impossible to predict weather systems beyond four or five days, there was a definable form to the chaotic behavior. Lorenz spotted this mathematically by graphing the constant changing and shifting of weather patterns. Though the patterns on the graph never repeated themselves or settled into any regularity, they did configure into a looping shape that observed certain boundaries. The continuous looping formed a figure-eight or "butterfly wing" pattern around two poles and to this figure was given the name "strange attractor."[13] It meant that the chaotic transitions of the system seemed to be pulled or attracted to a certain configuration. Thus, in spite of the chaotic unpredictability of the system, a certain regularity and order prevailed.

Chaos theorists in the years to come found other patterns of order in otherwise chaotic systems. One such was the biologist cum mathematician Robert May who was interested in what happened to the rate of growth of biological populations when a single parameter, say food availability, in a nonlinear mathematical equation was changed. May, in the course of his investigations, raised the parameter slowly, monitoring the consequent rise in population:

> Suddenly . . . as the parameter passed 3, the line broke in two. May's imaginary fish population refused to settle down to a single value, but oscillated between *two* points in alternating years. Starting at a low number, the population would rise and then fluctuate until it was steadily flipping back and forth. Turning up the knob a bit more . . . would split the oscillation again, producing a string of numbers that settled down to *four* different values, each returning every fourth year.[14] (emphasis added)

Continuing to increase the parameter produced continued bifurcations:

> As the parameter rose further, the number of points doubled again, then again, then again. It was dumbfounding—such complex behavior, and yet so tantalizingly regular. . . . The doublings themselves were bifurcations, and each bifurcation meant that the pattern of repetition was breaking down a step further . . .

[13] See Gleick, ibid., pp. 25-9 for a fuller discussion of strange attractors. See also Briggs J, *Fractals: The Patterns of Chaos*, New York: Touchstone, 1992, pp. 57-9.

[14] Gleick, op. cit., pp. 71-2.

> These bifurcations would come faster and faster—4,8, 16,32 . . . and suddenly break off. Beyond a certain point, the "point of accumulation," periodicity gives way to chaos, fluctuations that never settle down at all.[15]

May found that pushing the parameter upward through the chaos he had created produced another unexpected result: suddenly, without warning, the regularity would reappear again, showing periods of some stable number like 3 or 7. But continue to push the parameter, and these periods again start to bifurcate and re-bifurcate until they, in their turn, transformed into chaos. May's colleague James Yorke was one of the few, at first, to realize the implications of this work:

> James Yorke analyzed this behavior with mathematical rigor in his "Period Three Implies Chaos" paper. He proved that in any one-dimensional system, if a regular cycle of period three ever appears, then the same system will also display regular cycles of every other length, as well as completely chaotic cycles.[16]

The phenomenon of period-doubling and bifurcation governing the transition of May's population growth to a chaotic pattern is seen in other dynamical systems. In fact, it seems to govern most, if not all, observed transitions of dynamical systems from stability to chaos.[17]

Considering the whole of the period-doubling phenomena, its relationship to all possible numbers and to Chaos, it seems possible to resolve the whole thing down to the 2^n structure inherent in the Egyptian mathematical system. That is to say, the dynamical processes that change a regular, ordered system into a random, chaotic one seem to follow the progression 1,2,4,8, 16,32, etc., that is to say the 2^n system. We have seen that there are echoes of it in the Ishango Bone and it undergirds the entire mathematical system described in the Rhind Papyrus. Indeed, the method depicted in the Rhind Papyrus of multiplying by doubling seems to be an exact analogue of the phase transitions of dynamical systems as they transform into chaos.

It is pertinent to mention here that other African number systems are rooted in this 2^n system. One of the most conspicuous examples is the Ifa oracle of the Yoruba, a divination system employed by West African peoples from the Niger Delta to Ghana:

> The sixteen palm nuts are grasped in the right hand, leaving only one or two nuts in the left; if two nuts remain, a single mark is made on the divining tray; if one nut remains a double mark is made. Repeating this procedure four times will give

[15] Ibid., p. 72-3.
[16] Ibid., p. 73.
[17] Ibid., passim.

> one of sixteen basic figures . . . repeating it eight times gives a pair . . . of the basic figures, i.e., one of the 256 derivative figures.[18]

Each of these 256 possible figures has several proverbs associated with it that cover most life situations. A qualified diviner is therefore required to memorize a thousand or more of these proverbs. But it is clear that the manner in which the figures are generated and the divination carried out comes from the 2^n series (i.e., 1,2,4,8,16,32, etc.). In the case of Ifa, the number 4 has special power. The number 16, because it is 4^2, and 256, because it is 4^4, are also exceedingly powerful.[19]

It is worth mentioning here that apart from the 2^n series evident in RMP doubling operations for multiplication, it is also intrinsic to the so-called Horus-eye fractions. In this system, the parts of the right eye of Horus, emblematic of the sun, are assigned the fractional equivalents $^1/_2$, $^1/_4$, $^1/_8$, $^1/_{16}$, $^1/_{32}$, $^1/_{64}$ (see Fig. 19). As Gillings informs us, however, these Horus-eye fractions were used solely in the measurement of grain.[20] This series incorporates the reciprocals of 2 to its 6th power (and 4 to its 3rd). This numerical principle is inherent also in the Chinese *I Ching* divination system.

If all the fractional equivalents that compose the eye of Horus are added, the sum is $^{63}/_{64}$. There are some interesting features about this fraction, not the least is its accompanying mythology that tells us that it results from the duel between Horus and Set, where Set tore out Horus's right eye. Thoth restored the eye except for a small fragment, accounting for the missing $^1/_{64}$th. As already mentioned, it also incorporates the reciprocal 2^n series, that is to say $^1/_{2^n}$ out to the 6th power ($^1/_{64}$). Finally, $^{63}/_{64}$ is what is known as a complementary fraction, meaning that it is part of a series of fractions, beginning with $^2/_3$, that is found by substracting $^1/_n$ from 1 where n is any positive integer: $^2/_3$, $^3/_4$, $^4/_5$, $^5/_6$ . . . $^{63}/_{64}$, etc. Kurt Sethe voiced the opinion that the ancient Egyptians might have known of such a series, though the only fraction in this series they expressed with a numerator greater than 1 was the ubiquitous $^2/_3$ fraction.[21]

The Egyptian Fractional System

To the modern way of thinking, the Egyptian fraction system seems peculiar and awkward in the extreme, though it was adopted virtually without modification by the Greeks and was used up to the time of the European Renaissance. They created what is termed a "unit fraction" system meaning

---

18 Bascom W, *Ifa Divination*, Bloomington: Indiana University Press, 1969, p. 3.

19 Ibid., passim; see also Zaslavsky C, *Africa Counts*, Westport: Lawrence Hill & Co., 1979, pp. 214–8, for a discussion of the significance of 4 and powers of 4 in the Ifa system.

20 Gillings, op. cit., p. 210.

21 Cited in Chace AB, op. cit., p. 6.

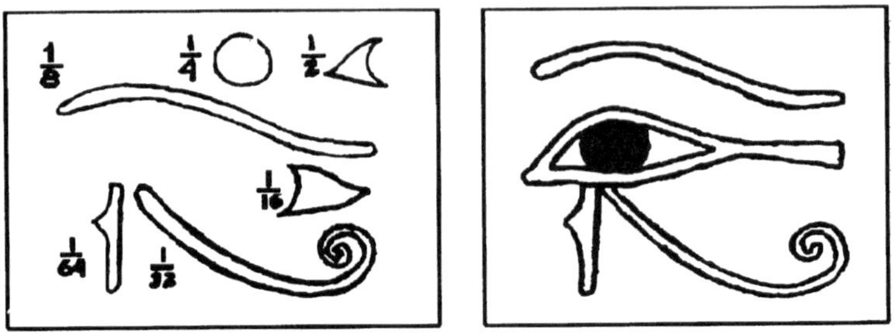

Fig. 19. The Horus-eye fractions used only for dry volume measurement. The sum total of the fractions equalled $63/64$ (James Laws).

that, with one exception, *all* fractions had to be expressed with 1 as the numerator. A fraction was indicated by the oval sign placed over any integer to represent that integer's reciprocal. The exception to this notation was the $2/3$ fraction, the *only* fraction without unity as a numerator and with its own special notation using the oval. Parenthetically, it is of interest that the oval sign transliterates as our "r" letter and sound and is also a hieroglyphic sign meaning "mouth, speech, words, opening, or entrance" (see Fig. 20). Why this particular sign should be used by the Egyptians as the indicator of the fraction (i.e., as a kind of numerator) is not clear.

The $2/3$ fraction is used so frequently in the operations of multiplication and division, that some mathematical historians assume that the scribes possessed a $2/3$ table of fractions. As Gillings has said, "This 'two-thirds' table would in general suffice for all the scribes' needs, and extended up to 100, it would enable them to find $2/3$ of any number, integral or fractional, because multiplication and division by 10 was a commonplace operation with them."[22]

The unit fraction system necessitated a notation that seems cumbersome by our standards. The fraction $2/5$, for example, would be written $1/3$ $1/15$; the fraction $5/9$ would be written $1/3$ $1/6$ $1/18$. As indicated earlier, fractions with even denominators could be handled with ease but odd-numbered denominators were another matter. The RMP reproduced a table of the unit fractions created from the division of 2 by all odd numbers between 3 and 101.[23] It is very evident that there was a careful method applied to this table because, given the unit fraction format, these $2/n$ fractions had to be decomposed into as many as four terms. For example, $1/89$ was notated as $1/60$ $1/356$ $1/534$ $1/890$. What has astonished historians, however, is that all of these fractions are decomposed in the most parsimonious form possible. That is, out

22 Ibid., p. 27.
23 Chace AB, op. cit., pp. 28–33.

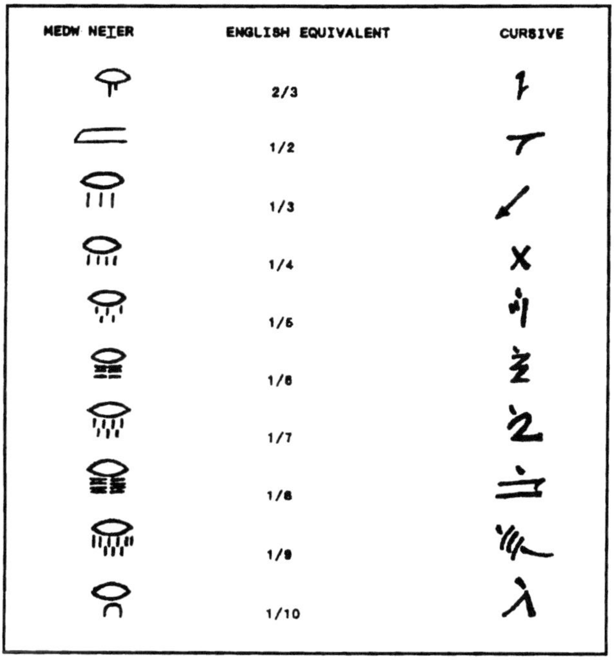

Fig. 20. The two-thirds and unit fractions of the ancient Nile Valley arithmetic system (James Laws).

of more than 22,000 possible combinations of terms to make up all the unit fractions, the creator of this table invariably chose the simplest set of terms for each fraction. This accomplishment has modern mathematicians scratching their heads:

> How was it possible for them, with only a knowledge of the twice-times table and an ability to find two-thirds of any integral or fractional number, to calculate unit fractional equivalents of $^{2}/_{5}$, $^{2}/_{7}$, $^{2}/_{9}$, . . . , $^{2}/_{101}$ without a single arithmetic error? And how did it come about that, of all the many thousands of possible answers to these decompositions, those recorded by the scribe of the RMP were in almost every case the simplest and best possible by his own prescribed standards?[24]

The parsimony is evident in the selection preference for unit fractions with the fewest terms possible. Given a choice, a two-term fraction is set down in preference to a three-term, a three-term in preference to a four-term, and there are *no* fractions with more than four terms. Moreover, no fraction is

---

[24] Gillings, op. cit., p. 47.

ever duplicated in a set, smaller denominators are preferred to larger ones, and even denominators are preferred over odd.[25]

It is not entirely clear why the Egyptians made such extensive use of the $^2/_3$ fraction though it is evident that for certain fractional terms, the $^2/_3$ fraction facilitated the parsimony that the Egyptians desired. For example, 5/7 could be written $^1/_2$ $^1/_7$ $^1/_{14}$ or $^2/_3$ $^1/_{21}$. Clearly, parsimony dictates selection of the second set of terms. On the other hand, $^4/_5$ can be written $^1/_2$ $^1/_5$ $^1/_{10}$ or $^2/_3$ $^1/_{10}$ $^1/_{30}$. The rules evident in the RMP dictate the selection of the first set because the denominators are smaller. Thus, it would seem that a fractional equivalent would express the $^2/_3$ term only if it reduced the number of terms in the set. Whatever the reason for performing operations by means of the $^2/_3$ fraction, it was clear that doing so became second nature to the mathematical scribes of ancient Egypt. In fact, whenever carrying out an operation to find the $^1/_3$ of any number, the scribes invariably found its $^2/_3$ first—very likely by a glance at a prepared table—and then halved the answer.

Many modern commentators have disparaged the Egyptians' unit fraction system and professed to be astonished that in the West, the system continued in use from Greek times until the eve of the Renaissance. Fibonacci is credited with introducing the so-called Hindu-Arabic numerals to Western mathematics in the middle of the 13th century but this notation did not supplant that of unit fractions for another 200 years. Richard Gillings offers one plausible rationale for the unit fraction system:

> The Egyptians' concern for the accurate dealing with fractions almost certainly originated from practical problems, such as the division of food supplies, and other things, either equally or in some specific ratio, among families, troops, members of working gangs, large crews, etc., in a country which had no metallic currency or money, and in which payments were made in kind.[26]

Harmonics, Complementary Fractions, and the Music of the Spheres

On another line of thinking, Schwaller de Lubicz has pointed out that when examining the ancient world system of the Egyptians, particularly their science, one encounters an entirely different mentality.[27] Science in the Nile Valley represented a complete interpenetration of the empirical, theoretical, and symbolical. In mathematics, therefore, number and numerical operations were not merely abstract creations of the human intellect but expressed the fundamental material and spiritual realities of nature.

---

[25] See Gillings, op. cit., p. 49, for a very useful discussion of the rules employed to create the 2/n table for all odd integers between 3 and 101.

[26] Ibid., p. 105

[27] Schwaller de Lubicz RA, *The Temple in Man*, translated by Robert Lawlor, Cambridge, Massachusetts: Autumn Press, 1977.

Pythagoras, who for 22 years underwent initiatic training in science, philosophy, and symbolism in a *per ankh* ("House of Life"), probably in Waset (Thebes), is the one responsible for passing the number symbolism gleaned from Temple teachings into Western culture. Indeed, through the fragments of Pythagorean writings that have survived we can learn much about the role of number and numerical proportions in the ancient system of Egypt. Thus, the discovery of the system of harmonic intervals that intertwines mathematics and music, and was attributed to Pythagoras by later historians, was never claimed by the philosopher himself. On this subject, Gardner Wilkinson states:

> Though it is impossible for us now to form any notion of the character or style of Egyptian music, we may be allowed to conjecture that it was studied on scientific principles; and, from the great attention paid to it by Pythagoras, many years of whose life were spent in learning 'the wisdom of the Egyptians' . . . [28]

He further cites Greek attestations that music was a sacred science among the Egyptians, vouchsafed to them by the gods, and therefore an ineluctable feature of the training of the higher priesthood. Of the many reasons that the Greeks after Pythagoras's time visited Egypt, improving and refining their musical knowledge was one of the most important.[29]

This being the case, the temple priests of the Nile had already mastered harmony centuries before Pythagoras was admitted among them:

> That they paid great attention to the study of music, and had arrived at a very accurate knowledge of the art, is evident, when we consider the nature of the instruments they used, and the perfect acquaintance they must have had with the principles of harmony . . . [30]

Wilkinson arrived at this conclusion based on his study of Egyptian wall paintings showing musicians using a variety of stringed instruments, especially the three-string guitar. Only skilled, well-trained musicians with a mastery of harmony can produce a variety of perfect melodies by foreshortening the strings of the instrument on a fingerboard as it is being played. The technique is familiar to anyone who has ever watched a contemporary violinist or guitarist and is represented on countless Egyptian drawings of musicians. Indeed, as we shall see in our discussions of the monochord below, the harmonic scale is achieved by repeatedly halving or shortening a single-stringed instrument. Based on the testimony of ancient writers and his knowledge of Egyptian instruments Wilkinson stated flatly: "It appears . . . that music was

---

[28] Wilkinson G, *The Manners and Customs of the Ancient Egyptians*, London: John Murray, 1878, pp. 431-2.
[29] Ibid., p. 432.
[30] Ibid., p. 433.

Fig. 21. Egyptian musician playing the 14-string harp.

studied by the Egyptian priests with other views than that of affording pleasure and entertainment, the *same science being borrowed by Pythagoras from Egypt . . .* "(emphasis added).[31]

We have already had cause to call attention to the series of complementary fractions, that is those fractions created by subtracting $^1/_n$ from 1 where n is any integer. As we have seen, this gives us the series $^1/_2$, $^2/_3$, $^3/_4$, $^5/_6$, $^6/_7$, $^7/_8$, $^8/_9$, etc. It just so happens that this series conforms to the harmonic scale whose properties were reputedly discovered by Pythagoras. Simply, these harmonic proportions can be ascertained by manipulation of a monochordal instrument. Upon examining the practical relation between number and harmony, one is led into a vocabulary that is familiar to us through the language of Chaos theory:

> The problem which the monochord presents is that the string can be divided at *any* point. The string represents an *indefinite continuum* of tonal flux which may be *infinitely* divided. How, then, is it possible to "create" a musical scale at all? The solution . . . resides in the *limiting power of Number*. (emphasis added)[32]

---

[31] Ibid., p. 444.

[32] Fideler DR, "Introduction," in Guthrie SG, *The Pythagorean Sourcebook and Library*, Grand Rapids: Phanes Press, p. 25.

The property that distinguishes a musical tone from mere noise is number or a numerical proportion. The doubling and halving that we find characterizing dynamical systems in transit to chaos finds yet another echo in harmonics:

> The overtone series provides . . . the architectural foundation of the musical scale, the basic "field" of which is the octave, 1:2, or the doubling of the vibrational frequency, which inversely correlates with a halving of the string.[33]

The musical scale can be constructed by a series of ratios obtained first by an arithmetic mean giving us, after the octave ratio 1:2, 2:3 or the perfect fifth. The formula for the arithmetic mean can be expressed as (A + B)/2. The second ratio on the scale is obtained by performing the operation of the *harmonic* mean, obtained by multiplying the two extremes of the octave, doubling the product, and then dividing this result by the sum of the two extremes. This can be expressed in the formula 2AB/A+B. We have before us two operative principles—the arithmetic mean and the harmonic mean—that are absolutely attested to in the RMP in Problems 74 and 76.[34] In any case, the harmonic mean gives a 3:4 ratio, i.e., the perfect fourth that is, harmonically, the inverse of the perfect fifth (2:3). We now have a series or scale of $^1/_2$, $^2/_3$, $^3/_4$ that are a series of complementary ratios or fractions and, in the words of David Fideler, "While we have not 'created' a complete musical scale, we have arrived at the architectural foundation on which it is based."[35]

In measuring the proportions that create overtones on the monochord, the C scale is often the starting point because it is the only one based on the eight primary tones that does not exhibit sharps or flats. The C scale thus contains a certain "purity" that makes it simpler to learn music on. Moreover, the C scale is created by successively halving the monochord thus creating vibrational frequencies that obey the properties of the familiar 2^n series, i.e., 1, 2, 4, 8. It is worth pointing out, however, that other scales, such as G, based on a string length of 1/3 and a frequency of 3, also creates its scale by halving and doubling.

If one starts with a C-note and wishes to create a D-note, the vibration of the C tone must be increased by a ratio of 8:9; from D to E, the vibration must be increased by the identical ratio of 8:9. However, the next tone, F, cannot be created by an E vibrational ratio of 8:9, but rather 243:256. The Greeks called this ratio the *leimma* or "left over" and it is the same as our semitone. This leimma is defined as the "excess of the fourth over the double tone: $^4/_3 : {}^9/_8 \times {}^9/_8 = {}^4/_3 \times {}^{64}/_{81} = {}^{256}/_{243}$."[36] However, starting from G, the

[33] Ibid.
[34] Gillings, op. cit., pp. 130–1.
[35] Ibid., p. 27.
[36] Ibid., p. 50.

same 8:9 ratio abides in creating tones until B is reached. The interval between B and C is, once again, a leimma.

No one who has even a passing acquaintance with the RMP can fail to appreciate the signficance of the proportions 8:9 and $^{256}/_{243}$. One of the most famous problems of RMP is the 48th that gives the formula for finding the area of a circle. The RMP formula can be rendered as $A = (^{8d}/_{9})^2$ or $^{64}/_{81}d^2$, where d is the diameter, and the RMP value for π derived from this formula is $^{256}/_{81}$ or (decimally) 3.1605. Note that the RMP's π is *exactly 3 times the leimma proportion* $^{256}/_{243}$ $(4^4/3^5)$ related to the semitones between E and F, and B and C. As we have seen above, this leimma is derived from the excess of the fourth over the double tone, i.e., $^4/_3$, and the RMP π figure equals $(^4/_3)^4$.

The $^8/_9$ in the RMP formula for the area of a circle is the exact proportion used to create different tones from C to D to E and then from F to G to A to B. There are simply too many correlations here to explain them away as coincidence. That Pythagoras studied geometry, number, and astronomy at Thebes in Egypt of the 6th century B.C. is one of the strongest arguments in favor of the interrelation of these ratios. In a citation by Gardner Wilkinson, this conclusion is underscored by the judgment of the explorer Bruce commenting on Egyptian harps:

> In noticing the harps of a tomb at Thebes, Bruce makes the following remark . . . [they] 'are altogether, in their form, ornaments, and compass, an incontestable proof, stronger than a thousand Greek quotations, that geometry, drawing, mechanics, and music were at the greatest perfection when this instrument was made . . . [37]

With this perspective, we can amplify our previous conclusion that it was in Egypt that Pythagoras studied the musical principles that gave rise to the theories linking number and harmony for which he is famous. Egypt taught him the "music [and the geometry] of the spheres."

These considerations all provide a clue why the RMP reproduced a π value that was close to, but less precise than, our modern value for π of 3.1416 or, in fractions, $3^1/_7$ (3.1428). Though we will discuss the Great Pyramid at greater length later, it has been known for at least 140 years that in several parts of its structure, the π relation of $3^1/_7$ was built into it.[38] The RMP π relation deviates from the modern value by nearly 1.9% whereas the Great Pyramid's $3^1/_7$ value deviates by only 0.1%. Clearly, the Egyptians knew of a more precise value for π than the RMP reveals, so why reproduce the less precise value at all?

---

[37] Wilkinson, op. cit., p. 435.

[38] See Petrie WF, *The Pyramids and Temples of Giza*, London: Histories & Mysteries of Man Ltd., 1990; also, Piazzi Smyth C, *The Great Pyramid*, New York: Bell Publishing Co., 1978.

We have already indicated that Egyptian mathematics, indeed all its science, was required to serve three equally important functions: (1) empirical, (2) theoretical, and (3) symbolical.[39] Which particular function predominated in any situation might vary according to the circumstances or purpose. One has to surmise that, while adhering as closely to empirical and theoretical requirements as possible, the $^{256}/_{81}$ (3.1605) π relation served preeminently a symbolical and "connective" function, relating it, and therefore geometry itself, to the harmonic proportions of the musical scale. Thus, number, geometry, and musical harmony were linked in a "wholistic" fashion that seems entirely foreign to the modern sensibility but powerfully informed the entire system that became known as "Pythagorean."

This line of argument receives support from Plato's famous dialogue, *Timaeus*. Alfred North Whitehead once opined that all Western philosophy was but a series of footnotes to Plato and in its turn it can be said that Plato's dialogues were but a series of footnotes to Pythagoras. That of course means that they are but commentaries, once removed, on the Temple wisdom of Egypt, more especially since Plato trained there for 13 years. Modern historians of philosophy despair of fathoming *Timaeus* because it seems but a hopeless mishmash of fable, metaphysics, symbolic geometry, and mythic cosmogony. Since they have not troubled to study the science and symbols of the ancient Nile Valley, it is no wonder that *Timaeus* seems impenetrable.

Plato's special terminology for the One and the Many or the Indivisible and Divisible was the *Same* and the *Different*. In describing *Existence*—the *World Soul*—he demonstrates the interrelation between the *Same* and the *Different* by a series of exponential numerical divisions or proportions of the World Soul which, starting from Unity, can be related as follows:[40]

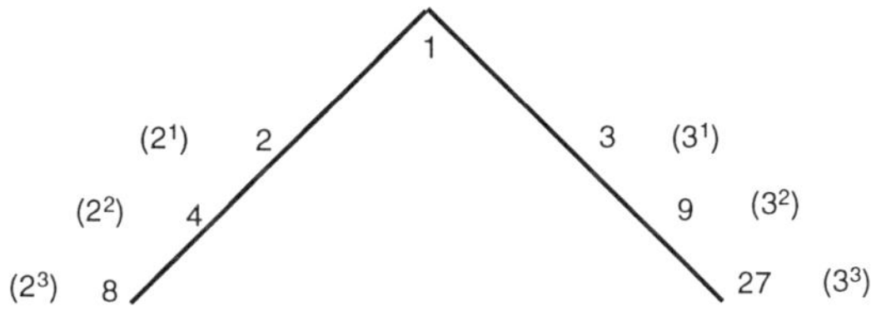

There are, as indicated, seven terms in this series. Then Plato describes the means by which the intervals created by these divisions were "filled in" by certain additional proportions:

[39] Some might say also a spiritual function as well.

[40] See Plato, *Timaeus and Critias*, translated by Desmond Lee, New York: Penguin Books, 1979, pp. 48–9, for further commentary on this section of the dialogue.

These links produced intervals of $3/4$ and $4/3$ and 9/8 within the previous intervals, and he went on to fill all intervals of $4/3$ with 9/8; this left, as a remainder in each, an interval whose terms bore the numerical ratio of 256 to 243.[41]

It is easy enough to see that these proportions are derived from the harmonic scale handed down from the Temple through Pythagoras, as shown previously. The leimma of $256/243$ is derived from the "excess of the fourth over the double tone," i.e., $4/3 \div (9/8 \times 9/8)$, or $4/3 \times 8/9 \times 8/9$. Now Plato gives us the solution to problem of relating geometrical constants to the harmonic scale:

He [God] then took the whole fabric and cut it down the middle into two strips, which he placed cross-wise at their middle points to form a shape like the letter *X*; he then bent the ends *round in a circle* and fastened them to each other opposite the point at which the strips crossed to make two circles, one inner and one outer. (emphasis added)[42]

The "fabric" to which Plato refers is the Fabric of Existence, i.e., the "World Soul," which has been divided into the "Pythagorean" proportions or ratios outlined above. Concerning Plato's scheme, Diop has this to say:

Plato seems to be inspired by the so-called Pythagorean theories, because the progression's mathematical intervals are now considered to be *musical* intervals that he fills in with mediators, meaning *arithmetic* and *harmonic* means, in such a way that one passes . . . from one extreme to the other: the mathematical relationship or ratio that unites musical tones is called . . . *logos*; it deals with a harmonizing operation; there is a symphony when the intervals are all in tune.[43]

What interests us here is the manner in which Plato turns these numbers, proportions, and harmonic intervals *into circles*. Here, then, is the connecting link between the RMP π constant of $256/81$ (3.1605), generated by the $8/9$ term in the formula for a circle, and the harmonic scale where tones are generated by the $8/9$ proportion and semitones by the leimma of $256/243$, which is exactly $1/3$ of π.

The two concentric, crossing circles described in Plato's scheme constitute the simplest form of an *armillary sphere*. Armillary spheres were constructed in pre-Christian antiquity to represent the great circles of the heavens. Plato's schema gives us the Circle of the Same and the Circle of the Different; the former represented the motion of the fixed stars, the latter the motion of the planets.[44] Other, more complex armillary spheres contained circles to repre-

---

[41] Ibid., p. 48.
[42] Ibid., p. 49.
[43] Diop, op. cit., p. 350.
[44] See Desmond Lee's footnote on page 49 in his edition of *Timaeus*.

sent the horizon, the median, the equator, the tropics, the poles, and the ecliptic. But what Plato has preserved for us is the Egyptian Temple schema for integrating and unifying number, geometry, musical harmony, and astronomy. No wonder the Pythagoreans and Platonists were profoundly moved by the "music of spheres." Geometry and harmony seemed to rule Heaven. Again, Gardner Wilkinson lends considerable weight to these suppositions:

> We may fairly conclude that he [Pythagoras] derived his knowledge of this subject [musical tones] from the same source [Egypt] as that of *the solar system*, which remained *unknown in Europe from his time to the days of Copernicus*, and with which Pythagoras, of all the Greeks, was alone acquainted. (emphasis added)[45]

We will have reason to return to this discussion of ancient Nilotic knowledge of the solar system in Chapter 5.

Chaos, "Self Similarity," and the 2/3 Fraction Table

We have already caught a glimpse of the way in which the 2^n series, upon which the mathematical system of ancient Kemit was founded, resonates with a definite mythic symbolism and yet also defines certain fundamental patterns in the transitions of natural dynamical systems. In modern science, Chaos theory seems to be the new paradigm connecting the symbolic essence of number to nature.

As Chaos theory, affecting a multiplicity of disciplines, was gathering steam in the mid-1970's, a Los Alamos physicist, Mitchell Feigenbaum, became immersed in studying the numerical properties of dynamical systems, from fluid turbulence to changes in biological population growth.[46] The doublings and splittings (halvings) of these dynamical systems en route to chaos seemed to occur in the "boundary conditions" separating stability and randomness, and Feigenbaum began by calculating the exact parameter values of these doublings and splittings. In effect, Feigenbaum was attempting to determine not only the speed with which the doublings occurred but the rate of change of the speed of the doublings. What he found was startling:

> There was an unexpected regularity hidden in this system: the numbers were converging geometrically, the way a line of identical telephone poles coverges toward the horizon in a perspective drawing. . . . The period-doublings were not just coming faster and faster, but they were coming faster and faster at a constant rate.[47]

---

[45] Wilkinson, op. cit., p. 448.
[46] Gleick, op. cit., p. 171.
[47] Ibid., p. 172.

Geometric convergence means that some number, some ratio is repeating itself on different scales. Feigenbaum calculated the ratio of convergence to three decimal places and came up with 4.669 or, in fractions, $4^2/3$. Feigenbaum then starting working on the speed and rate of period-doublings utilizing a much more complex trigonometric function and found, again, a convergence rate of $4^2/3$ (4.669):

> Incredibly, this trigonometric function was not just displaying a consistent, geometric regularity. It was displaying a regularity that was numerically *identical* to that of a much simpler function. No mathematical or physical theory existed to explain why two equations so different in form and meaning should lead to the same result.[48]

As James Gleick, author of *Chaos*, explains, Feigenbaum seemed to have stumbled on some scaling "law of nature," or, at the very least, a new "constant." The elegant, rigorous, complex equations and proofs so dear to a mathematician's heart seemed to be obeying an inner, hidden law with a defined, invariable numerical value. Looking below the surface of these equations revealed an identity of number—the same for a quadratic equation as for a trigonometric one—that made the equations themselves seem a bit irrelevant. In the end, Feigenbaum called his "new" reality "universality" and universality meant that, in the transition from order to chaos, different systems behaved the same.

What does any of this have to do with Egyptian mathematics? In this particular instance the invariable convergence rate of $4^2/3$ (4.669) instantly suggests the $^2/3$ table that Gillings insists must have existed among Egypt's mathematical scribes.[49] This $^2/3$ table was essential to the Egyptian method of handling fractions since $^2/3$ was *the only* fraction ever used by the Egyptians with a numerator greater than 1. Moreover, when we look at the number $4^2/3$, we find that it is exactly $^2/3$ of 7:

> To the ancients of all nations, the number 7 seems always to have held some ineffable fascination. . . . The number 7 was constantly before the Egyptian scribe, for in his multiplication, based upon the method of continued doubling (and halving in division), the sequence 1 + 2 + 4 = 7 appeared frequently. Furthermore, the Egyptian table of length was 4 digits = 1 palm, 7 palms = 1 royal cubit, hence 28 digits = 1 royal cubit.[50]

We find that this convergence number $4^2/3$—and therefore the $^2/3$ fraction itself—is related to the doublings of the 2^n series inherent in the change

---

[48] Ibid., p. 173.
[49] See Gillings, op. cit., p. 24.
[50] Gillings, op. cit., 100.

in dynamical systems. We have seen above that Yorke proved that "period 3 implies chaos." It is difficult not to think these various numerical relations are connected. Gillings tells us: "The scribe did not need an extensive one-third table [for fractions], because each entry [to make $^1/_3$] could be found by merely doubling the corresponding entries in the two-thirds table."[51]

There are other tantalizing clues. Claudia Zaslavsky has reprinted a photo of an Eygptian vase which represented a standard volume measure of $8^1/_6$ *hinu*.[52] The number $8^1/_6$ is exactly $1^1/_{16}$ of 7. The fraction $1^1/_6$ is exactly $^1/_4$ of $4^2/_3$ which itself is $^4/_7$ of $8^1/_6$. Looked at reciprocally, $4^2/_3$ is *4* times $1^1/_6$ and $8^1/_6$ is $1^3/_4$ times $4^2/_3$.

The Egyptian preoccupation with the number 7 is evident from the above quote and is explored in the Egyptian Mathematical Leather Roll and the RMP. From the various subunits of the cubit measure noted above we find that since 4 + 2 + 1 = 7 in digits, then $^1/_7$ $^1/_{14}$ $^1/_{28}$ = $^1/_4$ of the cubit equivalence in unit fractions.[53] In Problem 34 of RMP, we have the relation $4 \times (1^1/_2\ ^1/_4) = 7$; that is to say, $4 \times 1^3/_4 = 7$. There are other ways to establish to the unit fraction equality of $^1/_7$, $^1/_{14}$, $^1/_{28}$ = $^1/_4$, to wit, the multiplication of $^1/_4$ by 7:

1/	**1/4/**
2/	1/2/
4/	1/
7	1 1/2 1/4 (1 3/4)

Since $1^1/_2\ ^1/_4\ (1^3/_4) = {}^1/_4 \times 7$, then $1^1/_2\ ^1/_4 \div 7 = {}^1/_4$.

The point is that the technique of multiplication by doubling or dividing by halving helps establish the importance of 4 with its reciprocal $^1/_4$, and $1^3/_4$ with its reciprocal $^4/_7$, in Egyptian operations involving 7. We see, furthermore, that these numbers establish a direct correlation between $4^2/_3$ and a definable Egyptian measure (8 1/6 hinu). Thus, the number $4^2/_3$ participates in the $^2/_3$ series, the fractional operations of 7, and indirectly in the 2^n series, all of which are fundamental elements of Egyptian mathematics. Though there are those who would undoubtedly explain this proposed relationship between Feigenbaum's $4^2/_3$ (4.669)—a number that characterizes the rate of change of *all* dynamical systems, making it *the* number of dynamical universality—and the mathematics of the RMP as merely fortuitous, in light of what we have already seen, such a reflex dismissal hardly seems admissable. The impression grows that what is being called Chaos theory already has an old, old history in Africa.

The Egyptian preoccupation with number begins to become intelligible

---

[51] Ibid., p. 30. For example: 2/3 of 1/3 = 2/9; if 9 is doubled, 2/18 is the result, which = 1/9. 1/9 is in fact 1/3 of 1/3.

[52] Zaslavsky C, op. cit., p. 24.

[53] Gillings, op. cit., p. 100.

once we take Chaos theory into account. The Egyptians recognized simplicity, pattern, and order underlying all the phenomenal complexity of creation. All the apparent randomness and disorder that we perceive in nature hides a regularity that conforms to simple numerical relationships. It is why they insisted that the Rhind Papyrus represented a "correct method for enquiring into nature, and for knowing all that exists, every mystery, every secret."[54]

The Emergence of Algebra in the Nile Valley

Until the publication of Richard Gillings' *Mathematics in the Time of the Pharaohs* in 1972, there had existed a distinct tendency to belittle and minimize Egyptian mathematics. Otto Neugebauer typlified this attitude and he spent the best part of a generation, from the 1930s to the 1960s, proclaiming that ancient Egypt had produced no mathematics worthy of the name: "The role of Egyptian mathematics is probably best described as a retarding force upon numerical procedures."[55]

As if to close the door forever on the subject of ancient Egyptian science, Neugebauer pronounced that "Ancient science was the product of a very few men; and these few happened not to be Egyptians."[56] Nor was Neugebauer alone; T.E. Peet, whose reputation as a historian of science in the first half of this century equalled Neugebauer's, refused to acknowledge that problem 10 of the Moscow Mathematic Papyrus really contained a formula for finding the area of the curved surface of a hemisphere though his contemporary, Struve, had conclusively proved that it did. Later scholarship has proved Struve right and Peet not merely wrong but wrong-headed.[57]

As Diop points out, the Greeks were unanimous in their assertions that geometry was invented and reached its highest development among the Egyptians. Typically, Neugebauer refused to concede even this accomplishment to the Egyptians, but even where others did concede it, they tried to insist that the Egyptians never progressed to algebra: "We may note that there was very considerable development of geometry in Egypt, *but apparently no algebra*." (emphasis added)[58] This, like those of Neugebauer, is a remarkable statement considering that a translation of the RMP has been available since 1929 and no one who has studied that document with any care could possibly, in good faith, reach such a conclusion.

One of the clearest algebraic elements in Egyptian mathematics is seen in

---

[54] Title rubric from the Rhind Papyrus.

[55] Neugebauer, op. cit., p. 80.

[56] Ibid., p. 91.

[57] See Diop CA, *Civilization or Barbarism*, translated by Y.L.M. Ngemi, Brooklyn: Lawrence Hill Books, 1991, pp. 251-8 for commentary on problem 10 of the MMP and Peet's bizarre reasoning in his attempt to discredit Struve's reconstruction of the problem.

[58] Dr. Marshall Clagett quoted in Hapgood CH, *Maps of the Ancient Sea Kings*, New York: E.P. Dutton, 1979, p. 221.

the various number series present in the system. There are several examples of arithmetic series or, as they're called, *arithmetic progressions*. The most obvious such progression is the one obtained simply by counting by integers, i.e., 1, 2, 3, 4, 5, 6, 7, etc. which is called an *infinite progession* because it goes on to infinity. A *finite progression* contains a last *term*, a term simply being any number in the sequence. All number sequences, however, are not arithmetic progressions which must be obtained by adding a fixed number to *each* term after the first. In an arithmetic progression, the fixed numerical difference between adjacent terms is called the *common difference*.[59]

Problem 64 of the RMP is the outstanding example of an arithmetic progression in the document and it represents a problem of "distributing differences" (see Plate 11):

> Example of distributing difference. If it is said to thee, divide 10 hekats of barley for 10 men, the difference of each man to his neighbor $^1/_8$ this is, [what is the share of each]?
>
> The average share is 1 hekat. Take away 1 from 10; the remainder is 9. This makes $^1/_2$ the difference, that is $^1/_{16}$. Make this 9 times itself, becoming $^1/_2$ $^1/_{16}$ [$^9/_{16}$]. You remove $^1/_8$ hekat for each man until the last one. The doing as it occurs: $1^1/_2$ $^1/_{16}$ $1^1/_4$ $^1/_8$ $^1/_{16}$ $1^1/_4$ $^1/_{16}$ $1^1/_8$ $^1/_{16}$ $1^1/_{16}$ $^1/_2$ $^1/_4$ $^1/_8$ $^1/_{16}$ $^1/_2$ $^1/_4$ $^1/_{16}$ $^1/_2$ $^1/_{16}$ $^1/_4$ $^1/_8$ $^1/_{16}$; totalling 10 hekats.[60]

This can be put into modern terminology as follows: The sum of *10* terms of an arithmetic progression is *10* and the common difference is $^1/_8$. What are the terms of this series? Find the average value of the terms; $10 \div 10 = 1$. The number of differences is one less than the number of terms: $10 - 1 = 9$. Find half of the common difference: $^1/_8 \times {}^1/_2$ which equals $^1/_{16}$. Multiply 9 times $^1/_{16}$ which equals $^9/_{16}$. Add this to the average value of the terms, which equals $1^9/_{16}$. This is the highest term. Subtract the common difference of $^1/_8$ nine times until the lowest term is found. This gives the series $1^9/_{16}$, $1^7/_{16}$, $1^5/_{16}$, $1^3/_{16}$, $1^1/_{16}$, $^{15}/_{16}$, $^{13}/_{16}$, $^{11}/_{16}$, $^9/_{16}$, $^7/_{16}$.

Richard Gillings shows that this progression embodies the formula $S = {}^n/_2[2a + (n-l)d]$ where: a = first and lowest term; l = the last and highest term; d = common difference; n = the number of terms; S = the sum of *n* terms. This handling of an arithmetic progression in the RMP represents mathematical subtlety much more refined than the Egyptians are usually given credit for:

[59] Dolciani MP, Wooton W, Beckenbach EF, Sharron S, *Algebra 2 and Trigonometry*, Boston: Houghton Mifflin Company, 1983, 213–4.

[60] See Chace AB, op. cit., p. 126, for a translation of this problem.

> Hidden in a hieratic script of the Middle Kingdom, expressed in words, without the assistance of any form of algebraic notation at all, is the equivalent of a perfectly well-known modern formula for the sum to *n* terms of an arithmetic progression.[61]

Gillings, one of the most perceptive and unprejudiced of historians of ancient Egyptian mathematics, finds it difficult sometimes to hide his exasperation with some of the more disparaging evaluations of Egyptian mathematics vouchsafed by his colleagues. In reference to the above formula for the RMP arithmetic progression, Gillings was moved to comment on the derision evinced by the mathematical historian Morris Kline: "It is difficult to accept the recent pronouncement of Professor Morris Kline: 'The mathematics of the Egyptians . . . is the scrawling of children just learning how to write, as opposed to great literature.' Scrawling indeed!"[62]

Geometric progressions were also a familiar item on the menu of ancient Egyptian algebra. A geometric progression can be defined as any sequence in which each term succeeding the first is the product of the preceding term and a fixed number called the *common ratio*. The formula for this progression can be written as follows:

$a_{n+} = a_n \times r$ where the first term is a_1, the common ratio is r, and n = 1,2,3 . . .[63]

The familiar $2^{n,}$ series represents a geometric progression, the one upon which the mathematics of the RMP is built. Problem 79 of the RMP represents a geometric progression based on powers of 7 and not only illustrates the Egyptians' command of geometric progressions but also their preoccupation with 7 and all its permutations. Gillings offers one very reasonable explanation for the prominence of 7 in Egyptian mathematical operations; to wit, when 7 is multiplied by doubling, the first three multipliers are *always* 1,2,4 the sum of which *is* 7.[64] Problem 79 is stated rather perfunctorily in the RMP in two columns (see Plate 12):

A House Inventory[65]

	1	2,801	Houses	7
			Cats	49
	2	5,602	Mice	343
			Spelt	2,401
	4	11,204	Hekat	16,807
Total:	7	19,607		19,607

---

[61] Gillings, op. cit., pp. 173-5.
[62] Ibid.
[63] Ibid., p. 227.
[64] Gillings, op. cit., p. 168.
[65] Chace, op. cit., p. 136.

How the scribe arrived at a figure of 2,801 to begin the first inventory column is not set down. However, the sum of the first four terms in the *second* inventory column, 7+49+343+2,401 equals 2,800, and it also just so happens that the last digit of each number in the first inventory column corresponds to its positional column number. That is to say, the 1 in the far left positional column is repeated in 2,80*1*, 2 in the positional column is repeated in 5,60*2*, and 4 in the positional column is repeated in 11,20*4*. The scribe must have found the number 2,801 for the first inventory column by adding the first four terms shown in the second inventory column, getting 2,800, and then super-adding the 1 from the positional column on the far left. Then, it just so happens, that the doubling from there gives 5,60*2* and then 11,20*4*, corresponding to the respective positional numbers and, incidentally, obeying the 1+2+4 = 7 equality that governs all Egyptian computations by 7.

Another peculiar aspect of this progression is that in order to find the first figure (2,800 + 1) of a doubling computation that goes up to 4, the first *4* terms—or terms up to the *4th* power—of the geometric progression by 7's must be summed. It begins to be apparent that the ancient Egyptian scribes had explored and experimented with innumerable properties of number and numerical relationships and had formulated what Gillings calls a "theory of number." Still, everything appeared to harmonize with the 2^n series. The number 4 in this series seems very powerful possibly because of its "limiting" or "completing" capacity.[66] We find, for example, that no unit fraction in the 2/n table for the odd numbers between 3 and 101 in the RMP is ever composed of more than 4 terms. Moreover, we have seen that 4 completes the sum of the integers in the 2^n series that makes 7 (1+2+*4*). The number 4 as "completer" even figures in the decimal system since, as the Pythagoreans were aware, the number 10 is found by adding the first four integers, 1+2+3+*4*.

Gillings clarifies the nature of the geometrical progression by 7's in Problem 79 in the following manner: "Find the sum of 5 terms of the Geometric Progression whose first term is 7 and whose common ratio is 7."[67] Gillings and others have noted the near-identical likeness of the Geometric Progression of Problem 79 to the well-known nursery rhyme, "As I Was Going to St. Ives":[68]

---

[66] Carl Jung and his disciple Erich Neumann wrote extensively about 4 as mythically the number of "completion" or "totality." See Jung C, *Psychology and Religion: West and East*, translated by R.F.C. Hull, Princeton, N.J.: Princeton University Press, 1969, p. 115, and Neumann E, *The Great Mother: Analysis of an Archetype*, translated by Ralph Mannheim, Princeton, N.J.: Princeton University Press, 1955, 1974, passim.

[67] Gillings, op. cit., p. 170.

[68] See Newman JR, "The Rhind Papyrus," reprinted in *Scientific American's Mathematics: An Introduction Its Spirit And Use*, San Francisco: W.H. Freeman and Company, 1952, 1978, p. 13, for a discussion of Problem 79 of RMP and its relation to the Mother Goose Rhyme.

"As I was going to St. Ives	
I met a man with seven wives.	7
Each wife had seven sacks,	49
Each sack had seven cats,	343
Each cat had seven kits.	2,401
Kits, cats, sacks, wives,	
How many were going to St. Ives?	

As every school-child knows, the answer to this riddle-rhyme is "1" but with the exception of the absence of the last term recorded in the RMP, this bit of "child's play" represents a geometrical progression exactly the same as Problem 79. Numerous modern fables, fairy tales, and folkloric traditions have been shown to descend from ancient pre-Christian sources; there is no reason to rule out such a pedigree for this rather charming Mother Goose rhyme.[69]

As indicated above, anyone who insists that the Egyptians were entirely ignorant or barely cognizant of algebra simply has not examined the RMP. Fully 11 problems (i.e. Problems 24–34) are simple algebraic problems that solve equations of the first degree (i.e., those with one unknown). Problem 24, for example, best exemplifies what has come to be called Egyptian problems of the *aha* type (Problems 24–27) and is solved by the method of the "false postion"(see Plate 5).[70] The problem can be reconstructed as follows:[71]

> A quantity [aha] plus its 1/7th becomes 19. What is the quantity? Pose a false answer of 7 whose $1^1/_7$th = 8. As many times 8 multiplies to get 19 is the same number of times 7 must multiply to find the quantity.
>
> 8 divides into 19 to give $2^3/_8$; $2^3/_8$ times 7 equals $16^5/_8$. $16^5/_8$ = the quantity; $2^3/_8$ = its $^1/_7$th.

As several authors have noted, this statement of the problem can be rendered into modern notation as $X + {}^1/_7X = 19$; find X. In both formats, the task is to solve for a single unknown, the simplest form of algebra. In the RMP, the unknown is the *aha*, i.e., "the quantity"; in modern algebraic notation, it is "x."

Another pair of simple algebraic problems (Problems 28–29) constitute what Gillings calls "Think of a Number Problems." These problems are probably the earliest examples of mathematical "games" on record. Problem 28 can be reconstructed as follows (see Plate 6):[72]

---

[69] See Di Santillana G and Von Dechend H, *Hamlet's Mill*, Boston: David R. Godine, 1977, for a marvelously erudite treatment of the mythic sources of fairy tales.

[70] The term *aha* is translated literally as "quantity."

[71] For this and subsequent problems reproduced in this section, the fractions, where feasible, will be rendered in modern notation, instead of unit fractions, to make the operations easier to follow.

[72] Chace AB, op. cit., p. 98.

> [To a number], add its $^2/_3$; from the sum subtract its $^1/_3$. [Suppose] 10 remains. Make $^1/_{10}$, of this 10, to equal 1; by subtraction 9 is remaining (the number first thought of).
>
> The Demonstration: $^2/_3$ of 9, or 6, is to be added to it, making 15. $^1/_3$ of this is 5. Take away this 5 and the remainder is 10. This is the number in the making!

The number to be thought of in this problem is equivalent to the unknown, marking this problem as algebraic.

From the Berlin Papyrus (BMP) come two very similar solutions to quadratic equations, showing that Egyptians went beyond simple algebra. Gillings gives the first problem, exemplifying both of them, as follows:[73]

> If it is said . . . the area of a square of 100 is equal to that of two smaller squares and the side of one is $^3/_4$ the side of the other. Let me know the sides of the two unknown squares.
>
> Demonstration: Always take a square of the side; then the side of the other is $^3/_4$. Multiply this by $^3/_4$, giving $^9/_{16}$, the area of the small square. Together, these two squares have an area of $1^9/_{16}$. Take the square root of $1^9/_{16}$ which is $1^1/_4$. Then take the square root of 100 which is 10. Divide 10 by $1^1/_4$ which gives 8, the side of one square. Take $^3/_4$ of 8 which gives 6, the side of the other square.

This problem can be expressed in modern notation as follows:

$$x^2 + y^2 = 100$$
$$4x - 3y = 0 \qquad x = 6; y = 8$$

Thus, this problem proposes the solution of *two* unknowns from two simultaneous equations, one of them of the second degree.

The second problem from the BMP is identical in form, solution, and proof; only the equality in the first equation has changed:

$$x^2 + y^2 = 400$$
$$4x - 3y = 0 \qquad x = 12; y = 16$$

Thus, we find that the Egyptians had the capacity to solve quadratic equations. All the evidence taken together, the oft-repeated assertion that Egyptian algebra was, at worst nonexistent or at best vestigial, has to be abandoned.

---

73 Gillings, op. cit., p. 161.

Geometry: The Nilotic Science of "Earth Measurement"

It has been said of Plato that there was a sign over the door to his Academy that forbade entrance to anyone who was not a trained geometer.[74] From the time of Pythagoras, the Greeks considered geometry the "queen of sciences" and the starting point for all mathematics. And, the Land of Geometry *par excellence* was universally known to be Egypt.

The ancients were unanimous in their admiration for Egyptian geometry and even most modern historians give Egypt credit for exemplary skill in this branch of mathematics. However, these same modern historians like to point to the superiority of Euclid and other Greek geometers over their Egyptian counterparts, while neglecting to mention that Euclid and the other Greek mathematicians of note, virtually without exception, lived and studied in Alexandria, the "Hellenic" capital of Egypt. Few of them produced great mathematics on the soil of their native Greece and it is difficult to avoid the presumption that they became great mathematicians because Egypt made them so. The more honest chroniclers such as Herodotous, Diodorus, and Clement of Alexandria freely admitted this.

The Library of Alexandria was burned three times between 44 B.C. and 744 A.D., ultimately incinerating most of the scientific, scholarly, and literary heritage of the ancient world. Even the famous dialogues of Plato represent but a part of his written output and some think that these are merely the popular versions of his thought. Thus, it is very dangerous to assume that the Egyptians possessed no greater mathematics than can be gleaned from the four surviving mathematical papyri and the other odd mathematical fragments that have turned up. If the Greeks paid homage to Egyptian geometry, they had good reason.

The two surviving manuscripts that give a picture of the fundamentals of Egyptian geometry are the Moscow Papyrus (MMP) and the RMP. In demonstrating the principles of any science, the Egyptians quite naturally and logically started with the simple examples and progressed to the more complex. Problem 6 of the MMP and Problem 49 of the RMP seek to find the area of a rectangle and embody the formula $A = l \times b$, where l is the length and b is the base of the rectangle. Problem 4 of the MMP and 51 of the RMP to find the area of triangle can be resolved to the modern formula $A = {}^{1}/_{2}bh$, where b is the base and h is the height of a triangle (see Plate 9).

The RMP Problem 50 dealing with the area of the circle is of surpassing interest because it is here that we encounter the equivalent for π in an early (19th century B.C.) Egyptian mathematical document:[75]

---

[74] Fideler, op. cit., p. 34.
[75] Chace, op. cit., p. 93.

Example of a round field (circle) of diameter 9khet; what is the area?

Demonstration: Take away $^1/_9$ of the diameter, namely 1, and the remainder is 8. Multiply 8 times 8, making 64. Therefore it contains 64 setat of land.

Do it thus:	1	9
	1/9	1
this taken away leaves	8	
	1	8
	2	16
	4	32
	8/	64

Its area is 64 setat.

We can formulate this problem and its solution in modern notation as $\text{Area}_{\text{Circle}} = (^{8d}/_9)^2$, where *d* equals the diameter, with an equivalent π value of $^{256}/_{81}$ or 3.1605 in decimal notation (see Plate 8). We have already discussed this formula for an area of circle with its $^8/_9$ relation and π value of $^{256}/_{81}$ in connection to harmonics. We have also called attention to the lesser precision of this RMP π value compared to the π value imposed on the construction of the Great Pyramid. Any discussion about the mathematical properties built into the Great Pyramid is instantly controversial and has been for more than 140 years. But that most sober and exacting of "Egyptometricians," William Flinders Petrie, found not one but *several* examples of the π constant in the Great Pyramid.[76] He confirmed the discovery of Piazzi Smyth and others that the height of the pyramid bears a relation to the circumference of the base of the pyramid as a radius does to the perimeter of its circle. Thus, to find the perimeter of the base, one needs only to take $2 \times \pi \times \text{height}_{\text{pyramid}}$. Moreover, Petrie informs us: "The King's Chamber walls are determined by the same π proportion which rules the exterior of the Pyramid; the circuit of the side of the chamber being equal to a circle described by its width as a radius . . . "[77]

Generally, the π constant Petrie found in the Great Pyramid was based on the fraction $^1/_7$, i.e., $3^1/_7$. This fractional constant is useful because it contains a unit fraction that happens to be $^1/_7$, making it easier to manipulate and also fitting this all-important constant into the "7-system" intrinsic to Egyptian mathematics. Nonetheless, others have found additional architectural and mensural evidence that suggest the Egyptians knew of yet a closer value for π than $3^1/_7$, i.e., 3 177/1250 or 3.1416.[78] Earlier we offered one rationale, based on harmonics, for the $^{256}/_{81}$ π equivalent in the RMP which is less precise than the π values extractable from the Great Pyramid and other

---

[76] Petrie, op. cit., pp. 93–4.

[77] Ibid., p. 94.

[78] Tompkins P, *Secrets of the Great Pyramid*, New York: Harper & Row Publishers, 1971, pp. 195, 263.

structures. Doubtless there were other considerations factored into the ancient Egyptians' use of three different values for π.

We can leave this subject after considering another peculiar factor concerning geometry and harmonics. The word "chord" has two meanings: (1) a simultaneous playing of two or more notes (C,E, and G) to create a harmonious tone, and (2) a straight line that bisects a curve in two places. The term "chord" derives from "cord" and it is of some interest that when the Egyptians wished to geometrically establish the orientation and layout of a building, they did so by the well-known procedure of "stretching the cord." All this becomes more interesting when we recall that the harmonic intervals attributed to Pythagoras are created by dividing and subdividing a stretched (mono)cord.

That the Egyptians were able also to handle trigonometric functions is indicated by Problems 56–60 of the RMP. Problem 56 is indicative of all the problems of this group (see Plate 10):[79]

> If a pyramid is 250 cubits high and the side of its base 360 cubits long, what is its seked?
>
> Demonstration: Take $^1/_2$ of 360; it makes 180. Multiply 250 so as to get 180; it makes $^{36}/_{50}$ of a cubit. A cubit is 7 palms. Multiply 7 by $^1/_2$ $^1/_5$ $^1/_{50}$ [$^{36}/_{50}$].

1	7
$^1/_2$	$3^1/_2$
$^1/_5$	$1^2/_5$
$^1/_{50}$	$^1/_{10}$ $^1/_{25}$

> The seked is $5^1/_{25}$ palms.

This *seked* gives the slope of the triangular faces of this pyramid which rise 1 cubit for every $5^1/_{25}$ palms of horizontal length. This seked of $5^1/_{25}$ is the cotangent of the angle of slope of the faces of the pyramid. Cotangents take us, of course, into the realm of trigonometry.

It can be inferred that the author of the RMP was also in a position to obtain the tangent of the angle of the slope of the pyramidal face. This tangent is obtainable by dividing the height of 250 by the 180 that he arrived at in the Problem 56 (by multiplying the pyramid's base of 360 by $^1/_2$). What the RMP author did, in effect, was create a right triangle through the plane of the pyramid's height that allowed the determination of the cotangent and, by inference, the tangent of the angle of the slope of the pyramid's face. Furthermore, the face of the pyramid provides the hypoteneuse of such a right triangle and, since the hypoteneuse could be readily measured, the sine of this angle of slope, equal to the height/hypoteneuse, and the cosine of this angle of slope, equal to $^1/_2$ pyramid's base/hypoteneuse, were also readily

[79] Chace, op. cit., p. 97.

obtainable.[80] It is hard to imagine, given the advanced state of the ancient Egyptians' applied geometry, that they didn't know how to arrive at sine and cosine values.

It stands to reason that trigonometric geometry was invented by the Nile Valley mathematician-priests because without it, calculation of the regular, virtually nondeviant slope of the faces of the pyramids of Giza (c. 3,000 B.C.) and elsewhere obtained by the builders would not have been possible. As we will see later, the ancient priest-scribes carried trigonometric reckoning to the pinnacle of achievement, far beyond what the RMP or the MMP reveal.

One of the geometric achievements of the ancient Egyptians that has elicited the unqualified admiration of modern historians of science is found in Problem 14 of the MMP which expresses the formula for finding the volume of a truncated pyramid. The problem is set forth as follows:[81]

> Method of calculating a truncated pyramid. If it is said to thee, a truncated pyramid of 6 ellen in height, of 4 ellen of the base, by 2 of the top; reckon thou with this 4, squaring. Result is 16. Double thou this 4. Result is 8. Reckon thou with this 2, squaring. Result is 4. Add together this 16, with this 8, and with this 4. Result is 28. Calculate thou $^1/_3$ of 6. Result is 2. Calculate thou with 28 this 2 times. Result is 56. Thou has found correctly!

This demonstration can be rendered into modern notation as $V = {}^h/_3(a^2 + ab + b^2)$, where a = the base, b = length of the top, and h = the height. As Gillings observes, this is "the standard formula for the frustum of a pyramid." It has him and everyone else stumped:

> It has not been easy to establish how . . . they [the Egyptians] were able to deduce (in a most compact and far from obvious way) the formula for the frustum, which, in the words of Gunn and Peet, "*has not been improved on in 4,000 years.*"[82]

Though it is not specifically attested in the surviving literature, no one doubts that the Egyptians knew the equivalent of the far less daunting formula for the volume of a whole pyramid: $V = {}^1/_3ha^2$.

In modern eyes, the outstanding achievement of Egyptian geometry is found in Problem 10 of the MMP which incorporates the equivalent of the formula to find the area of the curved surface of a hemisphere. As noted earlier, interpretation of this problem has aroused much controversy, mostly because T.E. Peet 60 years ago sought to revise the translation of the prob-

---

[80] See Diop, op. cit., p. 261, for a discussion of the various trigonometric functions intrinsic to Egyptian geometry.

[81] Gillings, op. cit., p. 188.

[82] Ibid., p. 189.

lem accomplished by S. Struve, apparently because Peet found it impossible to credit the Egyptians with such sophisticated mathematics. More recent scholarship has upheld Struve, particularly since Peet's rendition involved inserting terms of his own invention into the statement of the problem. We will avoid the controversy here since Gillings and Diop have ably refuted all the objections Peet raised to Struve's translation. The problem runs as follows:[83]

> Method of calculating a basket [hemisphere].
> If it is said to thee, a basket with an opening
> of $4^1/2$ in its containing,
> Let me know its surface.
> Calculate thou $^1/9$ of 9, because the basket
> is half of an egg. There results 1.
> Calculate thou the remainder as 8.
> Calculate $^1/9$ of 8.
> There results $^2/3$ $^1/6$ $^1/18$ [$^8/9$].
> Calculate thou the remainder of these 8 left,
> after taking away these $^8/9$. There results $7^1/9$.
> Reckon thou with $7^1/9$, $4^1/2$ times.
> There results 32. This is its area.
> You have done it correctly.

This demonstration can be summarized in modern notation by the formula $A = 2\pi r^2$, where $\pi = {}^{256}/81$. As Gillings points out, this is the modern formula for the curved surface of a hemisphere and the ancient Egyptians apparently derived it 1,500 years before Archimedes! Diop was fond of saying, "he who can do more, can do less." Therefore, if the ancient scribes derived this formula for the area of the curved surface of a hemisphere, they ipso facto must have known the formula for the curved surface of a whole sphere of $4\pi r^2$, obtained merely by doubling the hemispheric formula.

Problems 41-43 in the RMP are concerned with finding the volume of a cylindrical granary (see Plate 7). The problems all exemplify the modern formula for finding the volume of a cylinder: $V = \pi r^2 h$ where *h* is the height. In these problems the task is to find the area of the circular end of the cylinder then multiply it by the cylinder's height. Problems 44-45 of the RMP are interesting as Diop shows us, because the first involves finding a cube of a number and the second, extracting a cube root.[84]

It is pertinent now to turn to the most famous theorem in all geometry, one involving the right triangle that took, erroneously it seems, Pythagoras's name. The Pythagorean Theorem is expressed as the sum of the squares of the two sides of a right triangle equals the square of the hypoteneuse, or A^2

[83] Gillings, op. cit., p. 195.
[84] Diop, op. cit., p. 269.

$+ B^2 = C^2$. Plato, taking his cue from the Pythagoreans, claimed that the building blocks of the cosmos were right triangles. Like so much in Egyptian mathematics, there is tremendous controversy over whether they were ever familiar with this "Pythagorean" equation. Even Gillings, generally admiring of all known aspects of their mathematics, refused to credit Egyptian knowledge of the Pythagorean Theorem. Generally his, like most scholars', is the argument from silence: no known Egyptian document contains a specific or general case of the Pythagorean Theorem. Unfortunately, none of the skeptics seems to have paid any attention to Flinders Petrie's influential monograph on the dimensions of the pyramids of Egypt:

> The use of angles of 4 on 3 (which has a hypoteneuse 5) in the Second Pyramid, and 20 on 21 (which has a hypoteneuse 29) in the N. Pyramid Dashur, seems to suggest that the square of the hypoteneuse being equal to the squares of the two sides may have been known; particularly as . . . the use of squared quantities is strongly indicated in the Great Pyramid.[85]

Thus, 4^2 (16) + 3^2 (9) = 5^2 (25) and 20^2 (400) + 21^2 (441) = 29^2 (841). Arguing that, because no Egyptian documentary evidence exists attesting to the Pythagorean Theorem, they had no knowledge of it is, in light of the angular dimensions indicated by Petrie, a little like arguing that the Great Pyramid could never have been built because no document exists describing its construction. The evidence is in the thing itself and if proportions pertaining to the Pythagorean Theorem are demonstrably present in the architecture of the pyramids, the only inference we can draw is that the theorem was known, in some form, by the ancient Egyptians 2,300 years before the time of Pythagoras. Since the priests were his teachers for 22 years, doubtless this revelation would not have come as any surprise to him.

Square Roots and Irrational Numbers

There are a number of instances where the scribe was obliged to take a square root of a number to solve his problem and these represented square roots of whole and fractional numbers. The Egyptians tended always to think geometrically and to extract a square root the mathematician was instructed to "make the (right) angle of a square." Now as Diop and others have shown, the so-called problem of the "double remen" is another indicator pointing to early Egyptian knowledge of the theorem attributed to Pythagoras.[86] The double remen was a fundamental unit of length which equalled the diagonal of a square whose side $a = 1$ cubit. The diagonal of a square is the common hypoteneuse of the two coterminous right triangles of the square. The

---

[85] Petrie, op. cit., p. 68

[86] Diop, op. cit., pp. 258–60.

length of the square's diagonal *c* is found by multiplying side *a* by the square root of two ($\sqrt{2}$). In a square, all sides equal *a*; thus we have $a^2 + a^2 = c^2$, *c* being *both* the diagonal of the square *and* the hypoteneuse of the right triangles created by the diagonal. Since $a = 1$ (cubit) in this case, $1^2 = 1$, giving us $1 + 1 = 2$. Thus, the c^2 here = 2 (and $c = \sqrt{2}$). The diagonal length *c* of the square is found by multiplying the side *a* of 1 cubit of 20.62 inches times the $\sqrt{2}$, giving 29.1325 inches for the length of a double remen since it is equal to *c*, the diagonal. A remen equalled *c*/2 or 14.6 inches.

The $\sqrt{2}$ is *the* irrational number par excellence. By definition, irrational numbers are not finite; the decimal points in these numbers continue nonrepetitively to infinity. This is in contradistinction to rational numbers such as $^1/_3$ or $^1/_6$ where the decimals *do* repeat themselves ad infinitum (0.33333 . . . or 0.166666 . . .). Pi (π) is another irrational number and, as such, very familiar to the ancient Egyptian mathematicians, as we have seen.[87]

Piazzi Smyth, Davidson, Schwaller de Lubicz, Tompkins, and Diop, among others, have asserted that the Egyptians constructed their sacred architecture and drew their human figures according to a proportion known as the Golden Section.[88] The Golden Section is defined by a line that is divided in such a way that the ratio of the whole line to its larger segment is the same as that of the larger segment to the smaller segment, giving a number called the Golden Number, also known as the *phi* (ϕ) proportion. This number, an irrational number, is rounded to 1.618 and was considered a sacred proportion by the ancients. It is said that the "architecture" of the cosmos conforms to the Golden Number.[89] This Golden Number is also related to the famous Fibonacci Series, deriving its name from a Venetian mathematician, Leonard of Pisa, who introduced it—and the Hindu/Arabic numerals—to Europe in 1202 after extensive mathematical studies in the Orient. The Fibonacci Series is created by a sequence of numbers in which each term is the sum of the previous two in the sequence giving, 1,2,3,5,8,13,21,34,55,89,etc. The ratios in this sequence approach the Golden Number, 1.618, by the ninth term ($^{55}/_{34} = 1.6176$), and this Golden Ratio characterizes the rest of the sequence. The perceptive reader will also note that the 10th term of the series, 89, is (rounded off) $^8/_9 \times 10^2$.

It has been found that certain natural structures such as snail shells and antelope horns can be mathematically modelled by the Fibonacci sequence and therefore by the ϕ proportion. There is no doubt that the Fibonacci Series was known in Hindu and Arabic mathematics and several authorities,

---

[87] Plain irrational numbers such as $\sqrt{2}$ are derived from algebraic equalities; π is *not* derived from algebra and so is called a "transcendental" irrational number.

[88] See Tompkins, op. cit., for an extensive discussion of the Golden Section in Egyptian architecture; see also, Diop, op. cit., pp. 299–305.

[89] West JA, *A Traveller's Key to Ancient Egypt*, London: Harrap Columbus, 1987, p. 95.

particularly the late historian of ancient measure, Livio C. Stecchini, claimed that the ancient Egyptians knew of it too.[90]

Finally, there is a relation between the ϕ and π such that $\pi = \phi^2 \times 6/5 = 3.1416$. This relation between ϕ and π was the tool that allowed the ancient Egyptian builders to translate spherical dimensions onto flat rectangular and triangular surfaces. It is for all the reasons adduced so far that the Egyptians incorporated the ϕ proportion, like the π proportion, into their stone architecture.

It is clear that the ancient Egyptians were thoroughly familiar with irrational numbers and performed geometric and algebraic operations with them. As we have stated several times, they were credited with a remarkable degree of mathematical refinement and subtlety by the Greeks. A clear-eyed study of such of their mathematics as has been preserved on the extant parchment *and* in stone resoundingly confirms this assessment. However, as in so many other areas of Egyptian life, one cannot escape the lingering impression that there was, mathematically, vastly more behind the papyric and architectural "documents" that have survived the ravages of time.

African Number and Symbol Beyond the Nile Valley

Relatively little is known about numbers and their use in Africa outside the Nile Valley. Only rarely have anthropologists investigated the subject so it has been universally presumed that sub-Saharan Africa possessed no math worthy of the name, another example of the argument from silence. The argument from silence only holds sway at all because only a fraction of the known archaeological sites south or west of the 1st cataract have been surveyed and an even smaller percentage have actually been excavated. In spite of this, a small body of literature offers intriguing insights into the manner in which sub-Saharan Africans handled number.

The Yoruba of Nigeria developed such an unusual method of numeration that Claudia Zaslavsky was moved to say that one must be a mathematician to learn the system.[91] The method of expressing their numbers can be exemplified as follows:

$$45 = (20 \times 3) - 10 - 5$$
$$106 = (20 \times 6) - 10 - 4$$
$$300 = 20 \times (20 - 5)$$
$$525 = (200 \times 3) - (20 \times 4) + 5$$

75[92]

---

[90] Cited in Tompkins, op. cit., p. 194.

[91] See Zaslavsky C, op. cit., p. 204. Most of this section relies heavily on Zaslavsky's book.

[92] Ibid.

As Zaslavsky points out, the very act of expressing one number constitutes a feat of arithmetic, involving addition, subtraction, and multiplication. From the above examples, it is evident that the Yoruba possess a base-20 or "vigesimal" system, one found among many other peoples in West Africa. Numbers 1–10 all have their unique terms, to wit:

ookan	=	1
eeji	=	2
eeta	=	3
eerin	=	4
aarun	=	5
eefa	=	6
eeje	=	7
eejo	=	8
eesan	=	9
eewaa	=	10

The numbers between 10 and 20 are created by one or more of three arithmetic operations, as we have seen: *ookan laa*, or 11, is "one in addition to 10"; *eeji laa*, or 12, is "two in addition to 10," and so on. The term for 20 is *ogun*. The next major unit after 20 that seems to be expressed "by itself" is *igba*, which is simply "200," and then *irinwo,* which is "400." According to Zaslavsky, the number used to calculate or express indefinite numbers is *egbaawaa* or 2,000 × 10, which is 20,000. The Yoruba do have an expression for one million, *egbeegberun*, which translates as "1,000 × 1,000." It is known that in Benin, one of the important Yoruba states, there is a term for one billion that idiomatically is expressed as "1,000 × 1,000 × 1,000."[93] It is said, moreover, that the idea of infinity is not unfamiliar to them.[94]

Several conclusions can be drawn from the adduced facts: (1) the Yoruba were quite capable of handling very large numbers, (2) the Yoruba possessed a concept of exponents or powers of numbers,[95] and (3) there is an underlying decimal character to this vigesimal system. And, though the number system is not, strictly speaking 2^n, it *is* founded on "decimal-doubling" parameters. Also, as we have already seen, the Yoruban Ifa oracle does in fact conform to the 2^n system.

Zaslavsky discovered in the course of her research that the Yoruba numerical system contained fractional quantities, particularly unit fractions, and

[93] Ibid., p. 211.

[94] Ibid. p. 206. For a valuable exposition of other traditional African numeration systems, see Obenga T, *La Philosophie Africaine de la Période Pharaonique*, Paris: L'Hammartan, 1990, pp. 392-6.

[95] The exponential operation is expressed as *lona* or "to be in [number of] places," for example, 4^2 = *erin lona meji* = "4 in 2 places;" 4^3 = *erin lona meta* = "4 in 3 places," etc. See Zaslavsky, p. 206.

the operation of doubling. In addition, she was able to say regarding the Yoruba (and other West African peoples): "Among these peoples certain numbers are held sacred, and the number *4* occupies the principal position among them."[96] This is undoubtedly the reason why numbers that are powers of 4, that is, 16 (4^2), 64 (4^3), and 256 (4^4) are the most important in the Ifa divination system.

Another symbolically significant number in the Yoruba system is 1,400, representing a multiple of 7 and 200. This number shows that 7 also possesses a semi-sacred quality among the Yoruba. We can see the significance of the numbers two, four, seven, sixteen, sixty-four, two hundred fifty-six, etc., in the Yoruba system. The Yoruba scholar J.O. Lucas insisted as far back as 1948, mostly on linguistic grounds, that there was a genetic connection between Yoruba culture and that of the ancient Nile Valley. When we look at the Yoruba number system in a certain way, Lucas does not appear to have been far off the mark.

Chapter 12 of Zaslavsky's most valuable monograph discusses the numerical constructions known as "magic squares" and a little known 18th-century Fulani mathematician of northern Nigeria, Muhammad ibn Muhammad, who mastered them.[97] This subject is of more than passing interest because so few names of individual African mathematicians come down to us at all, though, as Diop has observed, mathematics was one of many subjects taught at the University of Sankore at Timbuktu in the 16th and 17th centuries. Many of their scholars were known to possess private libraries considerably in excess of 1,600 books.[98]

Ibn Muhammad seems especially fascinating because he had earned quite a reputation as a mathematician and astronomer. He authored a manuscript in Arabic still extant in the the former Khedival Library in Cairo entitled, *A Treatise on the Magical Use of Letters of the Alphabet.* Zaslavsky was only able to obtain only five translated pages of a treatise nearly 200 pages long and the section in her possession dealt mostly with magic squares.

A magic square is created when numbers placed inside a square, arranged evenly in columns and rows, can be summed in any direction—vertically, horizontally, or diagonally—to obtain the same result. The creation of magic squares requires a command of number theory and they were deemed magical because they were used for divination purposes in the Orient. Chinese tradition mentions magic squares in existence more than 4,000 years ago. They became popular in the Arabian world about 1,200 years ago before being introduced to Europe in the 14th century.

A magic square is built up from the small square units filling a large one; a

---

[96] Ibid., p. 213.

[97] Ibid., pp. 137-151.

[98] See Dubois F, *Timbuktu the Mysterious*, for an account of the famous scholars of Timbuktu in the Middle Ages.

3rd-order square exhibits 3 square units on a side whose area, therefore, is 9 square units. A 5th-order square is 25 square units, and so on. The higher order the square, the more difficult and painstaking the process of putting in all the correct numbers to get an identical sum in the columns, rows, and diagonals. Ibn Muhammad worked with magic squares as high as 11th order.

Concerning the method of obtaining magic squares, Zaslavsky writes:

> A normal magic square of odd order n consists of a square array of the integers from 1 to n^2, arranged so that sum of each row, each column, and each of the two main diagonals is equal to the magic constant $n(n^2 + 1) \div 2$. By the use of the preceding formulas, we obtain squares in which the middle number in each array is $n^2 + 1 \div 2$.[99]

These formulas show the "algebraic" quality of magic squares and why sound knowledge of number theory is important in their creation. Ibn Muhammad's particular method of obtaining magic squares was admittedly and deliberately time-consuming. It appeared the student was to use these magic squares, especially at the higher orders, as exercises in patience, discipline, and concentration.

For a more extensive analysis of Ibn Muhammad's magic squares, a study of Zaslavsky's Chapter 12 is recommended.

The Zero-Concept and 2^n in the Dogon-Bambara Number Systems

We cannot leave this chapter without some discussion of the Dogon-Bambara number system. The Dogon and Bambara are ethnically related people that inhabit southeastern Mali. Some remarkable ethnographic research, spanning several decades, by Marcel Griaule and Germaine Dieterlen among the Dogon and Solange de Ganay among the Bambara have brought to light startlingly unified and coherent world-systems to rank with any.

C.A. Diop offers a brief analysis of Ganay's work in *Civilization or Barbarism* that reveals a symbolic number system entirely consistent with that of Pythagoras. Thus, in the Bambara system, according to Ganay, the number 1 represents the primordial thought of the creator that formed the world; 2, the splitting of the first principle; 3, the element of fire and masculinity; 4, the principle of water and femininity; 5, the synthesis of the creator and His work ("the quintessence"); 6, the male and female twinship; 7 (3 + 4), the unity of male and female and therefore fecundity, intelligence, and earth itself. The 8th number or sign in the Bambara system actually is defined as "the secret of nothing," i.e., the potentiality in the timeless void just prior to creation. Numerically it is equivalent to 0. Diop says of this system:

---

[99] Ibid., p. 143.

Here is a text that would be repudiated neither by the Pythagoreans, nor by the other Pre-Socratic philosophers, nor by Plato. The omnipotence of number in the Dogon and the Bambara systems, as well as in the Pythagorean one, is beyond question. In Pythagorism, the first ten numbers were endowed with secret properties comparable to those attributed to them by the Bambara.[100]

As for the symbolic number system of the Dogon, it can speak for itself from Dogon oral texts recorded by Griaule and Dieterlen:

In the beginning, before all things, was Amma, God, and he rested upon nothing. "Amma's egg in a ball" was closed, but made of *four* parts called "clavicles," themselves ovoid and attached. . . . the *four* clavicles are . . . the prefiguration of the *four* elements . . . ; likewise, the ideal bisectors which separate them will mark the collateral directions, . . . "angles *four*," that is to say, space. (emphasis added)[101]

We see ideas identical to those of the Bambara in Amma as the creative One resting on (the secret of) nothing, i.e., 0 ("egg in a ball"), containing the *4* primal constituents of space. The power of the number *4* is everywhere attested in Africa as we have already shown, thus it is no surprise to find it playing a formative role in Dogon metaphysics.

Still, all this is part of the ubiquitous 2^n system:

Amma's egg is represented in the form of an oblong picture covered with signs, called "womb of all world signs". . . . From the meeting point of the two axes extended two intersecting signs, forming the bisectors marking the four cardinal directions. Each of the four sectors thus formed contained originally eight drawings, each of which in turn, produced eight more. Thus, the oval (zero) contained $8 \times 8 \times 4$, that is, 256 outlines. . . . the 256 signs are "the complete signs of the world."[102]

That all of this derives from the doubling inherent in the 2^n system is explicitly attested in the Dogon text concerning the *po*, the source of the world: "The *po* will be *twin* because everything in this universe will have to be able to reproduce itself. The repetition of these two numbers in itself also connotes this essential duality."[103] The text goes on to say that the sign of Amma's womb from which everything will emerge is enclosed in a circle of 128 *yala* (the marks of potentiality), and that the sign of the egg itself is composed of 64 *yala*. All these numbers are familiar to us from the 2^n series.

As we might expect, the number 7 shows itself prominently in the Dogon system: "The number seven expresses the multiplication, for as we have seen,

---

100 Diop, op. cit. p. 318.
101 Griaule M and Dieterlen G, *The Pale Fox*, translated by Stephen C. Infantino, Chino Valley, Arizona: Continuum Foundation, 1986, p. 81.
102 Ibid., p. 84.
103 Ibid., p. 122.

it is the total of three, which symbolizes the male sex, and four, the female sex."[104]

Seven is intimately tied into Dogon astro-cosmology:

> When one invokes Amma during ceremonies . . . one says: "Amma, seven above, seven below, he spun fourteen worlds" . . . The number 14, attributed to the spiral stellar worlds produced by Amma, implies the concept of potential reproduction and multiplication. . . . Here it designates the seven earths and the seven heavens. Amma creates 7×2, that is to say, an infinity of worlds. The superposition of worlds and the concept of infinite multiplication of stellar universes are indicated by the fact that the number 28 is attributed to Amma as well; it is called "Amma's number" because one also says: "Amma fourteen above, fourteen below, etc."[105]

As in other African systems we have encountered, 7 and its multiples, 14 and 28, figure very prominently in the symbolic mathematics of the Dogon. We find a symbolic "decimal" component as well, for the Dogon oral texts tell us, "The number of man is ten, it is that of the Nommo (one of original Primordial Powers)."[106] And elsewhere we are told: "ten [is] the number attributed to the resurrected Nommo as the 'ancestor' of mankind."[107]

But there is more in Dogon number symbolism that, as Diop has already suggested, resonates closely with the Pythagorean system, itself a product of the Egyptian Temple. The Dogon, too, apparently had studied the relationship between harmonic tones and number:

> The [sacrificial] ceremonies are accompanied by calls sounded by someone playing the "Hogon's horn," *ogo buru*. . . . The sounds produced by the instrument, called voices or *mi* (eight in all), are the vibrations of the "word" that Amma placed in the seed. "The voice of the *ogo buru* is the image of the *po* spinning the world around.". . . . The "voices"—each of which has a name—are also those of the eight basic cereal grains.
>
> The seed is represented by a small circle from which emerge eight segments of increasing length, "the eight living things," that is the eight "articulations of the word" which . . . are sounds, the voices of the Hogon's horn.[108]

It simply defies reason to insist, as some will, that these correspondences between Pythagorean harmonics and the Dogon material just quoted are fortuitous. There is a climactic scene in the Berthold Brecht play *Galileo* where he is confronted by his inquisitors because of his "heretical" espousal of the

---

[104] Ibid., p. 128.
[105] Ibid., p. 194.
[106] Ibid., p. 497.
[107] Ibid., p. 497.
[108] Ibid., p. 441.

heliocentric model of the solar system. Three times Galileo points to his telescope, pleading with his accusers to look through it themselves to determine the truth or falsity of his theories. Three times, the inquisitors refuse to take the simple step of looking through the telescope, as if unconsciously realizing that such a step would be an irrevocable one and therefore not to be dared. African historiography—under which "Egyptology" must be subsumed—has been saturated with "authorities" who have dared not take the fateful step of looking at what is in front of their very eyes.

Conclusion

At no place or time in African cultural history did number, and the operations that comprise mathematics, ever exist for purely practical or theoretical reasons. Mathematics for its own sake, i.e., to indulge in abstract speculation, intellectual aesthetics, or mental gymnastics, was a usage to which it was seldom put. From the beginning, probably as far back as the time of the Ishango Bone, there was a symbolic and sacred character to mathematics that made it a tool for interpreting the Divine Plan. Moreover, there was an amazing continuity of mathematics in Africa: we can see the operations evident on the Ishango Bone regenerated in the Rhind Mathematical Papyrus; the unusual Yoruba number system seems to carry on traditions inherited from the Nile Valley; the Dogon may have preserved a category of Nile Valley thought that was of the same order as that vouchsafed to Pythagoras by his priest-mentors.

It seems axiomatic to historians that, outside of the Nile Valley, little native math in the formal sense existed. Even the work of Ibn Muhammad is subsumed under Arabic mathematics. But the door to indigenous African mathematics has barely been opened. One of the Great Projects awaiting Africa in the 21st century is to do her *own* historical spade-work. There are African civilizations still lying beneath the soil awaiting rediscovery. When the indigenous program of rediscovery begins in earnest, more of Africa's math will see the light of day. Then Africa will no longer have to be browbeaten by the unceasing drone of the argument from silence.

Appendix A

More on Myth and Musical Harmony in Ancient Egypt

Cheikh Anta Diop, in his typically polymathic way, provides an additional perspective on the rules of harmony that are ascribed to Pythagoras and analyzed by Plato in his *Timaeus*, but which actually originate in the Egyptian symbolico-mythical study of number. Diop points out that the series of harmonic intervals related to the $^8/_9$ proportion emanate from the Egyptian *ogdoad* and *ennead*. The *ogdoad* is the group of the original *eight neters*, representing 4 pairs of syzygies, which produced all living creation. The members of this sacred *ogdoad* changed in the course of Egyptian religious history but the *ogdoad* and its function always remained a potent feature of the cosmogony. The *ennead*, the group of *nine neters*, was created when the supreme One (*Ua*), be it Ptah or Ra, is added to the *ogdoad* of eight. As Diop states:

> . . . the Egyptian harp of eight or nine strings, reproduces the figures of the ogdoad or the ennead and . . . Pythagoras, who stayed twenty-two years in Egypt, became *very familiar with Egyptian musical theories*. (emphasis added)[109]

Thus, the $^8/_9$ proportion in the harmonic scale is directly analogous to the *ogdoad/ennead* relationship, where the *ennead* is generated by the *ogdoad*. As Diop indicates, a study of the 8-stringed and 9-stringed harps of ancient Egypt would undoubtedly reveal the harmonic/numerical properties at the root of all our modern notions about harmony.

Appendix B

Chaos, the 2^n Series, and Black American Music

For about the first 70 or so years of this century, Black music in America, particularly jazz, underwent an unprecedented evolution. At the turn of the century, the music that would become jazz was influenced by, among other things, the New Orleans funeral marches—themselves derived from Napoleonic marching music—in which the music was played in 4/4 time, equivalent to 1 or a full beat and in 2/4 time, equivalent to a 1/2 beat. Later, in the 1920s and 30s, innovators like Louis Armstrong and Duke Ellington perfected the musical language of Swing, melodically based on the quarter note. Beginning about 1940, and continuing for 15 years, bebop artists like Charlie Parker, Theolonious Monk, and Dizzy Gillespie created yet another jazz language constructed around extended instrumental improvisations on eighths. Finally, from 1955 until his death in 1967, John Coltrane ramified these developments to their nth degree by his improvisations on a dense series of sixteenth notes dubbed "the sheets of sound." Almost as soon as the sheets of sound technique emerged from Coltrane's tenor saxophone, Chaos descended on jazz in 1959 when Ornette Coleman, Cecil Taylor, and others sent forth atonal, aharmonic, rhythmically complex musical explorations upon the world that seemed to burst all received musical boundaries, earning the name "free jazz." Coltrane himself experimented on the boundaries of this musical Chaos, often descending into incredibly complicated textures of improvisation, driven by ever more complex polyrhythms, but never entirely forsaking the regularities of ordered improvisation.

Still, it cannot have escaped notice that the progression delineable in jazz

---

[109] Diop CA, *Civilization or Barbarism*, translated by Y.L.M. Ngemi, Brooklyn: Lawrence Hill Books, 1991, p. 350.

history follows a halving/doubling 2^n pattern that governs dynamical systems: 1, $^1/_2$, $^1/_4$, $^1/_8$, $^1/_{16}$ . . . Chaos. It has often been said that the artist tends to prefigure, indeed predict, in his art the cultural trends, currents, and forces that eventually engulf society.

4

The Sacred Craft: African Systems of Architecture

"Modern" architecture . . . started on a sound basis but has . . . , more often than not, ignored this human scale. Some architects have gone so far as to . . . ignore symmetry in religious architecture, which has consequently become grotesque.
—Alexander Badawy, *Architecture in Ancient Egypt and the Near East*, p. v

It has been much commented upon but poorly appreciated how much African architecture derives from its environment. In the Nile Valley, for example, the traveller may visit the monumental sites repeatedly for years before it dawns on him that those peerless builders drew their inspiration from nature. The mastabas, stepped pyramids, bent pyramids, "true" pyramids, not to mention the mounds and tumuli resting in silent profusion in the regions of the upper cataracts, did not emerge unbidden out of the imaginations of the early architects. This truth is brought home with startling clarity when flying over the desert just west of the Nile at Abu Simbel, where all the classic Nile Valley architectural styles repose stolidly in the landforms and rockscapes dotting that locality, etched into these geometrical shapes by the master hand of nature.

The civilizations and cultures of antiquity never lost their awe of nature, and the idea of "imposing" the will of man upon her was beyond conception. Thus, the edifices erected by the superlative Nile Valley architects of antiquity seem not merely to blend with the landscape but to emerge from it. The builders took extraordinary pains not to distort the landscape in any way; the material form of civilization was subsumed by nature. Hatshepsut's Deir el Bahari was carved out of and merges with the base of the massive cliffs rising straight up on the Nile's west side opposite Karnak Temple (see Plate 15). Moreover, it was possible as recently as a century ago to wander all over the Valley of the Kings without ever suspecting that one was surrounded by 62 extensive tombs. The tombs of the Nubian nobles at Aswan are, likewise, not immediately apparent to the eye unless one knows what one is looking at.

Similarly, the massive stone walls forming the perimeter around the hill enclosures at the Great Zimbabwe site incorporate the huge boulders and granite outcroppings that dot the region. Indeed, some enclosures conform

Fig. 22. The pyramids of the Old Kingdom, prior to the beginning of construction on the plateau of Giza. The Bent Pyramid and the other important pyramids at Darshur and Medum are shown.

so closely to the surrounding rockscapes as to be indistinguishable from them when viewed from afar. The complex of mud-brick dwellings and buildings, including the Great Mosque, at Timbuktu, again, seem to merge with the dry, dun-colored, Sahelian landscape that surrounds them.

Architecture, or for that matter art, as the *personal* or *individual* expression of the architect or artist is a concept alien to African thinking. With one exception, we do not know the name of a single pyramid architect in the Nile Valley.[1] The same rule applies to architectural constructions in Zimbabwe and Songhay. Only rarely do we find the African equivalents of Frank Lloyd Wright, Christopher Wren, or Le Corbusier; that is to say, architects whose distinctive style was translated into *individual* fame. Anonymity was

[1] This individual was none other than Imhotep himself, designer and builder of the Step Pyramid at Sakkara, whose name comes down as the "author" of this temple complex. Occasionally, favored architects of some of the non-pyramidal temples such as Senmut, builder of Hatshepsut's temple at Deir-el Bahari, were allowed to record their architectural accomplishments.

the rule; credit invariably went to the monarch who commissioned the project instead of the builder. The architect was recreating, as impeccably as possible, a divine ideal; he became the medium or conduit through which divine creative forces operated. His creation, veritably, was not his own. The king or pharaoh, however, could take credit because he partook of divinity as its visible representative on earth. Thus, the king, through his commission, was activating the divine will, realized through the technical skill and genius of the architect/artist.

The Pyramid: Divine Geometry in Stone

There is no geometric shape so completely identifiable with a single country as the pyramid is with Egypt. The ancient Egyptian name for the pyramid was *mer*, and therefore seems to have derived one of its names, *Ta-Meri*, meaning "Land of the Pyramid," from it. The mere sight of a pyramidal shape instantly calls Egypt to mind. As already mentioned, the Egyptians did not in any strict sense "invent" the pyramid; nature did that. But from early in its history, the master builders of the Old Kingdom appropriated the pyramid form and the pyramid became the architectural (and geometric) emblem of everything great and sublime in Egypt. It still is.

The stark, severely aesthestic symmetry of the pyramid embodied several mythic and symbolic functions. It was the stylized image of the primeval mound, rising from the formless void of *Nun*, giving the Demi-Urge a place to stand. It was solar geometry in stone, the pinnacle representing the source of the sun rays that fanned out toward earth in a "pyramidal" configuration. It was the architectural image of the Sacred 7, that is, the 3 upon 4 with four triangles conjoined on a square base. But as we shall see, the Great Pyramid in particular exhibited innumerable technical properties that show that it, like most features of Nile Valley science and technology, subserved functions that were at once empirical, theoretical, symbolic, and spiritual.

Pyramid literature is voluminous and since the middle of the last century has constituted a special discipline-within-a-discipline. In truth, most Egyptologists have eschewed pyramid studies, allegedly because of the bad name given to "pyramidology" by enthusiastic amateurs who devoted whole lifetimes to measuring and studying the Great Pyramid to prove or confirm, among other things, Biblical prophecies. But in truth, this is merely a smoke-screen. Highly qualified technicians and mensurists have studied every inch of the Great Pyramid and have published findings that orthodox Egyptology finds impossible to credit, not because the findings are in error but because acceptance means the complete overhaul of history.

William Flinders Petrie, who practically created modern Egyptology, making him the "spiritual father" of contemporary Egyptologists, was one of the more celebrated of the pyramid mensurists. But his semi-legendary standing among Egyptologists has not prevented his monograph on the pyramids from languishing in quasi-obscurity—except in one instance. Petrie's exacting measurements apparently refuted the previous findings of Charles Piazzi Smyth that one of the units of measure present in the pyramid structure is the "pyramid inch." Academic Egyptologists had scoffed at the "pyramid inch," especially because Smyth was one of the most insistent of the evangelists attempting to explain the building of the pyramid by Biblical inspiration. There was mass relief among Egyptologists when Petrie published his refutation for they then felt at liberty to discard *all* of Piazzi Smyth's pyramid work, even the considerable amount that was actually *confirmed* by Petrie's measurements.

Biblical prophecy aside, an acceptance of even a portion of the conclusions carefully amassed after more than a century of work would mean that our current notions of how and where advanced science and technology emerged bear no relation to historical reality. Indeed, we would be compelled to admit that, in some respects, the vaunted science and technology of modern times has not *yet* risen to the level of the ancient pyramid-builders. For those who prefer to insist that the Great Pyramid was merely an overgrown mausoleum, it is essential to ignore or dismiss pyramid studies.

It is the contention here that the various pyramid forms were inspired by natural formations that can be seen to best advantage in the region surrounding Abu Simbel. The conventional wisdom regarding the evolution of the pyramid commences with the *mastabas* (whose forms are also very abundant in the desert near Abu Simbel) which, in ascending layers, progressed to the true pyramidal form. This reasoning is sound enough, particularly when contemplating the earliest pyramid, the first great edifice in stone ever created by human hands, the Step Pyramid of Sakkara.

The mastaba became the first above-ground Old Kingdom tomb created. Simply described, mastabas were rectangular, single-story mud-brick structures with sloping sides and a flat top. They are called mastabas—the word coming from Arabic meaning "bench"—because when Napoleon's savants first caught site of them they were reminded of the benches used by the contemporary Egyptians. There are a profusion of remains of mastabas sitting in regular rows at Sakkara but these are mostly restored versions.[2] Virtually none of the superstructures of the original mastabas have survived. However, the underground portions have preserved markedly well and the walls are filled with drawings and paintings of geometrical design. From the layout of these interior portions, it has been possible to accurately reconstruct the mastabas as they must have originally appeared.

The mastabas ranged from 150 to 250 feet in length and surrounding the

---

[2] For a short but informative description of mastabas, consult John Anthony West's *The Traveller's Key to Ancient Egypt*, London: Harrap Columbus, 1987, p. 8.

central burial chamber within were smaller chambers to hold funeral offerings. A great deal about the state and condition of early Egyptian civilization can be gleaned from the large aggregation of varied artifacts—metal tools, unusual stone vases, fragments of tablets, and clay jar seals—that have been recovered from the mastaba-tombs. It is thought that the early rulers themselves lived in mastaba-like, mud-brick palaces and it is a curious fact that in all the 4,000-year history of dynastic Egypt, precious little remains of royal dwellings. Stone was used *only* to build tombs and sacred buildings, virtually never for the domiciles of mere mortals, not even for pharaohs.

It was not until the 3rd Dynasty, circa 3,500 B.C., that stone was put into architectural use on a grand scale. However that decision was arrived at, the initial result conveyed the impression that the Egyptians had been building in stone for ages. There is a conceit among many commentators—though the evidence, carefully considered, is clearly against it—that Egypt's dynastic civilization emerged on the world stage suddenly and ready-made, without clear-cut antecedents. *Deus ex machina*, as it were, ancient Egyptian civilization just appears. Though such an explanation is a distortion of the record, it has to be wondered how the ancient stone-builders architecturally achieved such supreme mastery of form and function on the first try.[3] Taken as a whole, John A. West is utterly correct in his assertion that there is no architectural edifice anywhere outside of Egypt to surpass Djoser's temple and funerary complex at Sakkara. Even within Egypt, the only true competitor is the Great Pyramid which represents an unsurpassed technical *tour de force*.

The step pyramid itself is unmistakably an elaboration of the mastaba-form (see Plate 13). One of the reasons why the structure is so unusual to study is that the original plan was revised during the course of construction at least four times at no apparent disadvantage to the final result. The original mastaba structure was about 69 square yards in area, encased in smooth white limestone above a square pit. Three additional terraces were superimposed on the base mastaba which itself was extended outward so that it encompassed a rectangular area of more than 14,000 square yards. Finally, the height of the entire edifice was raised by super-adding two more stepped terraces, giving six in all, the whole exterior lined, as originally, with fine white limestone. When completed, the Step Pyramid rose to a height of 197 feet and its slope angle was 72° 30'.

One of the best recent descriptions of the method of construction for the Step Pyramid was written by the late Peter Hodges, himself a master-builder.[4]

---

[3] John Anthony West points out that the hand of master stone masons is evident in some of the earlier mastabas at Sakkara but there simply was no stone-working on the scale evident in Djoser's funerary complex. One other possible clue comes from the revised age of the Sphinx which, if it holds up, does mean that large-scale (sculptural) stone-working prior to the 3rd Dynasty had occurred.

[4] Hodges P, *How the Pyramids Were Built*, edited by Julian Keable, Longmead, Shaftesbury, Dorset: Element Books Ltd., 1989.

In fact, Hodges preferred to call step pyramids "buttress" pyramids which, in his view, was more descriptive of the manner of their construction. The "true" pyramids, like those at Giza, he referred to as "coursed" pyramids because they were built up by laying squared blocks on level courses. By contradistinction, the buttress method could be characterized this way: " . . . the builders were doing no more than building walls, as their experience had taught them, but leaning them together so as to stabilise a mound."[5]

Buttress or step pyramids consisted of either six or seven concentric layers with the exposed faces limited to 30 feet in height. Stones in the lower layers weighed up to one and a half tons; in the upper layers, one-half ton. To begin, a center square and the concentric buttress lines were carefully outlined on a leveled area and then the actual construction would commence with the center core up to about chest height, according to Hodges. At that point work would begin on the first buttress that would serve as a stepping stone allowing the center to be continued upward to the prescribed height. Additional buttresses would be constructed around the inner ones and the first butresses or steps to actually be completed would be the outer two. After they had been raised to a certain height, scaffolds and/or building steps would be put in place to allow the building stones to be levered into position as the inner terraces were raised in height. The method employed made the most efficient use of the structure itself to economize on time and effort: "Every builder wants to gain as much height as possible before he has to add something temporary to his structure, such as a scaffold or a ramp."[6]

The pyramid itself is striking enough but there is a more powerful impression conveyed by the overall architectural ensemble that includes the forecourt on the south side of the pyramid, the entrance temple at its south end, and the massive wall, now almost entirely in ruins, enclosing the whole complex. The entrance temple is simply an amazing structure whose rectilinear design shows a cleanness of line with a refinement and grace of conception that became the hallmark of Old Kingdom architecture. The aesthetic perfection elicited by the elegant simplicity of Old Kingdom architecture, first realized at Sakkara, was never achieved again in Egypt and it is impossible not to admire the uncluttered genius of the architect.

The inside of the entrance temple is as admirable as the external formations. Here, for the first time in history, the "Doric"—some call it "proto-Doric"—column was erected. This architectural feature, the pride of the Parthenon, is no more Greek than the pyramid. We have already pointed out that, architecturally, the Egyptians' inspiration derived from the myriad natural forms surrounding them. These columns were explicitly modeled after papyrus reeds tied in a bundle. The columns in the entrance temple are not free-standing; a buttress attaches each of them to the wall.

---

[5] Ibid., p. 71.
[6] Ibid., p.. 66.

There is some quibbling about whether these marvelous columns in the entrance temple are truly "Doric" in style because they are "fasiculated" rather than "fluted,"[7] but there is no such nitpicking about the columns in the southern building attached to the entrance temple. So characteristically "Doric" are these that they admit of but one conclusion: the Greeks simply lifted this columnar style from Egypt (see Plate 14). And, many 19th and 20th century public buildings in Europe and America were done in a "neo-classic" style, prominently featuring the Doric column. Thus, we have the abiding paradox of the modern age: the more a Grecian style is striven for, the more are African-derived forms appropriated.

With the single exception of the Great Pyramid, the various pyramid structures of the Old Kingdom were royal tombs. This, however, did not prevent them serving other functions as well. One of the most successful of the pyramid surveyors, Moses B. Cotsworth, came to the conclusion that the terraced or stepped pyramid structure was devised to raise an obelisk higher off the ground so that it could cast a longer shadow.[8] Such an elongated shadow created a more precise sundial that enabled the ancients to measure the year more exactly. One generally finds, extending from the base of such a pyramid, a smooth, level court onto which the obelisk throws its shadow at noon throughout the year. As we have seen, a level, spacious, rectangular court extends southward from the base of the Step Pyramid at Sakkara. It is usually avowed that purely ceremonial and ritual functions were held on this court, but that does not rule out the distinct possibility that the part adjacent to the pyramid would have been constructed so that the shadow of the obelisk, functioning as a sun-dial, could be measured and followed throughout the year. Cotsworth also expressed the opinion that the evolution to the true pyramid form evident in the 3rd and 4th dynasties, culminating in the construction of Great Pyramid, was dictated by calendrical considerations:

> According to Cotsworth, the optimum design turned out to be the solidified Pyramid of Cheops, with its slope set for a particular latitude to swallow the equinoctal shadow. Once this method of establishing the precise length of the year had been found . . . there was no further need for enormous pyramids.[9]

Djoser's Step Pyramid temple complex at Sakkara, conceived, designed, and realized by the incomparable genius of the architect and polymath Imhotep, was humanity's first great architectural triumph. It established architectural forms, styles, and canons still in use today. The practical build-

---

[7] Ibid., p. 155.

[8] Cited by Tompkins P, *Secrets of the Great Pyramid*, New York: Harper and Row Publishers, 1971, p. 126. In this author's opinion Tompkins' book is simply the best compilation there is of the history of "pyramidology." Few books reveal as much about the exact sciences of ancient Egypt as this one.

[9] Ibid., p. 127.

ing technique and masonry evident in the entrance temple were never surpassed, though, as we have said, they were realized on a grander scale in the Great Pyramid at Giza. Furthermore, it is not unreasonable to suppose, after Cotsworth, that, along with its funerary function, the Step Pyramid furnished a platform to support an obeliskoid sundial, projecting its shadow on the southern court to provide a measure of the seasons. It is also pertinent to note that the famous Babylonian ziggurats were Mesopotamian versions of the Step Pyramid and that the form reappeared in the Meso-American pyramidal temples.[10]

Chronologically, the next important pyramid was the one constructed at Medum, 30 miles south of Sakkara, toward the end of the 3rd or near the beginning of the 4th Dynasty. Like Imhotep's Step Pyramid, Medum shows changes in ground plan as the work was being executed. It, too, developed from the basic mastaba form and was originally planned as a seven-tiered structure. After completion, it was decided to add an eighth terrace, which was done, then yet another decision was made to fill in the terrace spaces to complete a "true" pyramid, cased like all the pyramids in fine white limestone. At an undetermined later time, perhaps during the 19th dynasty, some of the outer casing stones were removed or extruded on their own, causing the outer packing and casing to fall away, leaving the original stepped core intact. Even today, it stands in the desert, a stepped tower surrounded by a huge agglomeration of rubble. It represents the only architectural "failure" of the Old Kingdom but it illustrates the step-by-step process by which the mastaba form, progressing through the terraced pyramid form, evolved gradually into the true pyramid. Moreover, there are those, like Peter Hodges, who think that the masonry of the Medum Pyramid, with facing stones perfectly laid and jointed and the external angles of the structure as precise and even as anything done today, represented an advance over the megalithic masonry of the Step Pyramid at Sakkara.[11] Thus, if it's a "failure," it is one that pushed Nilotic building technique forward.

The triumph of several generations of architectural experimentation, leading to the creation of true functioning pyramids, was finally achieved at the beginning of the 4th Dynasty, in the reign of Seneferu. This pharaoh, the founder of the dynasty, commissioned two pyramids to be built during his lifetime at Dashur, five miles south of Sakkara: the so-called Red Pyramid and

---

[10] See Tompkins P, *Mysteries of the Mexican Pyramids*, New York: Harper & Row Publishers, 1976, for an absorbing account of the history of Mesoamerican pyramidology. Like the Nile Valley pyramids, much scientific data was encoded in the structures of these edifices. The famous Sun Pyramid in Mexico City was constructed of six terraces and in this and other ways seems a replica of the Step Pyramid of Sakkara.

[11] Hodges, op. cit., p. 57. It is perhaps worth pointing out that the incredibly fine masonry of the entrance temple to Djoser's Pyramid is superior to that of the pyramid itself. This was the world's first attempt at construction of such a large stone edifice. The problems of translating the fine masonry of the temple complex to the pyramid structure itself would be worked out in succeeding generations.

the Bent Pyramid. The Bent Pyramid is so named because about halfway through the construction, the angle or slope of the pyramid was changed. This pyramid starts off at an inclination of 53°27' but, halfway to the pinnacle, the angle of inclination changes to 43°27'44". In every other respect, the masonry is just about perfect. There has been much speculation about the reason behind this shift in slope but no explanation has proved satisfactory.

With the Bent Pyramid, these Nilotic builders revealed a mastery of technique that would enable them to build structures on such a grand scale that their like would never be seen again. As Hodges described, the stones are squared and bonded together horizontally on successive courses.[12] Thus, the core stones of the pyramid could be bonded from the center to the faces. It is this method of bonding, in contrast to buttressing, that confers stability on this and others of the massive "coursed" or "true" pyramids.

Its sister pyramid, the Red Pyramid, in view of the subsequent collapse of the pyramidal exterior at Medum, carries the distinction of being the first true pyramid, one hardly less massive than the Great Pyramid. John Anthony West correctly points out that if Sneferu commissioned these two pyramids, then he presided over a construction effort that, in a period of 24 years, involved quarrying, transporting, and dressing *nine million tons of stone*![13] This exceeds the tonnage of the Great Pyramid by $2^1/_2$ million. Taken together, then, the sheer engineering and logistical effort behind the construction of Sneferu's two pyramids unfolded on a scale larger than that of the fabled Great Pyramid. It means that the necessary technical skill, knowledge, and logistical organization for such immense projects had already been thoroughly mastered by the time the building of the Great Pyramid commenced.

Before entering into a discussion of the Great Pyramid, it is worth asking the question, why build so many pyramids? By the time Egyptian pyramid-building was done at the end of the Middle Kingdom, a line of pyramids extended 94 miles south from Giza. The encyclopedic Peter Tompkins furnishes a clue:

> For right-angled trigonometry, the Australian engineer [R.T. Ballard] realized, true straight lines could be extended from the pyramids in given directions by direct observation, without the aid of other instruments, and that with the simplest of instruments, angles could be exactly observed from any point. . . . In a short time *anyone* might construct a table for himself answering to every degree or so in the circumference of a circle for which only forty or fifty triangles are required. . . . Such primary triangulation would be useful to men of almost every trade and profession in which tools or instruments were used.[14]

Taken with the pyramids of Giza, the string of pyramids extending southward meant that, by the method of triangulation, whole sections of the coun-

---

[12] Ibid., pp. 60–2.
[13] West, op. cit., p. 199.
[14] Tompkins, op. cit., p. 120.

try could be precisely measured and surveyed at any time. This capability was essential in a country where annual floods erased field boundaries.

The Great Pyramid: The Measure of the World

Which brings us to the Great Pyramid. There is simply no arguing that it is the most stupendous structure ever raised by human hands. It has never failed to strike observers with an overpowering awe and it has embroiled the imaginations of better than 150 human generations. Today, the fascination, indeed obession, with the Great Pyramid continues unabated, spawning a huge literature. Pyramid investigators fall into two camps: (1) orthodox academics who insist it is a tomb and nothing but; (2) a substantial coterie of scientists and surveyors who insist with equal adamancy that it is an extraordinary architectural and engineering marvel encoding all manner of mathematical, astronomical, and geodetic data.

Herodotus supplied the most complete classical description of the building of the Great Pyramid, relating that it took 30 years to build, employing shifts of 100,000 laborers working three months at a time. Herodotus informs us that the great causeway, over which presumably the blocks were moved as the edifice was being built, took 10 of those 30 years to build. The melodramatic 19th and 20th century imaginations—embellished repeatedly by Hollywood—painted lurid tableaus of ant-like hordes of slaves laboring and dying under the lash merely to create an oversized monument to the outsized ego of pharaoh. More sober minds have prevailed recently, particularly as the workmen's quarters surrounding the plateau of Giza have lately been unearthed, showing the laborers to have been *professional* artisans, masons, and stoneworkers who were full-fledged citizens of the country. Concerning this gargantuan construction effort, Julian Keable has this to say: "Subdued and fearful gangs of slaves have no place in such a scene—but a respect for repetitive manual work certainly has, a respect which is born out of the evident orderliness of the building operation, as well as its scale."[15] Furthermore, as Somers Clarke and R. Engelbach remarked, "We cannot help admitting that they were perhaps the best organizers of human labour the world has ever seen, and their method of carrying out a task always appears to be the most efficient and economical."[16]

The Great Pyramid rises to a height of 481 feet and 3 inches, composed, when it was completely intact, of 210 courses of masonry.[17] The angle of

---

[15] Keable J, "Foreward" to Peter Hodges' *How the Pyramids Were Built*, pp. ix-x.

[16] Clarke S and Engelbach R, *Ancient Egyptian Construction and Architecture*, New York: Dover Publications Inc., 1990, p. 3.

[17] There are numerous works available detailing the myriad measurements and dimensions of the Great Pyramid. Our discussion draws from the previously cited works by Petrie, Piazzi Smyth, and Tompkins in addition to A. Pochan's *Mysteries of the Great Pyramids*, New York: Avon Books, 1978.

the slope of each of the faces, determined from the remnants of casing on the northern base, was 51°51'. It is situated nearly 200 feet above sea level and nearly 130 feet above the ancient Nile flood crest. The estimate of the number of stones, cut and dressed, ranges from 2.3 to 2.8 million. The average weight of the interior blocks was $2^1/_2$ tons, with a range of 1 to 7 tons. However, some of the casing stones at the base weigh as much as 16 tons and granite blocks in the King's Chamber weigh 100 tons. More than 13 billion pounds of masonry went into the construction of the intact Pyramid, enough to build a wall four feet high two-thirds of the way around the globe.

The pyramid was built mostly of limestone cut from surrounding quarries, though the interior rooms such as the King's Chamber were constructed entirely of granite floated 500 miles down the Nile from Aswan. The outer casing, as for all Old Kingdom pyramids, was made entirely of fine white limestone that left a true and even pyramidal shape as precisely realized as any modern measuring apparatus could have achieved. Few of the casing stones remain, but those that do show a level of building technique that defies the imagination. The stones are fit together so evenly and finely that the joints would hardly admit a strand of hair between them (see Fig. 23). Never since has such perfection of masonry been seen:

> Though the stones were brought as close as $^1/_{500}$ inch, or, in fact, into contact, and the mean opening of the joint was but $^1/_{50}$ inch, yet the builders managed to fill the joint with cement, despite the great area of it, and the weight of the stone to be moved—some 16 tons. To merely place such stones in exact contact at the sides would be careful work; but to do so with cement in the joint seems almost impossible.[18]

Each side of the base of the pyramid measures 755 feet, 8 inches with an average error of $^1/_{10,000}$th from straightness. This incredible accuracy moved Flinders Petrie to say that here was precision such that only an optician could obtain, but this compares work on a scale of acres to that on a scale of millimeters.[19] The pyramid shape was the best one available to the ancient builders that permitted them to raise these edifices to such commanding heights using the least amount of material.[20]

The thorniest issue of pyramid studies concerns how the builders actually performed the task of raising such a massive structure using $2^1/_2$ million stone blocks. Even today, no one knows for sure, a climate of uncertainty that has given rise to an endless variety of fantastic speculations. The safest—and best attested—route is to assume that the builders of that time used only such instruments as were known to be available to them. With this

---

[18] Petrie WMF, *The Pyramids and Temples of Giza*, London: Histories and Mysteries of Man Ltd., 1990, p. 13.

[19] Ibid.

[20] Hodges, op. cit., pp. 2, 7.

Fig. 23. Remnant of the original casing stone of the Great Pyramid. Intact, the entire pyramid would have been covered with such casing of white limestone. The joints between the stones were so fine as to be barely visible.

premise firmly in hand, the possible building methods can be pared down to two or three choices. The theory favored by most academic Egyptologists involves the erection of ramps to move the stones into place by means of sleds and rollers. The other theory of consequence, one first described by Herodotus, involves the use of levering mechanisms.

Peter Hodges, in his book *How the Pyramids Were Built*, attacked the problem from the point of view of a builder who, with the available tools, would have had to master the challenges of moving 2½ million colossal stones to an increasing height that ultimately topped out at 40 storeys. What the Egyptians did *not* employ were wheeled vehicles or pulleys.[21] Moreover, Hodges pointed out that to have built ramps at a low enough gradient to allow stones to be pushed upward on a rising pyramid would have involved construction using *three times as much material as the pyramid itself*. All authorities agree that the Great Pyramid was completed within a generation (30 years); there is equal agreement that the Egyptians were the ancient world's greatest efficiency experts. These two facts alone make the ramp *modus operandi* implausible.

---

[21] In this the Egyptians were the same as the Incas who also raised great constructions of large, evenly fitted stones without the wheel. It is probably a mistake to assume that the Egyptians did not *know* of the wheel; for reasons that are obscure to us, they simply chose not to make use of it until the 18th Dynasty.

Having considered the problem, Hodges accepted the explanation of Herodotus, that the stones were levered into place, going so far as to say: "We should be able to recognise that such a method [levering] was the only one which could have been used to handle stones of such weight and in such numbers."[22]

The levering is best understood by thinking of the mechanism involved in jacking up a car.[23] In the absence of wheels and pulleys, levering enables the smallest amount of manpower to be used to lift the greatest amount of weight up a vertical height. As Hodges put it, one man can do the work of 12. Furthermore, levering precludes the use of ramps or scaffolding that would clutter up the work space. As Hodges noted, the only things the workmen would need to remove at the end of the project would be their own tools. There would be no ramps or scaffolds to dismantle or ramp material to dispose of. This "autostatic" method was also more efficient than the techniques used in raising the early step or buttress pyramids because of economies in time, labor, and material. Timber packing would have been inserted beneath the stone after each "jack" to lift it ever higher. By this method, a well-practiced team of laborers could raise a $2^1/_2$ ton stone step by step halfway up the pyramid within a day.

The task of raising and accurately setting in place hundreds of heavy stone blocks a day was itself a formidable challenge, but the pyramid-builders faced more subtle and refined tests of their ingenuity, care, and attention to detail. First, the ground surface had to be precisely leveled, achieved by creating small canals upon the construction site and filling them with water. Using a level water surface is still the most accurate method for ensuring a true horizontal plane. A similar use of water could be adapted on each course to ensure an evenly planar surface. Furthermore, the pyramid had to sit, course by course (210 in all) exactly true upon a perfectly square base, while the apex, at a height of nearly 500 feet, had to be set over the exact center of the base. The builders did not have a point 481+ feet high to aim at; they had to maintain the exact squareness and trueness of each course only by reference to the one below it. Thus, if there was even a small error in the setting out and laying of one concourse, it would magnify itself to eventually introduce a "twist" in the finished edifice.[24]

The most effective and practical method of creating consistently square and true ascending concourses would have been by raising the angles or corners first then laying in evenly the squared blocks. Continual testing of

---

[22] Hodges, op. cit., p. 19.

[23] Other authorities besides Peter Hodges accepted the levering explanation. Flinders Petrie, Somers Clarke, and R. Engelbach were also convinced that the lever system is the one that accounts for most of the facts known about pyramid construction.

[24] Clarke and Engelbach, op. cit., p. 125.

squareness is achieved by measuring the diagonals.[25] Through use of finely-calibrated calipers, it is possible even to test "cubeness" by measuring diagonals. In this stupendous structure, these Nilotic builders avoided even the slightest twist; clearly they had achieved a refinement of measurement equalling that of the most sensitive modern instruments. In the words of Peter Hodges: "The pyramid builders understood sighting, planes, squares, diagonals, and parallels, and that these skills could be enlarged for use on the building sites of the pyramids (see Fig. 24).[26] Hodges further insisted that the builders were so confident of their skill, so thoroughly in command of their techniques, that they knew, before they laid the first stone, that they could (and would) achieve a true and accurate pyramid shape, though several decades, millions of blocks of stone, and hundreds of thousands of masons, craftsmen, and laborers would be invested in the structure's completion.[27]

Finally, Hodges offered one of the more novel explanations of how the pyramids of Giza were finally brought to their characteristic shape. At the end of construction, the pyramids had a terraced look very similar to that of the step pyramids. Hodges suggested that the ledges or steps were then trimmed away by chisel from the top down to produce the final smooth pyramidal form.[28] Hodges admitted, however, that this explanation needed further verification.

The sides of the Great Pyramid are oriented perfectly to the four cardinal points. The first 19th-century astronomers to consider the problem of the pyramid's orientation, Sir John Herschel, Charles Piazzi Smith, and Richard A. Proctor, decided that the ancient builders had used the north pole star to correctly orient the pyramid.[29] The star closest to the polar position at the time the pyramid was built was α-Draconis, which at its closest point was 3°43' from the celestial pole. It was surmised that the Descending Passage, whose entrance was originally situated on the 16th course before the outside was dismantled, was used to align the northern face of the Pyramid to α-Draconis, and this accounts for the extraordinary straightness of the Passage that Petrie discovered after surveying it. In order for the Descending Passage to have been pointed straight at α-Draconis at its closest position to the North Pole, it would have to have been inclined at an angle of 26°17' and this is the *exact* angle of inclination of the Passage (see Fig. 25). Knowing this, we can determine the date for the construction of the Descending Passage, and therefore arrive at a close approximation of the date of completion of the pyramid itself.

---

[25] This discussion sheds light on the famous problem of the "double remen," obtained by calculating the diagonal of a square. It also shows how all "theoretical" Nilotic sciences were thoroughly grounded in practical, "everyday" realities.

[26] Hodges, op. cit., p. 36.

[27] Ibid., p. 33.

[28] Ibid., p. 91.

[29] See Tompkins, op. cit., pp. 149–50; also Piazzi Smyth C, *The Great Pyramid*, New York: Bell Publishing Co., 1978.

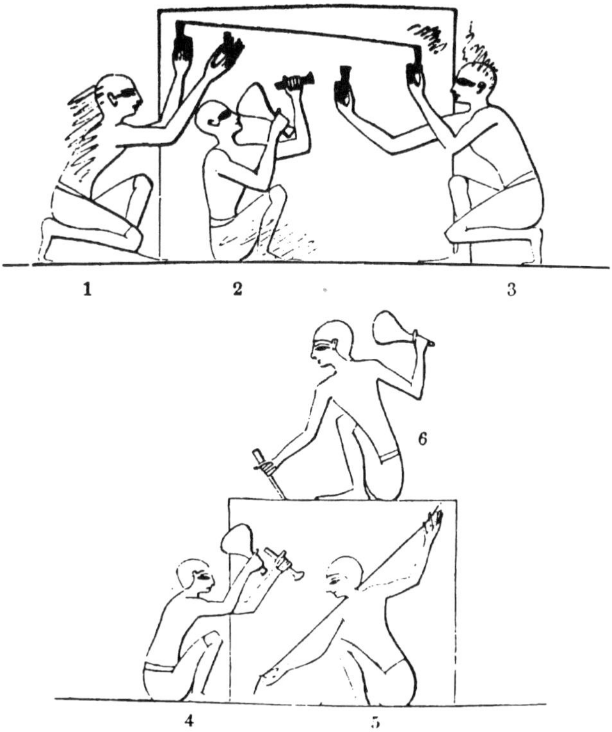

Fig. 24. ***Upper register:*** **Leveling a large building stone using chisel and mallet.** ***Lower register:*** **Squaring the stone by means of the diagonal. Here is the practical application of the problem of the "double remen."**

Alpha-Draconis occupied the position of 3°43' from the polar position twice in the last 5,500 years: the first time was 3,340 B.C. and the second time, 2,170 B.C. The second date, being of Middle Kingdom vintage, is entirely too late for an Old Kingdom structure like the Great Pyramid. Therefore, we are driven to conclude, on the basis of astronomical data, that the Descending Passage was excavated in alignment to α-Draconis 3,340 B.C. and the entire superstructure of the Pyramid was completed within 20 years after that. This tacks on about 800 years to the conventional age assigned to the Great Pyramid.

It is worth turning briefly to a consideration of some of the internal features of the Great Pyramid. The Descending Passage leads downward from the true entrance at an angle of 26°17' for a little over 69 feet to the entrance of the Ascending Passage, reflecting away from the Descending Passage and inclining upward toward the center of the pyramid at precisely the same angle of 26°17'. However, the Descending Passage continues down-

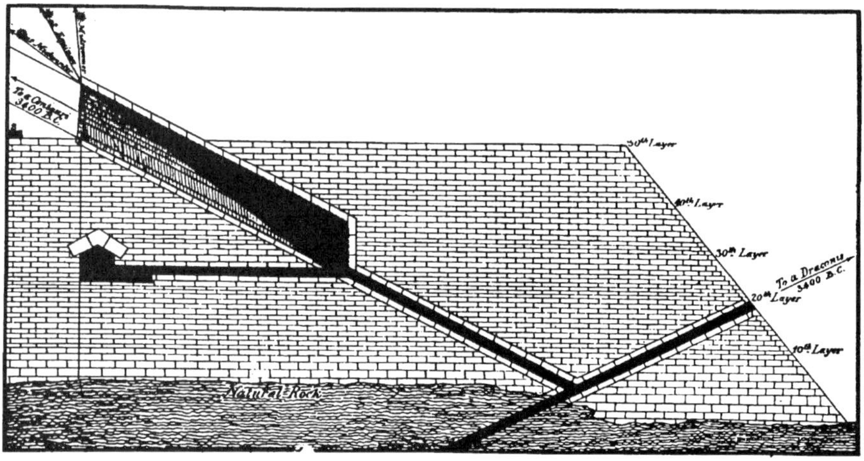

Fig. 25. Cross-sectional diagram of the Great Pyramid showing the alignment of the Descending Passage to a-Draconis 3,340 BC. The orientation of the Grand Gallery to the southern stars is also depicted.

ward past the Ascending Passage, narrowing to dimensions of 3'6" wide and 3'11" high, requiring anyone proceeding downward to bend forward at the waist for another 276 feet, giving 345 feet for the entire length of the Passage. At the end lies a subterranean chamber some 11 feet high whose function, ritualistic or otherwise, remains a complete mystery.

At the entrance to the Ascending Passage, a prospective entrant is confronted by a granite plug—actually three in all, lined up one behind the other—obstructing the opening to the Passage. To obtain admittance to the Passage requires crawling through a tunnel dug out much later around the three plugs to obtain ingress into the Passage. Its height is something over 4'4" which prevents most adults from climbing it standing erect.

At 129 feet from its opening into the Descending Passage, the Ascending Passage levels out and opens onto two other passages: a horizontal one, leading to the Queen's Chamber, and a continuation of the Ascending Passage, called the Grand Gallery, that is spacious enough to allow any adult-size individual to continue climbing erect. At the end of 157 feet, at the level of the 50th concourse, the Grand Gallery debouches into the King's Chamber.

The King's Chamber is unquestionably the most intriguing of the internal features of the pyramid for it is the only part of the Pyramid that is constructed entirely of granite blocks, some weighing more than 100 tons. Also, inside the King's Chamber is located the granite coffer, the only such accoutrement in the entire pyramid. It is too wide to have fitted through the entrance to the Chamber and is therefore thought to have been fashioned in place as the Chamber itself was being built.

Though it is evident that, to a large extent, symbolico-ritual considerations dictated their construction, the exact significance of these corridors, passageways, and chambers is still obscure. What is entirely lacking are *any* of the inscriptions, scenes, or paraphenalia associated with Egyptian mortuary practices. If the Great Pyramid was a tomb, it could not have been designed for mortal remains, though other pyramids undoubtedly were. That it served some grand, occult, initiatic purpose—along with its empirico-scientific functions—seems beyond question. Since initiation always involves symbolic rites of death, burial, and resurrection, it is only in this sense that the pyramid can be deemed a tomb.

It is time now to delve into the mathematic, scientific, and geodetic facts incorporated into the Great Pyramid's structure. As we have intimated, this topic has called forth reams of print, generating heated debate for and against acceptance of the ancient Egyptians' command of advanced scientific methods and concepts.

A word about the professional doubters—the "anti-pyramidologists"—whose numbers are legion. Certain mathematical "enigmas" such as the encoding of the π and ϕ constants into the dimensions of the pyramid have been repeatedly demonstrated, confirmed, and re-confirmed by scientific surveys from all over the world. Not being able to specifically refute these data, the skeptics have taken to explaining away the results by constantly invoking the threadbare cliche of "coincidence." Coincidence is the standard fallback position of those who cannot understand or refuse to accept facts that do not reconcile with their entrenched academic postures or prejudices. It is a way of explaining away inconvenient realities without actually having to disprove them. Thus, the invocation of "coincidence" means that the skeptic can't believe but he or she can't disprove either; the label of "coincidence" permits the skeptic to deny the "enigmas" academic currency. In academic parlance, in order for a theory, demonstration, or interpretation to be refuted, it must first be "falsified." One searches the literature in vain for a single academically rigorous falsification of the thesis that, built into the structure of the Great Pyramid, is a wealth of scientific data revealing the spirit of a civilization not less advanced than the modern one. This being so, the derision of the professional skeptics need not be taken seriously.

We have already seen that the Great Pyramid was aligned and oriented according to specific astronomical observations and measurements, and indeed it is clear that it was the technical and scientific requirements of the ancient Nile Valley astronomers that dictated the construction of parts of the Great Pyramid. There have been specific references to the astronomical functions of the pyramid by ancient Greek and Arabic writers.[30] In fact the Neo-Platonist Proclus tells us that the pyramid was used as an observatory prior

---

[30] Tompkins, op. cit., p. 147.

to its completion.[31] Now for making accurate observations of the movements of the heavenly bodies, it is essential to establish a heavenly *meridian* corresponding to an earthly one. A meridian is a great circle passing throught the poles of the heavenly sphere and the zenith of a fixed point on earth. Seen from earth, the sun, stars, moon, and planets appear to move in a circle around the earth and an exact earthly meridian allows the projection of a true meridian into the heavens (and vice-versa). Careful observation will allow a trained sky-watcher to detect the exact moment when any of the heavenly bodies move across or "transit" the meridian. This allows the observer to determine the body's rate of movement in its apparent circular orbit in the sky, providing the basis for charting and mapping the heavens, and, incidentally, charting and mapping the earth as well.

In Chapter 3, we made mention of the technique of laying and orienting the foundations of any sacred or official building in ancient Egypt by "stretching the cord." Part of this ceremony of stretching the cord included observing the culmination of a designated circumpolar star[32] to determine the north-south direction to be marked out precisely on the ground. The 19th-century astronomer Richard A. Proctor concluded that the ancient pyramid builders, having established a true earthly meridian, continued to use the polestar of the time to establish the angle of descent of the Descending Passage.[33] As seen above, this is how the Descending Passage came to be exactly aligned with α-Draconis in 3,340 B.C.

In Proctor's view, the ingenuity of the builders allowed them to devise an elegantly simple technique for establishing the angle of inclination for the Ascending Passage toward the center of the pyramid:

> . . . says Proctor, the builders hit upon the idea of creating an Ascending Passage at precisely the reflecting angle of 26°17'. By plugging the Descending Passage and filling it with water, they could reflect the polar star back up an Ascending Passage and continue to keep the passage truly aligned and the building level as it rose another score or more of courses.[34]

It is in the construction of the Ascending Passage that the original function of the Grand Gallery as the functional equivalent of a telescope comes into play. We have already seen that when the Ascending Passage reaches a level, it continues upward as the Grand Gallery, which is 28 feet in height, nearly 7 times the height of the Ascending Passage. The Ascending Passage and its continuation, the Grand Gallery, are aligned precisely to celestial

---

[31] Ibid.

[32] Circumpolar stars are those around either of the poles, which never set, making them ideal for orientation.

[33] Proctor RA, *The Great Pyramid*, London: Chatto & Windus, Piccadilly, 1883,pp. 78–183.

[34] Tompkins, op. cit., p. 151; see also Proctor, op. cit., pp. 118–21.

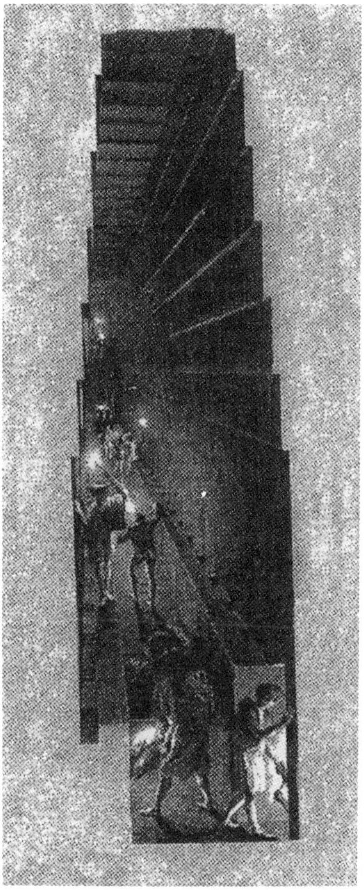

Fig. 26. Artist's rendition of the Grand Gallery as seen in Napoleon's time. The corbelled shape of the gallery is shown to good effect.

south. We have seen how this precisely southward alignment was achieved, that is, by reflecting the light of the north polestar back upward from the Descending Passage. Tompkins explains the initial function of the Grand Gallery in this way:

> Had an ancient astronomer wished for a large observation slot precisely bisected by a meridian through the north pole, so as to observe the transit of the heavenly bodies, what would he have requested of an architect? A very high slit with vertical walls, says Proctor, preferably narrower at the top, a gallery whose aperture, thanks to the reflected light of the pole star, could be designed so as to be exactly bisected by a true meridian.[35]

---

[35] Tompkins, op. cit., p. 152; see also Proctor, op. cit., pp. 127–8.

Considering that the construction of the pyramid was temporarily halted right at the completion of the Grand Gallery, giving any observer within it an unobstructed, clear, dry view of the southern sky along the polar meridian, the exact conditions described by the above quote are met in the Grand Gallery. As Tompkins notes, the entire southern circle of equinoctal constellations and their transits across a well-defined meridian could easily be observed. Every star in an 80° arc of the southern heaven could be observed and measured with great precision, and after repeated observations by trained observers at different levels of the Gallery over a certain span of time (perhaps a year), these ancient astronomers could have eventually developed an accurate sidereal clock. Then, when the period of observation was completed, all the sky measurements recorded, and accurate star-time developed, the builders could continue building the pyramid to completion.

Looked at in this way, it is hard to think of any other acceptable explanation for the manner in which the Grand Gallery was constructed. Concerning this, Proctor wrote: "We should find . . . in this double character of the Ascending Gallery, proof positive that it was intended for astronomical observations. *Only an astronomer would have set the architect such a problem*" (emphasis added) (see Fig. 27).[36] This interpretation fits in with verifiable data concerning the interior architecture of the pyramid, with information concerning the ancient Egyptians' formidable astronomical knowledge, and it resolves many enigmas. We have no shred of evidence suggesting that the Old Kingdom sky-watchers possessed telescopes, so the Grand Gallery, used as an observatory when it was still open to the sky, allowed them to glean the maximum celestial data from naked-eye astronomy.

What is also interesting about the Grand Gallery is that it debouches exactly at the 50th course of masonry of the pyramid, and at that storey the square platform is exactly *half* the area of the base of the pyramid (see Fig. 28).[37] Moreover, since the flat platform of this course, when it was still the truncated top of the pyramid under construction, was used to map all quadrants of the sky in preparation for the early zodiac, it is interesting to note that subsequent zodiacs, all the way into Renaissance times, took on a similar square configuration.[38] There is even a square zodiac at Denderah.

Finally, the observations made from within the Gallery and atop the platform onto which the Gallery opened would have permitted the accumulation of such data as to mathematically plot the Precession.[39] The Alexandrian as-

---

[36] Proctor, op. cit., p. 132.
[37] Ibid., p. 156.
[38] Ibid.
[39] The present writer has written about the Precession in several previous publications, so just a few words will be devoted to it in this chapter. We can say that though the tools for actually calculating the Precession precisely may have been made possible by observations taken in and on the pyramid, there are indications that the phenomenon was known some 7,000 years or perhaps even 30,000 years earlier.

Fig. 27. Proctor's conception of what the Grand Gallery would have looked like when it was open to the southern heaven. Trained sky-watchers taking numerous measurements of the southern stars transiting the open aperture would have been able to determine the heavenly (and hence earthly) meridian. They would also have been able to determine the time of the star's transit, thereby measuring the speed of rotation of the vault of heaven and thus of earth itself.

tronomer Hipparchus (c. 127 B.C.) is ordinarily credited with having discovered the Precession—though the first explanation of the mechanics of the phenomenon is attributed to Newton—but a number of modern commentators avow that the Precession was known long before Hipparchus's time.[40]

Few pyramidal debates are more heated than those centered around its geodetic functions. Geodesy is that branch of applied mathematics that exactly localizes specific points and areas over large expanses of the earth's surface and determines the shape and size of the earth itself. It is, of course, closely allied with geography and both of these sciences are fundamental to the creation of accurate maps. Specifically, geodesy involves finding the correct longitude and latitude of a given place on the earth's surface. Much of the reluctance to accept the possibility of advanced geodesy in high antiquity stems from the unwillingness to admit that the whole world had been traversed and mapped ages before Columbus.

---

[40] See De Santillana G and Von Dechend H, *Hamlet's Mill*, Boston: David R. Godine, 1977, for an extended discussion of ancient pre-Hipparchan knowledge of the Precession.

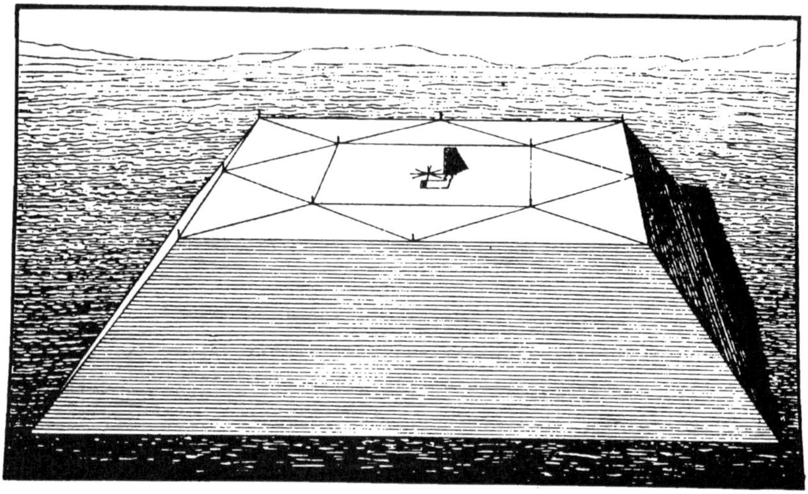

Fig. 28. The Neo-Platonist Proculus reported that the construction of the Great Pyramid was halted so the open platform could be used to make astronomical measurements. Proctor thought that the work was halted at exactly the 50th concourse whose area is exactly half the pyramid's base. The Grand Gallery terminates at and enters onto the 50th concourse.

Conventionally, it is assumed that longitude could not accurately be determined until the development of the sextant in the 18th century. However, it is *known* that Ptolemy of Alexandria, author of the famous *Almagest*, had apparently devised his own method of finding latitude and longitude, one based on astrononomical observation. Nonetheless, it is now clear, after some remarkably original research by Livio C. Stecchini, that both the Chaldeans (Babylonians) *and* the Egyptians had developed geodesy to a high level and that both cultures incorporated their knowledge in their architecture.[41] Nor is it too much to infer that Ptolemy's own work in geodesy owed considerably to his Egyptian predecessors since he was native to Alexandria.

Stecchini points out that Egypt was, in effect, re-mapped with the beginning of the dynastic period. Before that time, circa 4,000 B.C., Egypt had been divided into six zones; after that, it was divided into seven. This change may have been dictated by the need to reconcile Egypt's earthly geography with (1) the "geography" of the heavens which was septenary ("by 7's") and (2) the Egyptian harmonic scale, composed of 7 basic tones or notes. This change from the sextenary ("by 6's") to the septenary system is seen also in their units of measure. According to art historian Erik Iversen, the original

41. See Stecchini LC, "Notes on the Relation of Ancient Measures to the Great Pyramid," Appendix to Peter Tompkins', *Secrets of the Great Pyramid*, op. cit., pp. 287–382. Stecchini was one of this century's foremost experts on ancient measures. His dense, detailed 96-page essay at the end of Tompkins' volume must eventually revamp entirely the study of ancient history.

Egyptian cubit, the "short" cubit, was one based on 24 palms. The royal cubit was created out of this one by increasing it by 7/6 to 28 palms, reflecting the change from the sextenary to the septenary system.[42] Indeed, the first two pyramids, both constructed in the 3rd Dynasty, seem to indicate this transition from the sextenary to the septenary system upon which the geography of dynastic Egypt was constructed. Imhotep's Pyramid at Sakkara, the first of the pyramids, consisted of *six* steps or terraces. The one at Medum, built within two generations of Imhotep's, consisted of *seven* steps.

The first indication that the Great Pyramid was not randomly or haphazardly placed came from Napoleon's survey team in 1800. Seeking to accurately map northern Egypt, the surveyors quickly discovered that if a meridian was projected due north through the apex of the pyramid, the Delta was bisected neatly into two equal halves by area. Moreover, a right angle extending from the junction of the pyramid's northwest and northeast diagonals, completely and neatly enclosed the western and eastern boundaries of the Delta (see Fig. 29).[43] It was impossible to believe then, as it is now, that this was a fortuitous incident and even if no further information about the Great Pyramid had been forthcoming, this finding alone would have established uncontestably the ancient Egyptians' possession of advanced geodetic knowledge.

Stecchini tells us that the ancient Egyptians used three different latitudes to identify the Tropic of Cancer,[44] which was customarily taken as the southern limit of the kingdom. Now the Great Pyramid was situated at the "perfect" latitude of 30°, signifying it as the northern boundary of the southern portion of Egypt. The geographic shape of Egypt roughly conforms to that of the lotus flower, with the "stem" comprising the narrow, fertile strip hugging the Nile as it meanders northward, beginning at the First Cataract.[45] The open "bud" of the lotus-figure is the Delta itself; thus Upper Egypt comprised $^{6}/_{7}$ of the country by length. The northern (lower) edge of the First Cataract is situated at 24°06' north and the southern (upper) edge at 24° north. Thus,

---

[42] See Iversen E, *Canon and Proportions in Egyptian Art*, Warminster: Aris and Phillips Ltd., 1975, passim. Might this six-to-seven transition be the reason behind the colloquial phrase "at sixes and sevens," meaning "disordered" or "confused?"

[43] Tompkins op. cit., p. 46; Piazzi Smyth, op. cit., p. 226.

[44] The Tropics—Cancer north of the equator and Capricorn south of it—are defined as the highest and lowest limit of the sun's apparent movement on its ecliptic path. The sun's ecliptic is caused by the fact that the Earth is tilted as it revolves around the sun. On one side of the sun, the northern hemisphere is inclined to the sun; upon reaching the opposite side, the southern hemisphere is. At the summer solstice in the north (June 21) the sun is right over the Tropic of Cancer; at the winter solstice (December 21) it is right over the Tropic of Capricorn. Thus, when it is summer north of the equator it is winter south of it and vice-versa.

[45] Strictly speaking, the "First" Cataract is actually the *Sixth* Cataract because the Nile flows from south to north and the ancient Egyptians always deemed the south as the point of origin. Thus, of the six cataracts between Aswan and Khartoum, the one near Khartoum should be the first and the one at Aswan the sixth. However, we are tied to a 2,000-year convention that designates the cataract at Aswan as the "First."

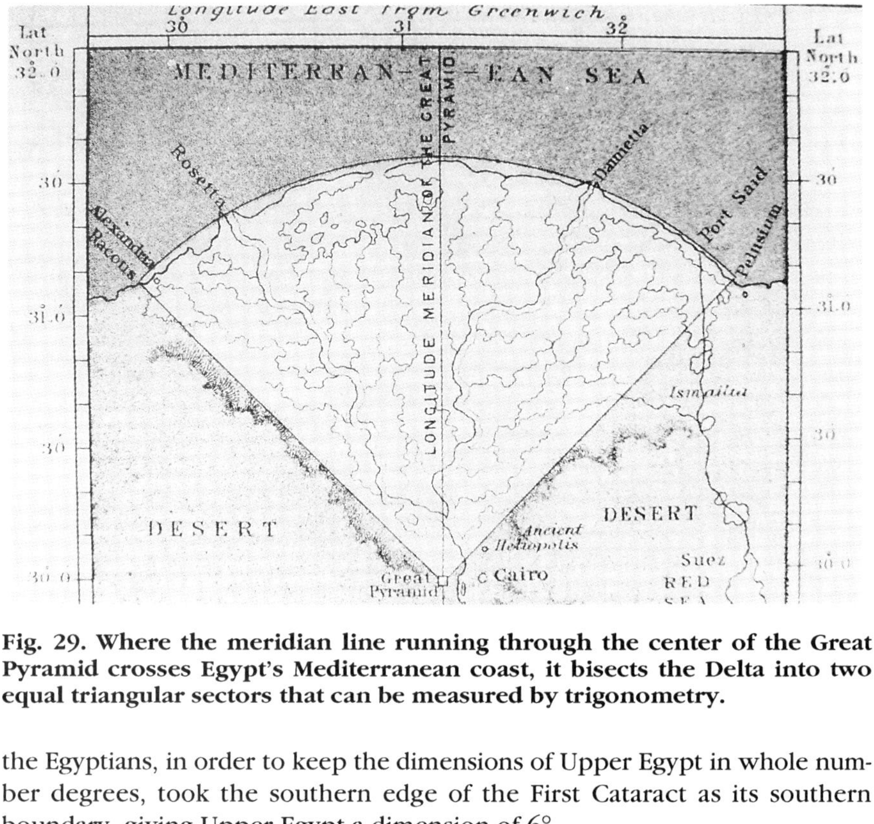

Fig. 29. Where the meridian line running through the center of the Great Pyramid crosses Egypt's Mediterranean coast, it bisects the Delta into two equal triangular sectors that can be measured by trigonometry.

the Egyptians, in order to keep the dimensions of Upper Egypt in whole number degrees, took the southern edge of the First Cataract as its southern boundary, giving Upper Egypt a dimension of 6°.

Aside from the pyramid itself, Stecchini states that there was another geodetic landmark that served as the meridian indicator bisecting the Delta, i.e., the Delta's apex at 30°06' north, geodetically 6 minutes north of the pyramid. This, too, could be considered Upper Egypt's northern limit while preserving the integrity of the 6° for the geodetic length of Upper Egypt since the southern boundary, in this instance, could be reckoned from the lower edge of the First Cataract at 24°06'.

In pre-dynastic times, the Tropic of Cancer was at or very near 24° north. Early in the dynastic period, the Tropic had shifted south to 23° 51' north and this change in the Tropic meant that there was no longer an even measure for the geodetic length of Upper Egypt based on either of the two northern landmarks: the Great Pyramid (30° N) or the Delta's apex (30°06' N). According to Stecchini, this discrepancy was resolved in a particularly ingenious way:

> The discrepancy was rationalized by considering the fact that when one follows the movement of the sun along the ecliptic by observing the shadow cast by a pointer, there must be introduced a correction of 15'. The position of the shadow

is not determined by the center of the sun, but by the upper limb of the disk. The apparent diameter of the sun is about half a degree [15'].[46]

If 15' is added to the Tropic of 23°51' at the time when the sun is exactly overhead on the first day of the summer solstice, the measuring point is actually at 24°06', already indicated as one of the southern boundaries in reckoning the length of Upper Egypt. It is for this reason that Stecchini insists that the Egyptians recognized three lines of the Tropic and this was inscribed in the tableau which depicts the neters—sometimes two forms of *Hapi*, sometimes *Heru* (Horus) and *Set*—uniting Upper and Lower Egypt by knotting the papyrus, symbol of Lower Egypt, and the lotus, symbol of Upper Egypt. At the bottom of the lotus register, three short horizontal lines are inscribed which, in Stecchini's view, represent the three values of the Tropic.

The above notwithstanding, the Egyptians still rationalized the geodetic dimensions of their country based on the actual tropic of 23°51' N by building Memphis, the capital city of the Old Kingdom, precisely at 29°51' N, i.e., 6° exactly north of tropic. Indeed, every major urban cult center was built at a distance that was in some fixed ratio to the Tropic. According to Stecchini the Egyptians went to extraordinary lengths to situate their cities at even proportional distances from their major geodetic landmarks. Such considerations even motivated religious revolutions and imperial enterprises.[47]

The Egyptian prime meridian was projected through Behdet, on the Mediterranean coast, because it was the coastal point, situated at 31°14' E and 31°30'N, that was aligned exactly with the meridian points through the pyramid and the apex bisecting the Delta. Behdet was apparently a pre-dynastic geodetic center, dedicated to Horus, that gave way in the dynastic era to the slightly more southerly latitude of 31°06' as the official northern boundary of the country, giving exactly 7° between it and the southern national border at the northern edge of the First Cataract. Thus was Egypt remade according to a septenary model.

The eastern and western borders were established at the corners of the Delta 1°24' East and West of the prime meridian passing through the apex at 31°14'E. Meridians were extended southward from these east and west boundary markers to reconcile the geography of Egypt as much as possible to a rectangle. The western meridian-border of Egypt, therefore, was considered to be 29°50'E. This is, in effect, the same meridian upon which sits Alexandria, located at the western corner of the Delta at the meridian 29°54'E, providing us with a perfect explanation of why Alexander the Great

---

[46] Stecchini, op. cit., p. 296.

[47] Thus, in Stecchini's view, the Amarma "revolution" of Akhenaten was, in part, dictated by his determination to reset the geodetic center of Egypt so as to re-establish the original predynastic geodetic system. Moreover, imperial land-grabbing in lower Nubia in the Middle and New Kingdoms also may have been partly motivated by the desire to institute a new geodetic system.

chose to establish his capital in that particular location. We know that Alexander's first undertaking after driving the Persians out of Egypt was a perilous 10-day expedition to the Oasis of Siwa in the Western or Libyan Desert, sacred to Amon. There, apparently, he became installed as the new "Son of Amon" and, since we find that he scrupulously followed age-old geodetic procedure in his adopted land, we can guess that it was the priests of Amon at Siwa who guided him in choosing the location of the new imperial capital Alexander was planning to build.

The eastern boundary, defined as a meridian situated at 32°38'E, created some discomfort for the ancient geodesists because a certain portion of the Nile, including the all-important First Cataract, actually lies east of this boundary by about 15'. The Egyptians rectified this anomaly by arbitrarily extending the region of the First Cataract, and therefore its southern border, about 50 or so miles southward to the point where the meridian of 32°38'E bisects the Nile right at the latitude of 23°N. This artifice added an extra degree to the geodetic length of Egypt (making 8 degrees total), and, in so doing, rationalized and justified the establishment of Thebes (Waset) as the new geodetic and cultic capital of Egypt. This switch may well have occurred as early as the onset of the Middle Kingdom around 2,200 B.C. and was to have serious geopolitical repercussions in Africa because it may have precipitated the centuries-long conflict between Egypt and Cush. In attaching this stretch of Lower Nubia in this fashion, it is evident that the Egyptian dynasts were clashing head-on with the interests of Cush who had exerted a hegemony in that region for a long time, if in fact it wasn't already a part of their country. In some ways, the reverberations of that rather imperious long-ago geopolitical act are still being felt today in those regions.

Stecchini insists that, because of its advanced geodesy and mensuration, it became convenient to consider Egypt's prime meridian of 31°14'E as the prime meridian for the whole world. He claims that Mesopotamian, Mediterranean, and Asian capitals were established in whole number or simple fractional relations to the Egyptian zero meridian. Thus, Egypt's prime meridian functioned exactly as the contemporary prime meridian, running through Greenwich, England, does. Moreover, Piazzi Smyth insists that the building of the Great Pyramid and the establishment of the zero meridian through its apex was also dictated by the fact that that meridian runs through more of the earth's north-south land surface than any other meridian. Complementarily, the pyramid's parallel of latitude runs through more of the earth's east-west land surface than any other latitude (see Fig. 30). Thus, the Great Pyramid can be said to be the center of the earth's land surface.[48] It is no wonder ancient nations formulated their own geodesy in relation to the Great Pyramid. It is no wonder that, in 1884, a strong case was made to reestab-

---

[48] Piazzi Smyth, op. cit., p. 89.

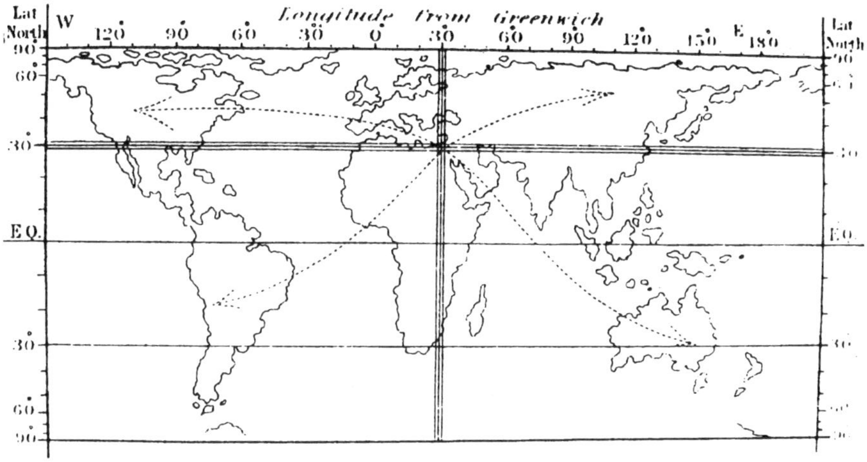

Fig. 30. The longitudinal line on which the Great Pyramid sits passes through more north-south land mass than any other global longitude; the latitudinal line of the Great Pyramid passes through more east-west land surface than any other. This makes the geodetic-geographic point of the Great Pyramid the center of the earth from the point of view of its surface land masses.

lish the modern prime meridian through the Great Pyramid until British and American pressure won the day for Greenwich, England.

All this is very heady stuff. It means, among other things, that the ancient priests of Egypt—indeed, the priests of all civilized countries—knew as much as we about the size and shape of the earth. It suggests further that ancient civilizations all over the globe were linked in a web of commercial, scientific, and diplomatic relations not at all unlike that of the modern era. Few indeed are they who are prepared, by their previous psychological conditioning or academic training, to accept any part of it. There is a dawning suspicion, once one grasps the implications of the work done on the pyramid, that at a certain point in historical time, a kind of correcting fluid, a veritable "white-out," was (metaphorically speaking) applied wholesale to the book of history, following which a whole different version of history was overwritten upon the blanked-out pages.

Ancient Nilotic Measure and Geodesy

If the Egyptians were in possession of advanced geodesy, as encapsulated in the Great Pyramid, certainly they must have devised measures of degrees of latitude and longitude along with the speed of the earth's rotation. Indeed, the Egyptians had developed a system of measures that coordinated length, volume, weight, and time.[49] Basic to this system was their unit of length, that is, the cubit. The cubit derives anciently from the length of pharaoh's arm between the elbow (*cubitum*) and the tip of the medial digit which was then formalized into a standard length. The smaller units of the cubit were subdivided into fingers, palms, and "fists." There were three forms of the cubit developed, beginning with the dynastic period and continuing to the Christian era. These were, in chronological order of appearance:

1. The "short" cubit of 24 fingers, 6 palms, 4½ fists.
2. The Royal Cubit of 28 fingers, 7 palms, 5¼ fists.
3. The Sacred Cubit of 34 fingers, 8½ palms, 6 3/8 fists.[50]

The short cubit measured 17.8 inches, the Royal Cubit 20.62 inches, and the Sacred Cubit 25.026 inches (see Fig. 31).

Other measures derived from the cubit were also in use. One such was the remen, composed of 20 fingers, 15 palms, 5 fists and spanning 14.83 inches.[51] Erik Iversen, author of *Canon and Proportion in Egyptian Art*, asserts that there was another Egyptian measure that he calls the "2/3 measure," consisting of 16 fingers, 4 palms, 3 fists and spanning 13.75 inches.[52]

Stecchini, in his appendix to Tompkins' book, has produced a lengthy discussion of the measurements of the degrees of arc of both latitude and longitude. The documentary evidence derives from an inscription discovered at Karnak Temple at Waset (Thebes). Ludwig Borchardt was the first in 1921 to discuss the geodetic implications of the inscription, though he ended by concluding that the possibility that the Egyptians measured their national dimensions by our equivalent of latitude must be absolutely excluded, apparently for no other reason than such advanced geodesy among them was considered intrinsically impossible. In brief, the inscription describes the exact distance between Behdet, located on the Mediterranean coast on Egypt's zero meridian, and Syene at the First Cataract:

> The inscription states that the distance between Behdet and Syene, the area of the First Cataract, is 106 *atur*, and divides this distance into 20 *atur* from Behdet to a place called Pi-Hapy, and 86 *atur* between Pi-Hapy and Syene.[53]

---

[49] Stecchini, op. cit., pp. 316–7.

[50] The existence of the "Sacred Cubit" is highly controversial since it is not specifically attested in the texts. Isaac Newton was the first to suggest its existence in the now-obscure pamphlet he wrote on the Great Pyramid. Others, such as Piazzi Smyth and Schwaller de Lubicz, accepted the historicity of the Sacred Cubit. Like other measures, it is said to be built into the dimensions of the Great Pyramid.

[51] See the problem of the "double remen" in Chapter 3.

[52] Mensurists such as Piazzi Smyth and Stecchini, among others, derive the modern inch, foot, and meter from these ancient measures.

[53] Stecchini, op. cit., p. 333.

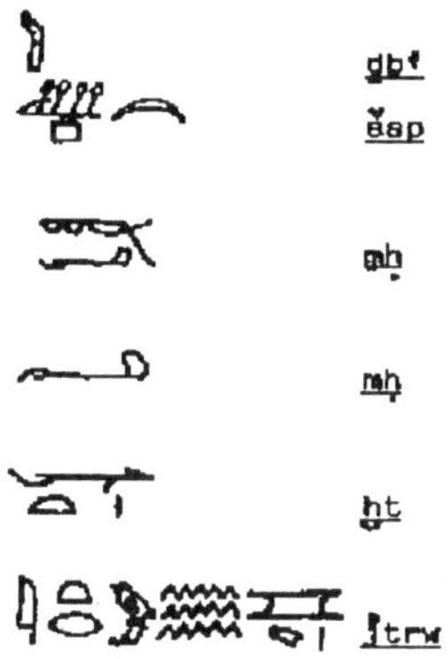

MEASUREMENT OF LENGTH

Measurement	Transliteration	Translation	Equivalent
	ḏbʿ	digit	0.73 inch
	šsp	4 digits or palm-breath	2.98 inches
	mḥ	28 digits or 1 cubit	20.64 inches
	mḥ	7 palms or 1 cubit	20.64 inches
	ḫt	100 cubits or 1 rod	171.8 feet
	ỉtrw	"river measure" 20,000 cubits	6.5 miles

Fig 31. Units and subunits of cubit measure showing both a septenary ("7s") and decimal character.

Apparently, it had crossed Borchardt's mind that these different distances reflected degrees of latitude but, as indicated above, he literally banished the thought. Stecchini exhibited no such qualms and determined the length of one atur to be 15,000 royal cubits. In using the inscription to support his figure, he points out that the difference in latitude between Behdet and the southern limit of the first Cataract is 7°30', and, on the basis of 15,000 cubits per one atur, the inscription distance of 106 atur equals 1,590,000 royal cubits. This distance of 1,590,000 royal cubits between Behdet and Syene equals 833,395.8 meters. The known geodetic difference of 7°30' between Behdet and Syene is equal to 831,091.6 meters so that the Egyptian distance, on the basis of 15,000 royal cubits per atur, exceeds the known geographic distance by less than 1 atur, $^2/_7$ atur (2300 meters) to be exact.

Stecchini makes the point that, in predynastic times, a measure between the short cubit of 17.8 inches and the royal cubit of 20.62 inches was used. Though this intermediate measure, like the short cubit, gave way to the royal cubit, it was used in geographical measurements because it yielded more precise results. Stecchini calls this measure the "geographic cubit" and it continued to be used, first by the Greeks, and then by the Arabs through the 19th century. The geographic cubit measured 18.15 inches. The reason why the geographic cubit was more suited to geodesy is that it gave whole, round numbers for the minutes of latitude and longitude respectively. We know, for example, that the apothem (the median line of the triangular face from the base to the apex) of the pyramid measured 352 *royal* cubits but exactly 400 *geographic* cubits; the side of the base of the pyramid measured 440 royal cubits but exactly 500 geographic cubits. It is known that the Cnidian philosopher Agartharchides reported that the pyramidal apothem of 400 geographic cubits represented 1/10th of a geodetic minute and the side of the base of 500 geographic cubits 1/8th of a geodetic minute; 4,000 geographic cubits thus equals exactly one geodetic minute (at the latitude/longitude of the Great Pyramid). Knowing this allows anyone to calculate the degree of latitude/longitude at the locale of the Great Pyramid. Moreover, it shows why the geographic cubit was more suitable for geodetic measurement than the royal cubit. However, the royal cubit maintained pride of place because it could be related to the septenary system governing Egyptian science and technology. All this being said, the 106 atur between Behdet and Syene could just as well be ramified, in Stecchini's view, to 1,802,000 geographic cubits (17,000 geocubits to an atur), giving a distance of 831,971.7, an excess of only 880.1 meters over the true distance. Thus, using the geocubit measure in round numbers gives greater precision than the royal cubit.

Above, we saw how the Egyptians obtained a meridian by using the Grand Gallery as an observatory to measure the passage of a certain star across the viewing aperture. This, then, would have been the method for obtaining longitude. Stecchini and Tompkins also note that once the observational apparatus of the Grand Gallery was in place, it also allowed the Egyptians to correlate length and time. The rotation of the vault in heaven, with the motion of the stars transiting across the meridian, gives an "absolute" time reference and, it just so happens, as Tompkins determined, the speed of rotation of the vault could be measured at *1,000 geographic cubits per second*.[54] This, too, allows more accurate determination of the longitudinal axes.

It struck this writer as he was reviewing the evidence showing substantial ancient Nilotic knowledge of the movements of the heavenly bodies through time and space that there was something very "Keplerian" about it all. Kepler was the 17th-century astronomer credited with the discovery of the laws of planetary motion that were directly responsible for Newton's discovery of gravity. The entire edifice of modern physics rests upon the labors of these two men. Be that as it may, this intuition concerning the Keplerian knowledge of ancient Nilotic astronomers was initially dismissed because there was no evident support for it until the author came across the work of Richard Anthony Proctor, who wrote:

> We know that [among the Egyptians] some harmonious relation between the distances and periods was supposed to exist. When Kepler discovered the actual law, he conceived that he had in reality found out the mystery of Egyptian astronomy, or, as he expressed it, that he had 'stolen the golden vases of the Egyptians.'[55]

---

[54] Ibid., p. 317.

[55] Proctor, op. cit., p. 262.

To this we might add that the *only* way time can be defined is as "movement through space," which underlies the Einsteinian conception of the universe as "space-time." Since the astro-architectural evidence shows that the ancient Nilotes had measured the relationship of time and space in exactly this way, it seems that more recent physicists have also recovered the "golden vases" of the Egyptians in the elaboration of their science.

The reader is invited to consult Tompkins's book, particularly Stecchini's treatise at the end, for more extensive details on these subjects. Suffice to say, the Egyptians had generated for themselves all of the analytical tools, tables, and data necessary to measure their country, their continent, *and* the world. That they actually did so is the subject of Chapter 6.

Before leaving the subject of the Great Pyramid, it is worth noting some other unusual features concerning it. Petrie was one of the first to call attention to an almost imperceptible hollowing effect in the Great Pyramid. This is to say that its triangular faces slant slightly inward toward their respective apothems. Up close, this effect is virtually impossible to detect on direct observation though it becomes evident on aerial photographs.[56] We should say more accurately that the effect is impossible to detect *except* at the spring equinox. As Pochan aptly demonstrates (with published photographs), at 6:00 a.m. on the morning of the spring equinox, a combination of light and shadow caused by the rising sun passes rapidly across the pyramid's southern face from left to right. This "flash" is a direct consequence of the pyramidal hollowing. At 6 p.m., 12 hours later, the flash reverses itself going from right to left. Thus, as Pochan notes, were the ancient Egyptians able to determine the equinox within 12 hours.[57] This, by the way, helped them keep track of the Precession since they would be able to mark the movement of the background of stars relative to the equinox year-by-year, century-by-century, millenium-by-millenium.

Another "calendrical" feature has been noted at the spring equinox, because at noon on that day, the pyramid "swallows its shadow." This means that with the sun overhead at noon on that day, the pyramid does not throw a shadow. Given these two means of determining the exact timing of the equinox, the pyramid, among its other functions, served as a gigantic almanac.

Even this summary treatment of the properties and functions built into the pyramid conveys the impression that something more than ordinary human genius went into its construction. Piazzi Smyth and other early pyramid enthusiasts thought they were contemplating divine Biblical revelation of the same nature as God's talk with Moses. Erik Von Daniken won a worldwide audience for his suggestion that only "space invaders" could have been

[56] See Pochan, op. cit., photos after p. 136.
[57] Ibid., pp. 231–2.

advanced enough to have built it. Yet we have nothing that concretely shows that any extraordinary machinery or supernatural technique went into its building. The pyramid appears to have been built with the tools of the time with an almost unfathomable ingenuity. More will undoubtedly come light about the Great Pyramid though it is hard to know if all of its essential mysteries will ever be revealed.

Some Facets of Nonpyramidal Nilotic Architecture

It is pertinent now to examine further some of the building principles underlaying Nilotic nonpyramidal architecture. In our discussions of pyramid building, we have already seen that limestone was the primary building material of the Nile Valley throughout the dynastic period. Limestone is a soft rock and the other soft rock used extensively for building purposes was sandstone, not significantly exploited for construction until the 18th dynasty (c. 1,580 B.C.). Egyptian alabaster was the only other soft rock used by the ancient Nilotic builder but on a much smaller scale than either limestone or sandstone.[58] Building-quality limestone can be found in the Nile Valley from Cairo to Nubia; quality sandstone is scarce above the 1st cataract. Quarrying methods, show, as Clarke and Engelbach indicated, the same orderliness and efficiency that characterized all other aspects of Nilotic building methods. Stone was cut away in roughly rectangular blocks, leaving nearly vertical quarry faces, instead of being wrenched from the quarry site. The standard view of the type of tools used to extract the stone characterizes them as made of well-tempered, constantly resharpened bronze or copper. And yet,

> . . . the tool was driven into the stone almost vertically, that small trenches were cut across the line of the separating trench at intervals of from one to two inches, and that the stone between these trenches was chipped or hammered away. It must be admitted that to do this would require a tool of great hardness.[59]

Since we now have evidence that iron was known and used in the Nile Valley from Old Kingdom times, it is worth posing the question of whether steel was also. As we have said in Chapter 2, iron conveys little advantage over bronze unless it is hardened by carburization to make steel. As Clarke and Engelbach suggest in the above quote, the kind of quarrying practiced by these ancient builders required hardened tools. Was ancient Nilotic bronze up to the task or was it steel that was employed for this purpose?

In addition to the soft rocks of limestone, sandstone, and alabaster, ancient Egyptians used the hard rocks granite, basalt, diorite, and quartzite, prin-

---

[58] See Clarke and Engelbach, op. cit., pp. 12–33, for an extended discussion of the building stones used in the ancient Nile Valley.

[59] Ibid., p. 18.

Fig. 32. Palatial home of a high Egyptian official.

cipally for statues, obelisks, and special interior structures of the temples. Basalt and quartzite outcroppings were found chiefly in the regions around Cairo; granite, almost always found as large boulders, was obtained almost exclusively from the east bank of the Nile at present-day Aswan. According to Clarke and Engelbach, the hard rocks, particularly granite, were extracted by wedging and pounding with dolerite balls. Again, the issue of the metal tools used has caused tremendous perplexity because all of the standard authorities have been constrained by conventional ideas about the beginnings of the Iron Age:

> Nowadays, the only metal used on the hard rocks is steel. To attempt to use copper, however much one may temper it by heating and hammering it, is to ask oneself whether the Egyptians did not use a tool of a hardness approaching tool-steel. It is tolerably certain that steel of this hardness was not known to the Egyptians.[60]

Thus, do the experts tie themselves into knots by their *a priori* reasoning. Be that as it may, it is known that these Nilotic technicians were able to quarry and *carry by ship* single blocks of granite weighing *two million pounds*.[61]

All the essential steps in the building process undertaken by modern architects are evident in the ancient Nile Valley. Pre-construction activities included drawing of plans and creating scale models, of which examples of both have survived. Indeed, such plans and blueprints were an indispensible

[60] Ibid., p. 25.
[61] Ibid., p. 34.

feature of temple archives since so many of them were rebuilt hundreds of years after initial construction along the same lines and in the same spot as originally. The architectural blueprint was always drawn on a squared papyrus surface, as were those of busts and statues. Virtually without exception, stone constructions and sculptures followed the well-known Canon of Proportion, most readily achieved by the technique of drawing on squared surfaces.

Clarke and Engelbach have criticized the ancient builders because, for most of dynastic history, the temples lack substantial foundations; some were built on no foundations at all.[62] It is only during the 25th or Ethiopian Dynasty that the builders of the Nile Valley began to establish solid foundations for their buildings:

> . . . under the princes of that country [Ethiopia] the buildings show a prodigality of materials in the foundations. Far from making a few trenches and holes to receive the walls and columns, the whole area of the temple was completely covered with carefully laid blocks three or four courses deep.[63]

Part of the reason why, prior to that time, seemingly lesser attention was paid to the foundation was because the walls and pillars of the temples tended to be so massive that strong foundations were somewhat superfluous.[64] It was the solidity and massiveness of the supporting structures that gave the buildings their stability. This same rationale applied to mortaring. Many commentators on ancient Nilotic building methods have remarked on the fineness of mortaring in Nile Valley buildings, allowing blocks to be apposed to within 1/50th of an inch of one another. But unlike building in later civilizations, mortar did not serve as a cementing material but a lubricating one, that is, a lubricant that allowed the blocks to be moved easily into place. Again it was the massiveness and solidity of the walls and columns that stabilized the building rather than the cohesion of the mortar. Thus, it can be seen that in the Nile Valley megalithic building was accomplished without mortar in the contemporary sense of it.

We have already touched on some of the methods employed by Nilotic builders to move huge, megalithic blocks. Granite blocks weighing hundreds of tons were floated on barges more than 500 miles down the Nile from their quarrying sites in Aswan. Exactly how this was accomplished is still somewhat mysterious. Sleds were also used and could haul heavy blocks of many tons gliding over surfaces made slick by lubricant applied ahead of the mov-

---

[62] Ibid., pp. 69–77.

[63] Ibid., p. 76.

[64] Clarke and Engelbach do point out that most of the important medieval buildings, many still standing, were erected without foundations. Again, for them it is a point of criticism whereas it may be that for buildings constructed in a certain way, foundations are not the most essential feature. Ibid., p. 77.

ing sled. Though, as we have seen, wheels were not used in the building projects of the Old and Middle Kingdoms, logs functioning as rollers were. Rockers, often in conjunction with levering implements, also served to appose the blocks tightly against one another.[65]

Though the Nile Valley civilizations were justly famed for their stone architecture, they were as skilled in the uses of brick. Indeed, most buildings, including the homes of the citizens of all classes, were built of brick. Even pharaonic palaces were of brick instead of stone, the latter material being reserved almost exclusively for sacred buildings. Bricks were made of Nile mud and, in the hot, dry climate, allowed to dry in the sun. Other cohesives added to the bricks consisted most commonly of wheat chaff or sand.

These bricks were fashioned by the millions and went into the construction of, among other things, walls attaining a height of 45 feet and a thickness of up to 80 feet. Few brick structures from dynastic times still stand, though, in this century, outstanding examples of brick masonry could still be seen at the ruins of the fortresses at Semna, Kumma, and Buhen in Nubia. Clarke and Engelbach noted also that grids of wooden beams were sometimes laid in every fifth concourse of a brick structure to stabilize it and prevent cracking.[66] This use of wooden beams is of more than casual interest because, as will be seen later in this chapter, the same technique was adopted in the mud-brick architecture of the Western Sudan, whose resemblance to Nilotic architectural forms has struck a number of observers.

The better homes and the palaces of the rich often attained great size, with more than a few able to accomodate 60 people. The royal palaces in particular were notable for their splendor, built with courtyards, gardens, and lakes. Indeed, every Egyptian of means maintained a garden where most of the leisure hours were whiled away. Walls and windows were constructed in such a way as to catch the northern breezes to keep the house well-ventilated. In the cooler months, particularly in the Delta, the thick brick walls also insulated the household within. Most of the better homes possessed indoor toilet facilities and, occasionally, running water. Mostly, water was brought in from wells or from the Nile (see Plate 16); waste water was carried out of the house and allowed to soak into the surrounding sand. However, in several cities, most especially Akhetaten and Buhen in Nubia, a well-planned and constructed public drainage system served the entire populace.[67]

---

[65] Ibid., pp. 88–93.

[66] Ibid., p. 210.

[67] See Badawy A, *Architecture in Ancient Egypt and the Near East*, Cambridge: M.I.T. Press, 1966, pp. 21–33.

The Astro-architecture of the Ancient Nile Valley

The astronomical relations governing the architecture of the Great Pyramid had begun to be recognized by the middle of the 19th century. However, it was not until the last decade of that century that the astronomical determinants of other Egyptian temples were established. This was almost entirely due to the work of Norman Lockyer who, like Piazzi Smyth, was an eminent astronomer in his own right. He had done some of the seminal work on sunspots and was the first to identify helium as one of the primary elements in the sun. Moreover, he was the founder and first editor of *Nature*, still one of the most influential scientific journals in the world. In spite of such gilt-edged credentials, his work on the astronomical orientations of Egyptian temples, set forth in the *Dawn of Astronomy* in 1894,[68] did nothing to endear him to either his colleagues or Egyptologists. Though not an evangelical Christian or Bibliolater like Piazzi Smyth, Lockyer did use the results of his investigations to launch into some highly speculative explanations concerning the dates of origin of Egyptian civilization. Nonetheless, his empirical work was and remains rock-solid. Furthermore, he applied his methods to an analysis of the astronomical orientations of Stonehenge, again to the derision of archaeologists. They stopped laughing, however, when Gerald S. Hawkins proved Lockyer right in 1963.

We will deal more fully with Lockyer's reconstruction of Egyptian astronomy in the next chapter, but we can introduce his work here since he was able to show the interconnection between Egyptian architecture and astronomy. Lockyer based his early work on the orientation of Egyptian temples on the temple surveys performed by the French under Napoleon in 1799 and the Prussians under Lepsius in 1844. That did not prevent him from travelling to Egypt on his own in 1890 to take his own measurements and sightings, particularly since neither the French nor the Prussians seemed to have drawn any conclusions relative to the astronomical implications of the alignments and orientations they had determined. But Lockyer found that temple after temple showed an axis of orientation precisely aligned to certain important stars. More astonishing from Lockyer's standpoint, temple axes seemed to have been deliberately altered by extended construction in order to "follow" a star as it, over a period of centuries, shifted away from the original alignment. In some cases, whole temples were completely abandoned and others erected in order to re-establish the correct alignment with a certain star. These axial shifts, among other things, convinced Lockyer—as others had been convinced from other data—that the Egyptians possessed a perfect knowledge of the Precession, of both the equinoxes and the polestars.

The Temple of Amon at Karnak, the largest of its kind in the world, was

[68] Lockyer JN, *The Dawn of Astronomy*, Cambridge: The M.I.T. Press, 1964.

constructed and oriented in such a way as to represent a perfect illustration of Lockyer's findings. The axis of the temple pointed toward the western hills across the Nile at a declination of 24°18'.[69] Lockyer points out that the obliquity, i.e., the tilt of the earth's axis, though 23°27' today, was, in 5,000 B.C., 24°22', meaning that the tilt has "straightened" by nearly 1° in 7,000 years. From these data, Lockyer concludes that the axis of the temple was originally aligned with the setting sun at the time of the summer solstice as early as 3,700 B.C. At that date, the sun at noon would have been poised directly overhead at the Tropic of Cancer at 24°18'N, corresponding to the declination of the axis of Karnak Temple.

As to the temple structure itself, Lockyer observed that the axial corridor, from the entrance pylons, stretched in a straight line back through a series of halls and portals of gradually diminishing widths, coming to an end at the Sanctuary. The entrance pylons were unroofed but the interior halls, i.e., the hypostyle hall, would have been covered. This led Lockyer to say:

> From one end of the temple to the other we find the axis marked out by narrow apertures in the various pylons, and many walls with doors crossing the axis. . . . In the temple of Amon-Ra there are 17 or 18 of these apertures, limiting the light which falls into the Holy of Holies or the Sanctuary.[70]

Lockyer further states:

> . . . every part of the temple was built . . . to limit the light which fell on its front into a narrow beam, and to carry it to the other extremity of the temple—into the sanctuary, so that once a year when the sun set at the solstice the light passed without interruption along the whole length of the temple, finally illuminating the Sanctuary.[71]

What is more, it occurred to Lockyer that these architectural arrangements, concentrating and focusing the light of the setting solstitial sun in a surrounding darkfield created by the temple's interior, produced exactly the same effect as a telescope. That is to say, the apertures created by the series of gradually narrowing portals, pylons, and separating walls functioned exactly the way diaphragms do in a telescope.[72] No one has yet found a telescope among the artifacts of ancient Egypt but it is evident that they possessed more than a passing understanding of optics. Whatever the case, at the precise moment of its setting at the solstice, the sun would send a thin shaft of light down

---

[69] Ibid., p. 119. The reader should consult Lockyer's book to understand the technical details of his method.

[70] Ibid., pp. 103–4.

[71] Ibid. pp. 105–6.

[72] Ibid., p. 108. Lockyer might have also said that these diaphragmatic principles are also intrinsic to the modern camera.

the axial corridor, brilliantly lighting the interior sanctuary wall for a period of two to five minutes, enabling the priests to measure the exact duration of the solar year.

Lockyer has found similar arrangements in most of the important cult temples of Upper Egypt, whether oriented to the sun or to other stars in their risings or settings. The star temples, therefore, were designed to direct starlight into the sanctuary in the same manner as the axial corridor of Karnak directed sunlight into its sanctuary. The orientations were exact, carried out by the familiar "stretching of the cord," usually performed by pharaoh himself. Several inscriptions describe the method by which pharaoh performed the stretching of the cord ceremony while sighting a specific star.[73]

It is also pertinent to mention the manner in which the Egyptians managed to paint scenes with brilliant colors and exact drawings in the deepest, darkest recesses of their tombs and temples without benefit of burning torches. No traces of soot or any combustion has *ever* been detected in the temples traceable to the time when the temples were built. Since no one seriously imagines that the Egyptians possessed electric lighting, it has been determined that they used a system of mirrors to direct sunlight from the surface through the twists and turns of the temple corridors. No matter how far removed from open air the inner rooms were situated, the sunlight conveyed into them by the mirrors caused them to be illuminated as brightly as electric light.

We will continue the discussion of Egyptian "astro-architecture" in the next chapter.

Harmony and Proportion in Ancient Egyptian Architecture

> Nowhere . . . is the belief in the existence of an aesthetics of numbers, an eurhythmy based on the harmonious coordination of proportions and ratios expressed in naked numerical terms such as manifest in any Egyptian work of art, professed with greater force and lucidity than in Philo Byzantius' laconic quotation . . . "beauty emerges imperceptibly from a profusion of numbers."[74]

This axiom applies equally well to Egyptian architecture. As we have seen, most of what we know concerning the interconnections among number, geometry, music, and astronomy in ancient Egypt comes to us indirectly through the traditions passed on by the Pythagoreans and the dialogues of Plato. The governing principle of *all* these disciplines from the Egyptian standpoint was harmony. Indeed, one might say that harmony and all its correlatives—proportion, symmetry, and eurhythymy[75]—comprised the true defi-

---

[73] Ibid., pp. 173-9.

[74] Iversen, op. cit., p. 13.

[75] Iversen defines *proportion* as "the correct adjustment of the parts to anthropomorphic standardization;" *symmetry* as "the commensurability of the parts;" *eurhythymy* as "the coordination of proportion and symmetry into a harmonious unity in accordance with the established aesthetics of numbers"; ibid.

nition of *Maat*. Harmony, of necessity, governed the creation of art and architecture as well.

There is a theme that recurs through several of Plato's dialogues, particularly *Timaeus*, *Laws*, and the *Republic*, that refers to the building of the ideal society. As John Michell points out, in Platonic society, only certain kinds of music are permitted, i.e., those emanating from the immutable laws of harmony which, in the playing, promote psychic stability and balance in the citizens of that society.[76] In *Laws*, one of the protagonists, an Athenian, laments the tendency of the nations of that time to allow musicians to compose and play any kind of music that whim and fashion dictated, without regard to their effects and consequences.[77] What interests us is that the Athenian commentator specifically relates that Egypt represents the sole exception to this state of affairs, that is to say, deharmonizing, "destructive" musical forms were expressly forbidden in that country:

> One could doubtless find things to criticize in other Egyptian institutions but in the matter of music it is a very noteworthy fact that it has proved possible to canonize those forms of music which are naturally correct and establish them by law.[78]

As Michell tells us:

> By Plato's time the very idea of a canon of music had been forgotten everywhere except in the academies of Egypt, but he himself had evidently studied and learned it, for the number code behind it is at the root of all his mathematical allegories.[79]

The evidence is clear: it is in Egypt that the concept of music providing a harmonic guide to govern society was elaborated. Furthermore, that harmony was definable by number is also an Egyptian concept. Since society, in the form of a sacred city, had to be established according to the rules of harmony and numerical proportion, it followed that art and architecture in the ideal society had to conform to harmony and numerical proportion as well.

Plato espoused some definitive if lofty ideas on how to construct the earthly version of the cosmic city. As Michell, Schwaller de Lubicz, and others reveal, the Temple was the central institution around which this pre-Christian "City of God" had to be constructed. Thus, Michell tells us:

> Like all products of the canon, the temple was also a world-image, synthesizing in its proportions the measurements of the human frame with those of the cos-

---

[76] Michell J, *The Dimensions of Paradise*, London: Thames and Hudson, 1988, pp. 12-3.

[77] It will not escape the attentive reader that this passage has implications that spill over into the modern day where a kind of "anti-music" has evolved that both mirrors and impels the social deconstruction everywhere evident.

[78] Cited in Michell, op. cit., p. 13.

[79] Ibid.

> mos. It was designed on the principle that "like attracts like," on the understanding . . . that if one wishes to attract any aspect of the universal spirit one must create a receptacle in its image.[80]

In earlier parts of this book, we have alluded to the numerical proportions and constants such as π and ϕ built into the Great Pyramid. Other numerical constants and values dictated the architectural design of pyramids and temples alike. One such was the $^4/_3$ proportion noted by Bent Hansen in the Step Pyramid, the Great Pyramid, and other temples in Egypt.[81] Now the $^4/_3$ proportion takes us once again back to the "Pythagorean" harmonic scale. Proceeding up the scale from C, the tones are created by the 8:9 proportion (of string length) until E (and later B) is reached. The tone ratio composing E, of string length $^{64}/_{81}$, is created by the operation $^9/_8 \times {}^9/_8$ which = 81/64.[82] However, to go from E to F can only be accomplished by producing a semitone, obtained by multiplying $^4/_3 \times {}^8/_9 \times {}^8/_9$, that is to say multiplying $^4/_3$ by E's string length of $^{64}/_{81}$, giving the leimma of $^{256}/_{243}$ (see Chapter 3).

It would therefore seem that a key constant of the harmonic scale was built into the pyramids and temples of ancient Egypt. We have already mentioned that the leimma of $^{256}/_{243}$ is exactly one-third of the Rhind value for π of $^{256}/_{243}$. And, in fact, if the E ratio of $^{81}/_{64}$—from where the leimma is generated—is multiplied by the Rhind π of $^{256}/_{243}$, one obtains the double-octave, defined by a string length of 1/4 and a tone ratio of 4. Since the double-octave was considered the limit of the ordinary human voice, Pythagorean harmonics ended there. The reader will notice, however, that at the end of all of this stands the all-powerful number 4 which anchors so much of African number symbolism.

One notices that this all-important $^4/_3$ constant that we continue to encounter in any discussion concerning Nilotic harmonics, geometry, and architecture is exactly the *double* of $^2/_3$. We have already seen how important the $^2/_3$ fraction is in the ancient Egypt arithmetic system and we are entitled to surmise that it is derivable from this harmonic constant of $^4/_3$. Since the ancient Nilotes kept everything perfectly consistent, we see the doubling and halving that dictated their mathematic operations is perfectly preserved here with $^4/_3$ being *double* $^2/_3$ and $^2/_3$ being *half* of $^4/_3$.

Gradually we see everything in the Egyptian cosmos—earthly and heavenly, human and divine, material and spiritual—integrated into a whole, and

---

80 Ibid., p. 16.

81 See Hansen BH, "An Architectural Analysis of the Great Pyramid," in *Proceedings of the First International Symposium on the Application of Modern Technology to Archaeological Explorations at the Giza Necropolis*, Cairo: Egyptian Antiquities Organization, December 14–17, 1987, pp. 206–219.

82 The tone ratio is *always* a reciprocal of the string length (of the monocord). Thus, a string length of 1/2 gives a tone ratio of 2, and so on.

Plate 1. A profile of the Sphinx in its quarry showing the erosive patterns on the Sphinx's body and the surrounding quarry wall caused by rainfall. (Photo: V.D. Floyd)

Plate 2. Bronze offering stand from Napatan Cush, circa 750 B.C., at the outset of the 25th Dynasty. (Photo: Museum of Fine Arts)

Plate 3. Copper bust of an Ife ruler from 500 to 800 years ago showing a perfection of technique in a metallic medium particularly difficult to cast. (Photo: D. Bakker)

Plate 4. Head of an Ife ruler cast in a zinc-brass alloy, possibly as far back as 800 years ago. The face may be that of a woman. (Photo: D. Bakker)

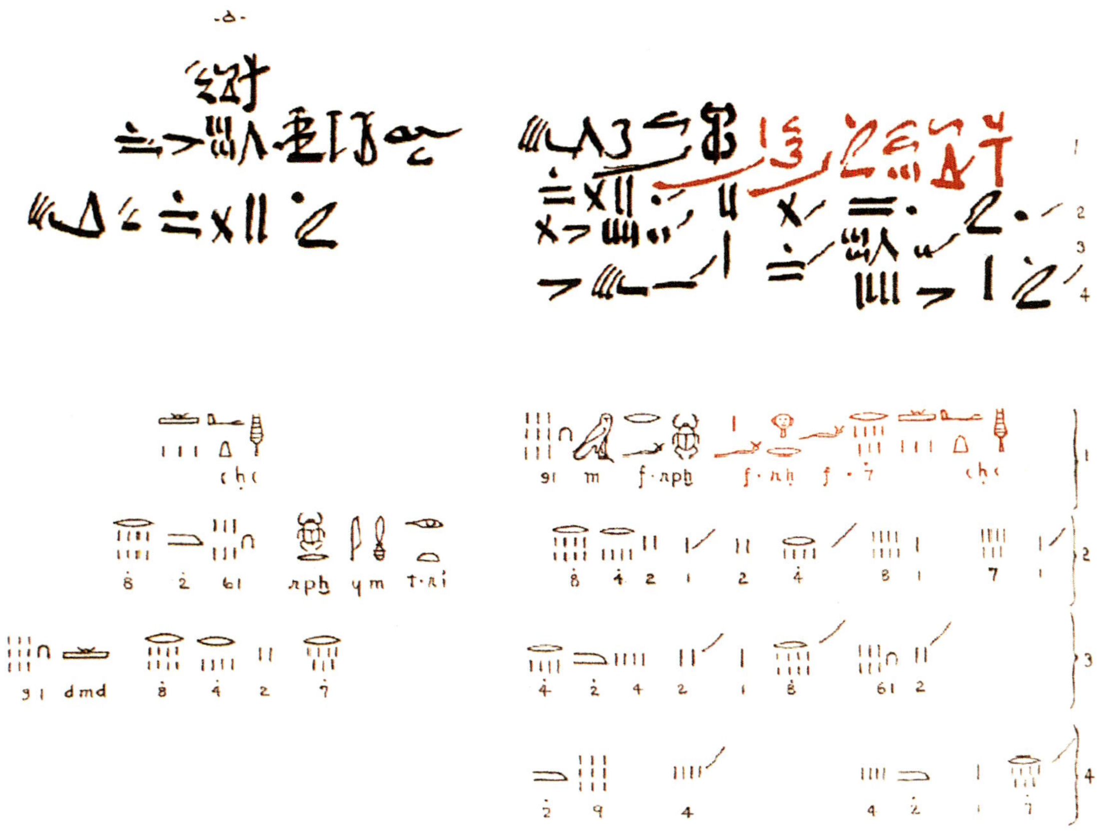

Plate 5. Problem 24 in the Rhind Papyrus, by which the problem is solved by means of an unknown quantity (*aha*). This is the earliest example of algebraic procedure.

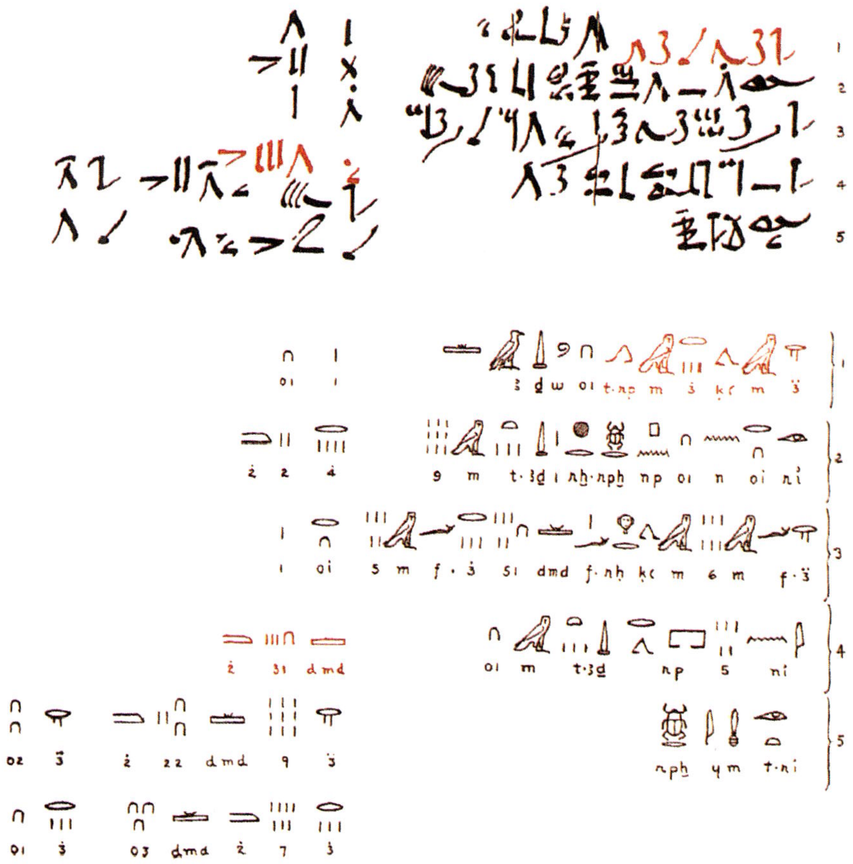

Plate 6. Problems 28–29 in Rhind Papyrus representing the first "think of a number" problems; also algebraic in character.

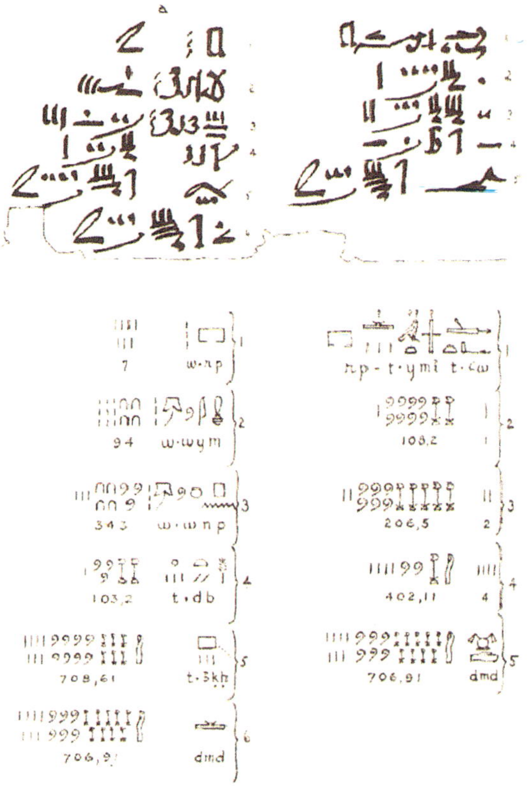

Plate 12. Problem 79 of the Rhind Papyrus giving a geometric progression.

Plate 13. Step Pyramid, with its six terraces, as seen from the forecourt of the complex. The structure, with its temple complex (not seen), was designed and built by Imhotep during the 3rd Dynasty reign of Djoser. (Photo: C. Finch)

Plate 14. "Proto-Doric" columns in the Step Pyramid temple complex at Sakkara. These columns were erected during the 3rd Dynasty, circa 3,600 B.C. (Photo: C. Finch)

Plate 15. Hatshepsut's temple at Deir el-Bahari, on the west bank of the Nile opposite and aligned to Karnak Temple on the east bank. The temple was carved out of the cliff by Senmut. The style of recessing terraces makes the temple completely unique; 18th Dynasty, circa 1,480 B.C. (Photo: C. Finch)

Plate 16. Temple conduit at Kom Ombo, engineered to filter and purify water conveyed to the temple from the Nile. (Photo: C. Finch)

Plate 17. The "palace" in the coastal Swahili city of Gedi (Kenya), circa 1,000 A.D. The city is 20 square miles in area of which only one square mile has been excavated. (Photo: A. Akers)

Plate 18. Aletha Akers in front of the "Pillar Tomb" at Gedi. An obeliskoid structure rises in the background. (Photo: A. Akers)

Plate 19. Temple or mosque used for Islamic prayer, the largest excavated structure at Gedi. At the entryway, on the other side of the standing figures, were troughs to wash hands and feet. The water trickled out of the trough, passed through a limestone filter, and then flowed back to the well from which it was drawn to be recycled. (Photo: A. Akers)

Plate 20. Aperture in the roof of the sanctuary in the chapel of Isis behind the Hathor temple at Denderah, showing sunbeams etched into stone as waves. Western science did not "prove" that light travelled as waves until the 17th century. (Photo: C. Finch)

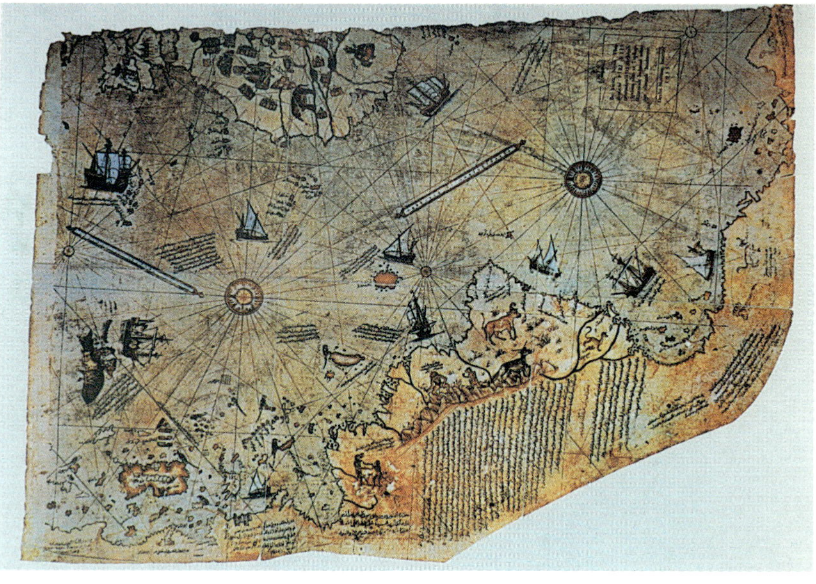

Plate 21. The Piri Reis Map (1513) drawn in typical portolano style. The Straits of Gibraltar, western Africa, eastern South America, Antarctica, and the Caribbean are all shown in correct latitudinal and longitudinal relationship to one another. A curved line of compass roses in the Atlantic define the left rim of a circle whose center is off the map. The surviving fragment represents only half of the original map whose center was located 100 miles west of Aswan in Upper Egypt. (Photo: Library of Congress)

Plate 22. Professor Cheikh Anta Diop with the author outside Dr. Diop's radiocarbon lab at IFAN, Dakar University, August 1983.

Plate 23. Maam Adji Fatou Seck, 110-year-old Senegalese *ndeppkat* (healer) with Dr. Erick Gbodossou, medical doctor and *vodoun* initiate at the Coumba Lamba USA healing ceremony, Penn Center, St. Helena Island, South Carolina, August 1996. (Photo: L. Foster)

harmony was the glue. More fundamentally, the numbers and numerical proportions underlying harmony were the "chemical bonds" uniting everything: music, math, art, architecture, astronomy, *ad infinitum*. That which was Good and Beautiful was that which was Harmonious; the Evil and the Ugly were, by definition, Unharmonious. Moreover, Harmony was not a vague, personalized, subjective sensation or apperception; it was a definite and measurable quality by the ineffable properties of number.

On his first trip to Egypt, this author, as he, in a fog of dazed wonderment, viewed a seemingly unending vista of architectural masterpieces, was soon struck by a daunting thought: these people built their civilization as if they were reading the Mind of God.

Southern Nile Valley Architecture

It bears repeating that, south of the Egyptian border, Africa, archaeologically, is practically untouched. There are, along the stretch of the Nile running the length of the Sudan, thousands of sites awaiting the spade. Therefore, we have only a fraction of the data on Nubio-Cushitic architecture, spanning at least 3,000 years to 300 A.D., as we do for Egyptian architecture.

Nubia and Egypt in pre-dynastic times belonged, in effect, to the same cultural horizon, as most pre-dynasticists will attest, and this broad cultural unity persisted until the complete demise of Nile Valley civilization with the collapse of Meroe around 330 A.D. Some of the most distinctive early architectural structures found in Nubia-Cush are the funerary tumuli, examples of which also abound in Upper Egypt of the same era (circa 4,000–3,000 B.C.), consisting of a burial pit, shaft, and mound or tumulus overlying the grave. As an aside, the wealth and abundance of even the common Nubian graves continue to astonish archaeologists since graves of persons of comparable status in Egypt do not exhibit such testaments of generalized affluence.

The tumulus style of funerary architecture underwent considerable development in the early centuries of Nubio-Cushitic history. W.Y. Adams gives a concise description of the so-called C-group tumuli (circa 2200 B.C.):

> A ring of dry-laid stone masonry, generally about three feet high and five to fifteen feet in diameter, was built around the top of the grave shaft. The interior of the structure was filled with rubble or sand, and sometimes covered at the top with flat slabs, so as to form a solid tumulus.[83]

However, following a style similar to developments realized with the adoption of the mastaba type of funerary architecture in Egypt, the tomb within the tumulus changed form:

---

[83] Adams WY, *Nubia: Corridor to Africa*, Princeton, N.J.: Princeton University Press, 1977, p. 156.

> Grave shafts gradually evolved from a round to a rectangular shape. By the middle of the C Horizon they were sometimes lined with upright slabs or mud brick. . . . The idea was evidently to preserve the body within an air chamber, as had long been the practice in Egypt and elsewhere. This method of interment reached its fullest development at the close of the C Horizon, when the body was laid within a rectangular brick chamber covered over with a brick vault.[84]

The tumulus funerary structure achieved its greatest development during the Kerma culture, between 2,000 and 1,500 B.C. So called Tumulus X, the largest of the surviving "great tumuli," measured 300 feet in diameter with an interior structure far more extensive and complex than any of the more famous Egyptian pyramids. These great tumuli were generally bisected by a long, straight brick corridor. From the northwestern edge of this central corridor emanated a series of 25 or so straight brick walls extending all the way to the perimeter of the tumulus, creating a like number of smaller corridors. On the southeast side of the central corridor, 30 or so straight dividers extending to the perimeter created a similar series of parallel corridors. This system of walls and dividers is thought by W.Y. Adams to have served as the support structure for the overlying tumulus.[85]

These great tumuli though have a sinister aspect: in the largest of them, Tumulus X, the remains of 322 individuals, in various stages of contortion, were discovered. These individuals seemed to have been sacrificed or buried alive with the royal occupant.

Perhaps the most unusual edifices that have been discovered at Kerma, indeed the whole Nile Valley, are three massive mud-brick structures that are called by their Nubian term, *deffufa*. We have already seen that mud-brick was the standard building material everywhere in the Nile Valley, and, as in the cognate Egyptian culture to the north, stone was reserved for sacred buildings and tombs. In fact there seems to be some ambiguity as to the exact function the deffufas served. Given the considerable amounts of jewelry, pottery, and various trade and luxury items strewn about in their vicinity, Adams considers them trade depots or warehouses.[86] However, on the basis of the columns, faience tiles, and the large quartz altar incorporated into the largest of these structures, other authorities indicate that the deffufas were temples.[87]

The largest deffufa measured 60 feet in height, 75 feet in width, and 150 feet in length. The walls on this one spanned 12 feet of thickness but a morturary deffufa described by Adams possessed walls of brick exceeding 30 feet in thickness. Adams, as Reisner before him, was impressed with the

---

[84] Ibid., p. 157.
[85] Ibid., p. 204.
[86] Ibid., pp. 199-202.
[87] Haynes JL, *Nubia: Ancient Kingdoms of Africa*, Boston: Museum of Fine Arts, 1992, p. 34.

straightness and evenness of the walls, the high quality of the masonry, and the capability of building such a large structure with mud-brick. Consequently, he assumes it was of Egyptian design and supervision. This reasoning is consistent with the pronounced tendency to assume that everything showing skill and mastery in ancient Nubia had to have come from its northern neighbor. At all times in Nile Valley history, there was a two-way flow of ideas, techniques, and influences between Nubia-Cush and Egypt consistent with a definable cultural unity that spanned more than 4,000 years. This cultural unity abided even where distinct regional and local flavors are evident.

Where Egyptian motifs exerted a special influence in Nubia-Cush there were nonetheless present local craftsmen, artisans, and builders with the know-how and technical skill to produce them with their own materials. Whether or not it was possible to call a certain building method distinctly "Egyptian," the architecture and the masonry assumed their characteristic forms under the hands of local craftsmen.

From about 1,500 to 1,000 B.C., Nubia-Cush was essentially part of Egypt, bearing much the same relationship to their northern neighbor that Scotland (or Wales) does to England. It was the era of the most intense Egyptianization of Nubia-Cush, though, as intimated above, the influence was not nearly so one-sided as customarily asserted. Cushitic words, hairstyles, religious beliefs, deities, and handicrafts found their way north. The Egyptian soldiery and constabulary became ever more Nubianized and more than one pharaoh was born of a Cushite mother. Some of the 18th Dynasty and 19th Dynasty pharaohs actually seemed to show a greater partiality toward Nubia-Cush than toward their own country. They built as many temples in Nubia as they did in Egypt and often elected one of their own royal sons to serve as viceroy of Cush.[88] Rameses reserved his finest temple-building effort for Abu Simbel in Nubia. And the supreme favor Nefertari, his Nubian consort, enjoyed is evidenced by the temples and statues—the most outstanding also at Abu Simbel—he erected to her alone of all his 60-odd wives and concubines. It is also of interest that in the declining centuries after the New Kingdom, increasing numbers of priests, aristocrats, and soldiers emigrated to Nubia-Cush from Egypt, where they evidently found a congenial environment. The point to be made is that Nubia and Egypt were not so culturally, artistically, and technically distinct as historians like to insist.

Around the 8th century B.C., when Cushite power finally eclipsed that of Egypt, the Cushites revived an architectural form long extinct in Egypt: the pyramid. These pyramids were invariably tombs—or more correctly, superstructures for tombs—and none attained anything like the size or scope of the pyramids of the Old Kingdom. Moreover, these pyramids were con-

---

[88] The title "Royal Son of Cush" given to the Viceroy of Nubia reminds us of the title given to the heir-apparent to the British throne: "Prince of Wales."

Fig 33. Cushite pyramids near Meroe. Pyramids in the Sudan are smaller and built with sharper slopes than in Egypt, but there are many more of them.

structed at steeper slopes than those in the north and, by comparison to the Egyptian pyramids, the stone blocks were much smaller. The earlier Nubian pyramids are all solidly built structures, though, by later Meroitic times, more effort seemed to have been invested in the adjoining chapels and temples than in the pyramids themselves. However, there are more surviving pyramids in the Sudan than in any country in the world, including Egypt.

Some of the earlier, and best, Nubian pyramids are of the terraced or stepped type, similar to Djoser's pyramid at Sakkara. Furthermore, Ivan Van Sertima points out that it is at this very time (circa 750 B.C.) that pyramidal structures begin to appear in Meso-America in the Olmec culture, and these too are of the terraced type.[89] We will deal at length with African navigation, transoceanic voyages, and map-making in Chapter 6 but we can at least point here to Van Sertima's inference that the wave of pyramid-building in pre-Columbian Meso-America reflects a Nubian influence.

The rulers of Napatan Cush during the period of the 25th Dynasty of Egypt (750–660 B.C.) were assiduous builders, none more so than Taharqa, usually regarded as the most outstanding pharaoh of the dynasty. His monuments are abundant throughout the length of the Nile Valley but perhaps the most unusual technical feat of his reign was accomplished at the sacred mountain of Jebel Barkal, sacred to Amon. At the base of Jebel Barkal, an important temple dedicated to Mut, the Consort of Amon, had been cut into the mountain. On the east face of the mountain there is a pinnacle jutting away from the main body that rises to 260 feet. A team led recently by Timothy Kendall of Boston's Museum of Fine Arts climbed this precarious pinnacle in order to study the

[89] Van Sertima I, *They Came Before Columbus*, New York: Random House, 1977, pp. 155–56.

Fig. 34. A stone ram, sacred to Amon, at the temple at Naga, 40 miles south of Meroe, circa 300 B.C.

inscriptions and the remains of an ancient statue.[90] The way the pinnacle juts out from the mountain makes it look like a rearing cobra, very similar to a uraeus diadem adorning a royal crown. Kendall rightly supposed that this cobra-shape made a powerful impression on the Napatans and explains the extraordinary effort expended to place a statue on the pinnacle.

Kendall and his collaborators found sockets cut into the rocks of the pinnacle that very likely held the logs of a scaffold. Two shadoof devices, one on the pinnacle and one on the opposite cliff, served to hoist men and materials up to the pinnacle. Kendall also found remains of bronze nails still in place at the summit of the pinnacle, leading him to believe that the area was covered over by a gold sheet that could reflect sunlight for miles. A statue seems also to have been set there in Napatan times.

The striking thing is the marvelous ingenuity employed to emplace a statue in an otherwise inaccessible location. The materials and methods were simple though utilized with such cleverness as to solve the most difficult engineering problems.

---

[90] See Kendall T, "Kingdom of Kush," *National Geographic*, (178)5, November 1990, pp. 96-125.

It is during the Meroitic period (c. 300 B.C. to 300 A.D.) that the most distinctive style of Nubio-Cushitic architecture emerges. By this time, the Nubio-Cushitic domain over which Meroe ruled was a great and wealthy empire, participating in world affairs on a par with Rome, Persia, and India. Diverse influences from other parts of the world show up in her monuments and architecture. The remains of a Roman bath were uncovered at the present site of Meroe.[91] At the cult center of Naga, 40 miles south of Meroe, "Indio-Brahmanic" motifs are evident. But none of these influences can account for the remarkable urban site located at the present-day Musawwarat es-Sufra, 10 miles north of Meroe. There lies a fabulous complex known as the Great Enclosure which, as Adams relates, has no architectural parallel anywhere in Egypt, Nubia, or the world.[92]

The Great Enclosure consists of a series of edifices and walled enclosures surrounding a temple sitting on a raised platform. The temple is encircled in its immediate vicinity by a colonnade, outside of which was built an ingenious series of corridors and ramps connecting all parts of the enclosure. The other remarkable feature of the Great Enclosure is the abundance of elephant motifs. The ubiquity of the elephant images is so striking that it is assumed that a major function of the complex was to tame and train elephants for work or war. This would have been no mean feat since African elephants are not nearly so pliable as their Indian cousins. Thus, having to contend with the fiercest member of the pachyderm family meant that the Meroitic tamers must have been of a caliber far overshadowing their modern circus counterparts.

War elephants were used by the Ptolemies of Egypt and, of course, by Hannibal himself in his Italian campaigns. It may therefore be logically presumed that the Carthaginians obtained both the elephants and their handlers from Meroe. Thus, indirectly, Meroe may have participated in one of the most famous military campaigns in history.

The other notable architectural feature of both Musawwarat and Naga is the presence of huge artificial lakes that held the cities' water supplies during the nine-month dry season. Adams also mentions more than a dozen other Meroitic urban sites, some situated in the drylands as much as 60 miles from the banks of the Nile, that were also served by large man made water basins, without which settled living in these locales would have been impossible:

> These colonies, made possible by considerable feats of hydraulic engineering, are the only permanent settlements which the Butana Steppe has ever supported; they offer a unique sidelight on the vigour prosperity of Nubian civilization in the last centuries BC and the first century AD.[93]

---

[91] Adams, op. cit., p. 315.
[92] Ibid., o. 319
[93] Ibid., p. 321.

Nubio-Cush absorbed diverse influences in their building methods, and in comparatively late times, i.e., after 300 B.C., architectural forms apparently unique to Nubia sprung up. For example, hollow-wall masonry is never seen in Nilotic architecture until this late period where it appears at Qalabsha in Nubia. Its construction may have been dictated by economic factors since, if the outer walls are sturdily built as the ones at Qalabsha were, this style conserves both material and labor. The other interesting style that seems to have been peculiar to Nubia in the late period was the laying of courses of stone masonry on beds concave to the horizon. Again, the best examples of this technique were to be found at Qalabsha.

Also in Nubia and Cush there were a number of pre-planned towns built. Public drains ran alongside the streets and issued out of the gateways carrying wastewater away from the habitations. Though pre-planned towns with drainage systems are known in Egypt—Aketaten being the outstanding example—most Egyptian towns lacked drainage facilities. The waste water was usually allowed to seep down into the sand outside buildings.[94]

As we have noted, there has been a definite tendency to view Cush merely as an Egyptian epiphenomenon. There were indeed periods of pervasive Egyptian influence but, as suggested earlier, there existed a two-way flow of influence. Much Cushitic architecture shows the unmistakable stamp of Egypt but there is an equal amount that derives from more or less local developments. Like their northern counterparts, the Cushites architecturally mastered the use of stone though they seemed to have been rather more dependent on sun-dried mud-bricks and fired-bricks for building materials. At all times, the Cushites exhibited great architectural and engineering ingenuity and skill. The Great Enclosure at Musawwarat and the mounting of a statue on the pinnacle of Jebel Barkal are but two examples. Thus, we can say that master-builders flourished in the Nile Valley, from the Delta to the Sixth Cataract, for more than 4,000 continuous years.

Great Zimbabwe: The Stone Architecture of Southern Africa

Outside the Nile Valley, the region of Africa evincing the most skilled use of stone as building material for massive construction was Zimbabwe. For a few decades after the discovery of the first of more than 500 villages constructed in stone (*zimbabwe* means "stone house") in what was then Rhodesia late in the last century,[95] some highly colored theories concerning migrating colonies of Phoenicians or Sabaeans dominated explanations of the presence of these massive stone structures in Africa's interior. Beginning in

[94] Badawy, op. cit., p. 6.

[95] There have now been counted more than 2,000 zimbabwes, all smaller and less extensive than Great Zimbabwe, that dot the countryside of modern Zimbabwe and extend into South Africa.

the 1920s, the excavations of Elizabeth Caton-Thompson, and those succeeding her, put such far-fetched theories to rest, showing that the stone ruins of Zimbabwe were creations of local peoples beginning 800 years ago.

The stone architecture of Great Zimbabwe, the most magnificent of the zimbabwe ruins, not infrequently calls forth a sense of awe not unlike that of Nile Valley monuments:

> The main features represented by the ruins of Rhodesia . . . are those of massive strength, enormous stolidity, monotonous plainness and similarity, complete symmetry, and most excellent workmanship, inspiring in their grand silence . . . a far deeper spirit of awe in the mind of the beholder than do the more modern styles of either Greece or Rome. The walls, in some instances are so massive that on their present summits might be placed an ox-wagon and team of sixteen oxen, with room to spare.[96]

Two features of Great Zimbabwe stand out: (1) the so-called Acropolis, a kind of citadel sitting on the hilltop contained within the walls of the complex (see Fig. 35), and (2) an elliptical building styled the "temple" situated lower down on the plain beneath the Acropolis.[97] Like all the constructions, everything within these complexes has been fashioned out of granite blocks. The overall effect is described by Basil Davidson: "The whole complex of buildings . . . has a dignity and strength of purpose in this lonely place that is irresistibly impressive."[98]

What makes the building of these massive stone structures all the more astonishing is that millions of cut stones have been laid to build high walls, towers, and buildings without any mortar or cement. Davidson, echoing the conclusions of more recent scholarship, insists that it is an entirely local genius that is responsible for this astounding architecture:

> The more one ponders on these buildings, the more they appear to have sprung from the native craftsmanship and ingenuity of peoples who worked here without any outside architectural influences to guide or help them. Everywhere these structures are marked with an originality which seems to owe nothing to the rest of the world.[99]

---

[96] Hall RN and Neal WG, *The Ancient Ruins of Rhodesia*, London: Methuen & Co., 1904, p. 159.

[97] The term "acropolis" was one first used by R.N. Hall and W.G. Neal in 1902. More recent investigators shun the term and refer to these hilltop ruins simply as the "Hill Complex." What Hall and Neal referred to as "the temple" is today called "The Great Enclosure." See Chipunza KT, *A Diachronic Analysis of the Architecture of the Hill Complex at Great Zimbabwe*, Uppsala, Sweden: Societas Archaeologia Upsaliensis, 1994, pp. 27–9.

[98] Davidson B, *The Lost Cities of Africa*, Boston: Little Brown & Company, 1970, p. 247.

[99] Ibid.

Fig. 35. A view from below the Hill Enclosure at Great Zimbabwe. The walls are set upon the granite massif and the right lower corner of the wall in the photo incorporates a granite boulder in typical style. At the bottom of the hill, the remains of an ancient dry wall is evident. (Photo: C. Finch)

Indeed, radiocarbon analysis marks the Hill Complex as the oldest of the datable ruins, going back to 1085 A.D. The Great Enclosure, situated on the plain below the Hill Complex and exhibiting the finest of Zimbabwean masonry, was constructed some 250 years later.[100] Given these dates, it is impossible to view Great Zimbabwe as anything but a local creation.

Two of the earlier investigators of the Zimbabwe stone ruins, Hall and Neal, remarked on the geometric quality of the complexes. The elliptical walls and curved passageways seemed to them to reflect a geometric sensibility in the builders.[101] They also point out that numerous buildings were oriented to the solstitial positions of the sun and to the stars. Hall and Neal claimed

---

[100] See Garlake P, *Great Zimbabwe: Described and Explained*, Harare, Zimbabwe: Zimbabwe Publishing House, 1992, p. 9.
[101] Hall and Neal, op. cit., p. 161.

Fig. 36. Ascending staircase to the Hill Enclosure. The narrowness of this winding passageway would have served defensive purposes admirably. (Photo: C. Finch)

ages of 3,000 to 4,000 years for the earliest sequence of ruins.[102] As seen above, modern chronology has effectively refuted these presumed dates by establishing the beginnings of Zimbabwean architecture in the 11th century A.D. Thus, while certain aspects of this stone architecture invite comparison to Nile Valley monumental masonry, a firm chronology makes it impossible to invoke a Nilotic influence.

There are four perceptible phases of evolution of this distinctive architecture. The early writers on Zimbabwe, consistent with their theory of Phoenician influence, insisted that the most skillful construction was the earliest and the inferior masonry arose later, suffering noticeably at the hands of local natives who could never master the Phoenicain techniques. It is now

[102] Hall and Neal's chronology was dictated by their assumption that the Zimbabwe complexes were originally erected by Phoenicians. Therefore, until additional datable material is discovered, the lower chronology of modern scholarship is the most tenable one.

Fig. 37. Conical towers perched on top of the outer walls of the Hill Enclosure. This view conveys some sense of the panorama commanded by the Hill Enclosure. (Photo: C. Finch)

well-established that the more primitive constructions are the earliest and, with a rapid improvement in methods over 250 years, the fine "classic" style of architecture was achieved. The later ruins revealed cut stones laid in level concourses without mortar but with such skill and effectiveness that walls reaching 35 feet in height and 12 feet in breadth at the top were completed at the Great Enclosure. However, within 100 years after this signal achievement, Great Zimbabwe underwent an irreversible decline and no further refinement in building is evident. Upon observing the evolving styles and stages of Zimbabwean building, it is to be wondered if these people, given a little more historical "space," might have been able to elaborate an architecture built up with stones as precisely planed, cut, finished, and laid as Nilotic architecture. They were already very close to it in the Great Enclosure.

Early investigators of Great Zimbabwe like Neal and Hall commented on the defensive nature of the redoubts of the Hill Complex, so that whatever

Fig. 38. High, elliptical inside walls typical of Great Zimbabwe in the Great Enclosure. The courses of fitted, non-mortared stones can be seen in this photo. (Photo: C. Finch)

other purpose the Hill structures served, there was a distinct military function to them. Garlake rejected this notion, with particular reference to the Great Enclosure, and other modern writers appear to have accepted this dictum.[103] However, personal observation, at least of the Hill Complex, conveys an impression closer to that of Neal and Hall. In the Hill Complex, the massive walls, narrow entrances, rounded gateways, stepped recesses, and twisting and turning passageways protected by buttresses and transverses give it the character of a fortress (see Fig. 36). Moreover, many of the outer walls of the Hill Complex incorporate huge boulders into their structure in such a way as to enhance their defensive position. Also, the Hill Complex commands such a panoramic view of the surrounding country that it is just about perfect for sentinel duty (see Fig. 37). All things considered, the Hill Complex is so admirably suited for defense that it is difficult to imagine that the effect was achieved by chance.

---

[103] Garlake, op. cit., p. 13.

It is also pertinent to mention here that modern writers assert that there was no unified ground plan, no "blueprint" developed prior to the construction of any part of the Great Zimbabwe complex. Again, personal observation makes it difficult to believe that Great Zimbabwe, particularly the Hill Complex, was built haphazardly, without some kind of guiding idea. The shape and placement of scores of curvilinear walls and passageways reveal purpose and foresight (see Fig. 39). If the builders lacked an overall unifying plan, they still must have had a sense of what they were doing and what they trying to achieve.

Hall and Neal have provided some of the most thorough descriptions of the architectural features of Great Zimbabwe, exemplifying the archictecture in its most developed phase. To wit:[104]

- Massive solidity and symmetry of the buildings, the walls being on the broadest foundations averaging 5 to 15 feet, with tops averaging $3^1/2$ to 7 feet in width, the summits paved with granite slabs.
- Main and divisional entrance wall-ends are rounded, often supported by buttresses, with intricate interior passages.
- Walls are elliptical in design.
- Foundations of the main walls rest on the rock formation and follow the surface outline.
- Main walls are all curved or elliptical though divisional walls extend from all angles from inner sides of main walls.
- Building craft and masonry of very superior quality, with chevron motifs running the length of the top of some of the main walls.
- Fine *dagha* clay, finished to a beautifully smooth glaze, was used to finish the floors and also to construct steps, tables, seats, and fireplaces. Walls were usually covered over with highly polished dagha.
- Drains for fluid run-off were installed.
- Conical towers with platforms on their summits are unique to the latest phase of the architecture.
- The later stone structures are all situated on knolls rising from the plain of the country.

The architecture of the Great Enclosure at Great Zimbabwe shows building technique and architectural conception at an extraordinarily high level of development, comparable to the best dry stone stone masonry anywhere. As stated earlier, it represents the culmination of an architecture evolving over a period of 250 years. In the earlier architecture the walls are not so massive or solidly built, the foundations rest on soil instead of bedrock, there are no drains present, and the curvilinear style is not nearly so prominent as

---

[104] Hall & Neal, op. cit., pp. 168–9. See also Garlake P, *Life at Great Zimbabwe*, Gweru, Zimbabwe: Mambo Press, 1982.

Fig. 39. Another view of the inside walls of the Great Enclosure showing their height (35 feet). Also shown is the style of terminating the walls in ellipses. (Photo: C. Finch)

later. Rectilinear styles are somewhat more obvious; the cutting of stone and overall masonry are of lesser quality.[105]

In terms of materials and methods, in the vast majority of constructions, granite blocks were used although occasionally diorite was employed. The style and quality of workmanship is described by Hall and Neal:

> The buildings show the most elaborate care in their construction. . . . The planes of the walls are even and smooth; the courses are regular, each stone fitting closely into the next, while one stone always overlaps the joint between the two stones below it, and the courses in the . . . buildings are bonded from front to back. . . . In carrying a wall over a boulder the joints are cleverly made.[106]

---

[105] Ibid., 169-71. While Neal and Hall are invaluable for their descriptions of Great Zimbabwe, it must be remembered that what they call "first-period" architecture is actually the latest phase of Zimbabwean architecture.

[106] Ibid., p. 173.

Numerous iron (steel?) tools such as chisels, wedges, hammers, and trowels have been discovered among the ruins and undoubtedly belonged to the original masons.

Most of the entrances face north or west to provide protection against prevailing easterly and southeasterly winds. On the Hill Complex, the width of the main wall entrances vary from $1^1/_2$ to $3^1/_2$ feet, and the narrower entrances may have been constructed for better defense against attackers. There are generally parallel walls inside the main walls, often of the same height, that created somewhat labyrinthine passageways and, when divided, as many as a half-dozen (or more) interior enclosures. Again, the maze-like configurations appear to have been designed to confuse and thwart unwanted intruders. The conical towers, with platformed summits and accessible by stairways, may have served a religious-ritualistic function, though this is only a surmise (see Fig. 40). They, too, could have served a military function since it is possible to survey the whole complex and its surroundings from them. In some of the earlier constructions, granite cellars were built; no cellars have been discovered in the later buildings.

Certain ornamentions were constructed into the tops of the walls of the complexes on the plains below the Hill Complex. Five ornamental styles have been identified: (1) dentelle, (2) chevron, (3) herring bone, (4) sloping block, and (5) check. Dentelle and chevron patterns are most prominent at the Great Enclosure; the other styles have been found at other zimbabwes in the region (see Fig. 42). Another distinctive feature of the Great Enclosure is the manner in which steps were constructed: "They [steps] were set in the thickness of the lower courses of the walls and the face of each sucessive step was constructed in a system of multiple curves of progressively decreasing radius, an extremely ingenious and beautiful design."[107] This is one of many features that make the dry stone archictecture of Great Zimbabwe so distinctive and aesthetically pleasing to the eye, even in it its ruined state.

It should also be noted that the buildings of Great Zimbabwe, roofed with thatch over the stone walls, were, as noted above, floored with a polished, enamel-like clay substance called *dagha*. This is another unique element in the buildings of region, completely different from the hard-packed dirt floors typically associated with African rural dwellings. Polished dagha was also used to coat walls, fashion tables, and make bedsteads. Garlake makes the point that it is difficult from the state of the ruins today to picture what Great Zimbabwe looked like in its heyday. Floors and walkways were laid in beautifully polished red dagha and dagha-covered walls were painted in striking colors. The effect must have been dazzling.

Finally, it is important to call attention again to the construction of the walls along curvilinear and ellipitical lines, particularly during the most refined phase

---

[107] Garlake, *Great Zimbabwe*, op. cit., p. 21.

Fig. 40. A cylindrical edifice inside the Great Enclosure sometimes styled "The Temple." (Photo: C. Finch)

of the architecture embodied at Great Zimbabwe. It is not easy to ascertain the practical or symbolical motivation behind this stylistic device but it sets Great Zimbabwe apart from almost all other monumental stone architecture. The graceful curves of the walls, towers, and buildings immeasurably enhance the visual impact of the place, invoking the sense of wonderment ordinarily reserved for the great Nilotic edifices. Also, the architecture is intrinsically *of* the place and this may offer a clue to the elliptical building style. Particularly on the Hill Complex, the ingenious manner in which walls were carried over the boulders to incorporate them into the wall, shows how the architecture was adapted to local environmental conditions instead of being imposed upon on the landscape. The kind of living space that flowed from the building style bonded the inhabitants to the land, the ecology, and nature.

Most historians date the end of Zimbabwe stone-building to the 15th century. At its height, Zimbabwe—purely on the evidentiary strength of its fabulous architecture—must have been a powerful, centrally organized kingdom,

Fig. 41. Curvilinear steps mark an entrance to the Great Enclosure. This style of steps seems altogether unique to Great Zimbabwe. (Photo: C. Finch)

indeed the hub of a veritable empire. The country was awash in gold and copper and, though located 500 miles from the coast of the Indian Ocean, participated in long-distance trade that brought it commercial items from Arabia, Persia, India, and China. Why it declined, collapsed, and disappeared can only be guessed at. By the 16th century, its greatness was but a memory, the splendid stone ruins standing as mute testaments to Zimbabwe's ancient glory.

The Maritime Architecture of Swahili Civilization

The history of the Swahili city-states is almost unique in Africa. These were thriving commercial entrepots that, at their height, were sending giraffes to China and welcoming traders and ambassadors from all the lands bounded by the Indian Ocean.[108] Moreover, their civilization, though firmly rooted in Bantu culture, represented perhaps the most complete cultural synthesis achieved by African and Arab in the history of the continent.

Naval expeditions from Mediterranean lands had been sailing the waters off the coast of eastern and southern Africa at least since 600 B.C. when a Nilo-Phoenician fleet, under a commission from Pharaoh Necho, circumnavi-

---

[108] See Davidson B, *The Lost Cities of Africa*, Boston: Little, Brown and Company, 1970, pp. 171-210, for a succinct and readable historical survey of the Swahili cities.

Fig. 42. Characteristic chevron motif running along the top of the wall of the Great Enclosure. (Photo: C. Finch)

gated Africa from east to west. Earlier Egyptian voyages to Punt, dating as far back as 1,500 B.C., may well have reached the coastal waters of what would later become the Swahili cities. Pre-Islamic Sabaeans formally opened trade with Africa's east coast—rich in gold, ebony, ivory, copper, and ostrich feathers—and after the time of the Prophet, Arab merchants intensified this trade by establishing entrepots in the emerging cities. For nearly 1,000 years the development of the East African trading cities of the coast was shaped by: (1) an ever-greater expansion of commerce, (2) increasing numbers of Arab merchants establishing residence in the cities, and (3) a more or less seamless grafting of Arab Islamic culture onto a Bantu African base. This syncretism was especially noticeable in the language Ki-Swahili, which emerged as Arabized Bantu. It gave its name to the entire maritime civilization of East Africa's coast, and, typical of trade languages, became—and remains still—the *lingua franca* of East Africa.

Wealth flowed through such East African ports as Mombasa, Kilwa, Malindi, Zanzibar, and Sofala and, on the basis of this wealth, a highly cultured society developed. In the 13th and 14th centuries, Arab travellers rated Kilwa one of the most beautiful cities in the world.[109] It is therefore not surprising that a distinct, highly aesthetic architecture developed in these Swahili cities, showing marked Islamic influences to be sure, but whose fundamental features evolved out of a native East African context.

The two primary materials used for building in Swahili cities were coral

[109] See Davidson B, *African Kingdoms*, New York: Time-Life Books, 1971, p. 81, for a quotation from Ibn Batuta on the beauty of Kilwa.

and stone.[110] In the earlier phases of Swahili history, most buildings were constructed of coral blocks laid with mortar. As the architecture developed, stone was more frequently employed and the skill with which it was used showed noticeable improvement over time. At Kilwa, where some of the finest examples of Swahili architecture were to be found, many edifices sported spherical and pointed cupolas, semi-cylindrical vaults, and stone columns. Late in the 14th century, the walls, cupolas, and vaults were decorated by in-laying Chinese and Persian porcelain dishes into the cement. At its best, Swahili architecture excited breathless admiration. The domed ceiling of Kilwa's great mosque invited comparison to Spanish Cordoba's world-famous mosque by one Arab writer.[111]

The wealthy inhabitants of these cities exhibited no hesitation in building their private residences in stone, and the care they lavished on these dwellings frequently showed Swahili architecture in its best light. These homes were usually oriented north or east with a spacious courtyard in front. Inside the house were long, narrow rooms usually in parallel; the front room opened into the courtyard. In the rear were situated indoor toilet facilities. As time wore on, two- and three-storey houses made their appearance. There was only one window installed in the front part of the house overlooking the courtyard so that the remainder of the house was characteristically dark. However, ceramic lamps were plentiful in the wealthier dwellings and produced enough illumination to allow, among other things, reading and writing.

Like so many other sites in Africa, most of the Swahili archaeological remains are still untouched. However, the ruins at Gedi, a Swahili settlement just 3–4 miles from Malindi on the coast of Kenya, shows some of the more unusual features of the Swahili style, though only one square mile of the 20 square-mile site has been excavated (see Plates 17–19). The testimony of a modern traveller amplifies our appreciation of Gedi's main Islamic center of worship:[112]

> [The largest structure] was the temple [sic] used for Islamic prayers. Outside of the temple just beyond the doorway . . . and to the left were troughs to wash your hands and feet before entering the temple. Water was taken from a basin and poured over the hands and feet. It was recycled because the trough was slanted and ran over limestone before returning to the source.

Further observation disclosed a well into which the recycled water returned that connected, at its base 600 feet down, with an underground stream sup-

---

[110] See Matveiev VV, "The Development of Swahili Civilization," in *Africa from the Twelfth to the Sixteenth Century*, edited by D.T. Niane, Berkeley: Heineman-UNESCO, 1984, pp. 469–75, for a concise description of Swahili architecture.

[111] Davidson, op. cit., *African Kingdoms*, p. 181.

[112] The author is indebted to the written communications of Aletha Akers, a Watson fellow travelling in East Africa (1994) at the time the acute observations quoted here were made.

plying the settlement with water.[113] The most striking feature to note is the recycling of water through a limestone purifier. The homes of the more affluent inhabitants of Gedi also boasted indoor toilets and bathing facilities complete with running water, apparently controlled by faucets.[114] Thus it is that the water portage, sanitary, and drainage facilities of this old East African city—and it was hardly atypical—were surprisingly "modern." In Europe of the same time, i.e., 500–1,000 years ago, all human and animal waste products were unceremoniously dumped into the street with scant consideration for the hapless passersby.

Through the Swahili trading cities, East Africa was in touch with the world. Through them, East Africa's riches flowed outward and exotic luxury items from faraway lands percolated deep into the African interior. Swahili architecture, though largely of local inspiration, reflected this cosmopolitan spirit.

The Western Sudanic Architectural Style of the Medieval Period

Western Sudanic architecture, like the people who created it, though intrinsically rooted in the region, seems to echo another time and place. This can be attributed to the mounting body of evidence that shows that all the major groups in the lands that comprise the once-great empires of the Western Sudan (the Sahel) originated in the lands drained by the Nile. It is beyond our present scope to examine this evidence in detail, much of it hinging on complex comparative linguistics. Still, a few nuggets can be considered. Cheikh Anta Diop, the pioneer in this field, observed that his clan name "Diop" still exists as a clan name in the Nilotic Sudan. The people known as the Lebu, a subgroup of the Wolof of Senegal, claim that they originated in what is now Libya. Now the ancient Egyptians called the peoples inhabiting the lands west of the Nile "Lebu," the very word from which the Greek name "Libya" was derived. Moreover, this word also meant "lion" and it is a curious fact that the present-day Lebu were among the earliest inhabitants of the coast of Senegal *and* that the lion is present-day Senegal's national symbol.

Finally, a word concerning a startling piece of research by Professor Aboubacry Moussa Lam of C.A. Diop University of Dakar, Senegal, that came to light in 1991.[115] Professor Lam discovered an obscure historical tract written nearly 80 years ago by a French-educated local chieftain named Yoro Dyâo in which he reported oral traditions of the Senegambia that attest to at least six migrations from the Nile Valley in ancient times. Lam finds it possible to identify the pharaohs and the time-periods connected to these migrations,

---

[113] Personal correspondence, Aletha Akers, 1994.

[114] Personal communication, Aletha Akers, 1994.

[115] The relevant information is contained in Lam AM, *Le Sahara ou la Vallée du Nil,* Dakar, Senegal: IFAN/Ch. A. Diop, pp. 59–79, 1991. The amazing thing is that this material was apparently unknown to Cheikh Anta Diop himself, the teacher of A.M. Lam.

the first of which seems to have taken place at the time of the Persian invasion of Egypt in the 6th century B.C.[116] The material goes far to confirm the theories of Cheikh Anta Diop in striking fashion and helps establish with greater certitude the genetic connection between the ancient civilizations of the Nile and West Africa.

Even a cursory acquaintance with the cosmomythology of the Dogon, now living along the Bandiagara Cliffs in southeastern Mali, will reveal pronounced similarities between their system and that of ancient Nile Valley religion. Not surprisingly, the Dogon relate that their original home was somewhere to the northeast, from where they began their migrations 1,400 years ago. These and other examples could be multiplied indefinitely to show that originally the peoples of the Western Sudan were Nilotic or "peri-Nilotic" peoples. They had been, in some way, intimate and active participants in Nile Valley culture. Thus, it stands to reason that their architecture would show marked affinities to that familiar to us in the Nile Valley.

The history of the Western Sudan was decisively impacted by the rise and fall of three vigorous empires: Ghana, Mali, and Songhay.[117] Ghana, the oldest and longest-lived of the empires, can be dated at least back to the 3rd century and lasted until the 11th century when sacked by the Almoravids. Its successor state, Mali, incorporated a larger territory, one equal to the size of Europe, and became so fabled for its seemingly inexhaustible gold wealth that it found a place on Medieval European maps.[118] The historian D.T. Niani estimates its total population at its height at 50 million.[119] Finally, in the 15th century, Songhay, the most powerful and expansive of the Western Sudanic empires, emerged. Songhay, larger than Mali, controlled the trans-Sahara salt-and-gold trade almost until the turn of the 17th century when it was brought down by a Moroccan invasion force equipped with firearms. Thus, for 1,400 years, the Western Sudan was dominated by large imperial states whose wealth caught the world's attention.

These Western Sudanic kingdoms contained many cities whose prominence was owed to their positions astride the salt-and-gold trade. The most renowned of these were Gao, Jenne, and Timbuktu. Gao, the Songhay capital, was home to 100,000 inhabitants.[120] Timbuktu was a bustling metropolis of 80,000 souls. And Jenne, whom many consider the real "jewel" of the Western Sudan, contained a population of 60,000 persons. Jenne, a walled

---

[116] Ibid., pp. 62–70.

[117] This era of African history is among the best-documented so there is a comparative wealth of historical surveys of this region during the medieval period. Good introductory sources include E.W. Bovill's *Golden Trade of the Moors*, Oxford University Press, 1958, and J.S. Trimingham's *A History of Islam in West Africa*, Oxford University Press, 1962.

[118] Bovill, op. cit., p.

[119] Niane DT, "Mali and the Second Mandingo Expansion," in *Africa from the Twelfth to the Sixteenth Century*, op. cit., p. 156.

[120] Ibid.

city situated on an island in a mid-Niger delta, was virtually immune to military conquest. Even Sonni Ali, the great Songhay conqueror, could not reduce the city after a seige of nearly eight years; the impasse was resolved only when the city voluntarily opened its gates in return for guarantees of its sovereignty.[121] Jenne was always home to the finest specimens of Western Sudanic architecture.

Recent archaeology has uncovered conclusive evidence that Jenne can trace its existence back to before the time of Christ.[122] The present city arose around 1400 A.D., but its predecessor, known as Jenne-Jeno and situated less than two miles away, could be traced back to 250 B.C. It had become a thriving pre-Medieval commercial center of 20,000 persons by 800 A.D. The old city had been girdled by a wall 9-feet thick and was a center of long-distance trade involving gold, iron, copper, salt, ceramics, and foodstuffs long before Islamic penetration into the region. Thus, Jenne's urban lifestyle and traditions are perhaps the most venerable in the Western Sudan.

The western Sudan, unlike the Nile Valley or Zimbabwe, is not blessed with abundant building stone such as granite or limestone. Thus, the building material of choice is a particular clay native to the Sudan that allows the fashioning of flat, long, regular bricks or, for simpler dwellings, irregular ball-shaped bricks that were sun-baked, set in mortar, and the finished off with a rough casting.[123] These edifices, simple though they may have been, provided protection against storms and could endure for centuries.

With the finer cut long bricks, the architects of Jenne (and throughout the Western Sudan) were able to construct more massive structures whose characteristic features make this architecture among the most identifiable in Africa. However, more than one observer has been struck by the evident affinities to Nile Valley architecture:

> The summits of their [Songhay] dwellings are ornamented by those triangular battlements which may be seen on the palaces of Rameses Meiamoun. The pylon, which is another characteristic of Egyptian architecture, gives access to the dwellings of Jenne, and forms, too, a motive of decoration, the facades of the houses being adorned with great buttresses of pylonic form.[124]

The most remarkable standing example is the great mosque of contemporary Jenne where the Sudanic architectural features, with their Nile Valley "flavor," are shown to marvelous effect (see Fig. 43). This, above all other Sahelian monuments, exemplifies most perfectly the following comments of

---

[121] See Dubois F, *Timbuctoo the Mysterious*, translated by Diana White, New York: Negro Universities Press, 1969, p. 147.

[122] See McIntosh S and McIntosh R, "Finding West Africa's Oldest City," *National Geographic* (162)3, September 1982, pp. 396–418.

[123] DuBois, op. cit., pp. 148–9.

[124] Ibid., p. 150.

Fig. 43. Entrance to the Great Mosque at Djenne, one of the finest examples of Western Sudanic architecture from the medieval period.

Felix Dubois: "In short, the effect of the whole, its harmonious proportions, the symmetrical distribution of its ornamental motive, and its massiveness, unmistakably proclaim the art of Egypt."[125]

Dubois's reference to harmony and symmetry can be well appreciated by anyone who sees even photos of the mosque. As yet, we have no concrete evidence that the architects of the Western Sudan had reduced harmony to number, impressing itself on architectural forms, as was the case in the Nile Valley. Nonetheless, their constructions are unerringly pleasing to the eye in a manner entirely consistent with the harmonic canons of Nile Valley architecture. Thus, their work seemed to be guided by an interior "canon" of harmony.

The original mosque, built on a square plan, measured 183 feet on a side and rose to a height of 39 feet. The building was oriented to the four cardinal points and each side showed the "pylonic" features and rows of buttresses built into the walls that gave the whole its "Nilotic" sensibility. It was said that the original structure had stood for 800 years until destroyed in the course of religious wars in 1830.

It is worth reiterating that though stone buildings were not unknown in the Western Sudan, its architecture was mostly carried out in sun-baked clay

[125] Ibid.

Fig. 44. Nineteenth-century artist's rendition of Timbuktu, dominated by the Great Mosque.

brick. A Sudanic house usually consisted of two storeys, the upper one surmounted by a terrace. The wealthier houses showed exterior facades ornamentally adorned with obeliskoid pillars. The walls inside and out were plastered over to give a smooth overlay to the brick skeleton. The flat roof was constructed by means of a grid of wood planks atop the upper walls that were allowed to extend through the walls to the exterior, giving the studded appearance so peculiar to Western Sudanese buildings. Over this grid of planks was laid a final layer of clay to serve as the roof.

A hundred years ago, the French traveller Felix Dubois was shown houses that had stood for 400 years, testifying to the great stability of the materials and construction methods. Some of the homes possessed inner courtyards, and strategically dispersed throughout the dwellings were openings for air and light, ornamentally embellished with terra cottas. All of the more substantial dwellings boasted indoor drainage facilities: "A system of baked pipes is established in every dwelling to carry away household water, and latrines, with perfectly constructed drainage, are established on all the terraces."[126]

Like elsewhere in Africa, Western Sudanic architecture conformed to its ecology. It made the most of the materials at hand, creating functional, durable structures that possesed their own unique aesthetic qualities. This is so despite the architecture's apparent stylistic pedigree derived from Nilotic prototypes. Once again it showed that Africa's architecture could be as vivid and diverse as the people themselves. Once again it showed that the indigenous architecture was seldom imposed on the environment but defined by it.

---

[126] Ibid., p. 152.

Conclusion

At all times and places, Africans showed themselves wonderfully capable of building with whatever materials the environment supplied. The ubiquitous round, thatch-roof mud hut, enountered in "the bush" the length and breadth of the continent, has created a stereotype of African building, fueling assumptions that, beyond the Nile Valley, Africa possessed no proper architecture. Careful archaeological work in recent years has laid such assumptions to rest.

All over the continent, the people tended to lavish their greatest care on their religious edifices, and most often the buildings that have survived over the centuries are such. More subtlely, Zimbabwean architecture leaves an impression of a geometric sensibility redolent of Nilotic architecture; Western Sudanic architecture expresses a sense of harmony and symmetry that is also Nilotic. Thus, even where observable styles are distinct and varied, it is sometimes possible to detect a unified ethos that wafted all over Africa. And architecture never quite lost its designation as the sacred craft.

5

As Above, So Below: African Investigations into Space and Time

Lens, film, and amplifier circuit had not yet come to the aid of the old sky-clerks. Doubtless most of them lacked even the easy elegeance of algebra and geometry of the last millenium or two, but one resource they had in plenty, which modern astronomy cannot replace. They had time. Time for super-novas, time for the stars to shift, time for all the unusual arrangements of the planets, for the apparent close approach of Venus, Saturn, Jupiter, time for everything.
—Philip Morrison, *Astronomy of the Ancients*

It is impossible to overestimate the importance of the above passage for understanding ancient astronomy. If we can appreciate our remote ancestors' seemingly unlimited capacity for time, then some of their astronomical accomplishments do not seem quite so unbelievable to our modern incredulity. In our instantaneous, "real-time" world, the idea of unceasingly pursuing a particular goal for 10,000, 20,000, or 30,000 years appears to be utterly beyond ordinary human capacity.

Our modern technological contrivances have effectively speeded up history. Cultural changes that formerly evolved over centuries or even millenia are now compressed into years, or sometimes months. Wars cannot even be fought anymore except under the glare of CNN's camera lights. During the Persian Gulf conflict, one was often assailed with the eerie, Kafkaesque impression that some invisible, off-camera director was about to yell "Cut!" as the missiles were falling because he was not satisfied with the camera angles. Nor would one have been unduly surprised if, obligingly, the missiles had stopped abruptly on their downward course to wait for the television crew to set up for a better shot.

Ours has become an instant gratification, drive-thru, convenience store culture. We barely tolerate standing five minutes in line at the post office; we cannot conceive of cultures that may have studied and charted the heavens continuously for 26,000 years (or longer). And yet the evidence, carefully considered, tells us that this is exactly what happened.

The "received" opinion of historians is that "true" astronomy begins with the Babylonians, though refined and perfected as a science by the Greeks. By this opinion, Egyptian astronomy was of a lesser standard; but evidence will be brought forward in this chapter to challenge that conclusion. In this connection, it is worth remembering that Thales, at the close of the 7th century B.C., was the first to bring scientific astronomic concepts to the Greeks *from Egypt*.[1] It was Thales who first gave the name *Ursa Minor* ("Little Bear") to the northern constellation that now bears that name. He seems to have been the first in the Greek world to attach different names to constellations already known in Egypt. We know, for example, that prior to Thales, *Ursa Minor* was the constellation of the Jackal *Wep-uat* ("opener of the way") in the Egyptian planisphere. He must have known the constellation by that name but decided on "Little Bear" for the Greeks since the jackal was not an animal native to their country.[2]

Concerning another famous Greek savant, Eudoxus, who lived in the 4th century B.C. and was a friend of Plato, Richard Hinckley Allen says of him: "Eudoxus, who, according to common story, brought the constellations from Egypt, and about 366 B.C., was the first to publish them in the original prose *Phainomena* . . . "[3] Further on, Allen informs us, " . . . indeed the Greek astronomers always acknowledged their indebtedness to Chaldea and Egypt, but gave most of the credit to the latter."[4] And finally: "Many have maintained that Egypt was the first to give shapes and names to the star-groups; Dupuis, perhaps inspired by Macrobius of our 5th century, tracing the present solar zodiac to that country and placing its date 13,000 years anterior to our era. . . . The little that we know of Egypt's early constellations indicates that they apparently were of native origin, and in no respect like those of Greece, which, if adopted at all, were so at a very late time in that history . . . "[5]

In the 4th century B.C., the Egyptians were known to have recorded their observations of 832 lunar eclipses.[6] A lunar eclipse is observed during a full moon and occurs when the earth comes between the sun and the moon, causing the moon to fall within the earth's shadow. The average number of lunar eclipses per year is about two, which can be observed from any part of the earth where the full moon is above the horizon. Thus, the 832 lunar eclipses known to have been observed by the Egyptians in Alexandrian times mean that this group of eclipses had been observed for over 400 years, tak-

---

[1] Thales, though the first acknowledged "Greek" scientist and philosopher, was actually Phoenician.

[2] However, in late antiquity, *Ursa Minor* was sometimes referred to as *Cynosuris*, meaning "tail of the dog." This is the same as our word *cynosure* which means "to lead or guide," a reflection of the ancient dog/jackal's role as the guide-by-night in the Nilotic underworld. See Allen, RA, *Star Names: Their Lore and Meaning*, Dover Publications, Inc., 1963, p. 456.

[3] Allen RA, op.cit., p. 18.

[4] Ibid.

[5] Ibid., p. 19. This is an undoubted reference to the Great Year or precessional cycle of 25,900 years whose current cycle began in 10,858 BC in the sign of Leo. See Finch CS, *Echoes of the Old Dark Land*, Decatur, Georgia: Khenti Inc., 1991, pp. for an extended discussion of the Great Year.

[6] Obenga T, *La Philosophie Africaine de la Periode Pharoanique*, Paris: L'Harmattan, 1990, p. 273.

ing the period of observation back to around 750 B.C.[7] Plutarch verified their ability to predict eclipses so they may well have known the cycle of Saros.[8]

Theophile Obenga further informs us that, according to Diogenes Laertius, these Nilotic astronomers recognized that a lunar eclipse was caused by the full moon falling within the earth's shadow.[9] This fact corroborates other evidence pointing to ancient Nilotic knowledge of the sphericity of the earth, its revolution around the sun, and its relation to its orbiting lunar satellite. Thus, it comes as no surprise that Achilles Tatius, confirming evidence brought out in previous chapters, asserted that the Egyptians had been the first to measure the earth.[10]

It seems not too much to infer that the ancient Nilotes had, in effect, discovered the solar system. Macrobius tells us that they knew that Mercury and Venus revolved around the sun.[11] When we recall that ancient writers asserted that they had deduced earth's orbit around the sun (and the moon's around the earth), it might be reasonably inferred that they had discovered the solar orbits also of Mars, Saturn, and Jupiter. They called these planets "Travellers," clearly distinguishing them because of their more rapid movement, from the stars outside the solar system. If, therefore, they had discovered the revolutionary movements of the first three planets of the solar system, the inference that they knew the revolutionary movements of the second three is not out of place.[12]

A physicist and historian of science, John Pappademos, has compiled a list of contributions to astronomy attributed to the Egyptians. Below is a partial list of these contributions (ones not otherwise mentioned in this chapter):[13]

1. Identification and naming of 43 stellar constellations.
2. Writing of astronomy texts, of which Clement of Alexandria listed four, including (a) *On the Disposition of Fixed Stars and Stellar Phenomena*, (b) *On the Disposition of the Sun, Moon, and Five Planets*, (c) *On the Syzygies and the Phases of the Sun and Moon*, and (d) *On Risings*.

---

[7] It is interesting to note that this date marks the beginning of the 25th or Ethiopian Dynasty of Egyptian history.

[8] Cited in Obenga, op. cit., p. 273. The Saros is a cycle—usually counted a Babylonian discovery—occurring every 18 years, 11 1/3 days, where the Sun, Earth, and Moon return to the exact same geometric aligment. Knowledge of the Saros enabled ancient astronomers to predict lunar and solar eclipses.

[9] Obenga, op. cit., p. 273.

[10] Ibid.

[11] Ibid.

[12] As cited in Chapter 3, Gardner Wilkinson remarked that, alone of the Greeks, Pythagoras knew the solar system and had learned of it from the Egyptian priests.

[13] See Pappademos J, "The Newtonian Synthesis in Physical Science and Its Roots in the Nile Valley," in *Egypt: Child of Africa*, edited by Ivan Van Sertima, Highland Park, N.J.: Journal of African Civilizations Ltd., 1994, pp. 318–9.

3. Compilation of star culminations and risings.
4. Discovery of the occultations of the stars and planets by the dark side of the half-moon.
5. First proof that the angular diameters of the sun and moon are unequal.
6. Discovery of the conjunction of the planets with each other as well as with the fixed stars.

Pappademos likewise points out that several famous Alexandrian astronomers, including Sosigenes and Ptolemy, though known to us through their Greek names, were actually ethnic Egyptians. Sosigenes reformed the calendar on behalf of Julius Caesar and thus is the true author of the "Julian" calendar and Ptolemy was the author of the famous *Almagest*, *the* standard astronomical text in the Western world for 1,000 years. Other celebrated Greek astronomers, i.e., Oenopides, Eudoxus, Straton, Eratosthenes, and Hipparchus all studied in Egypt, the latter three at Alexandria. Greek astronomy was, in effect, refurbished Nilotic astronomy, a truth that *no* Greek ever denied.

J. Norman Lockyer, whose book *Dawn of Astronomy* constitutes one of the major sources for this chapter, has to be considered the most important authority on ancient Nilotic astronomy. Not all his conclusions can be accepted unmodified but no one did more than he to elucidate the foundations, evolution, and development of Nile Valley astronomy. He stood practically alone in his time in showing the manner in which temple architecture was used to chart stellar movements. Lockyer says this about the observational continuity of Nilotic astronomy through time:

> In Egypt the change of cult might depend upon astronomical change—upon the precession of the equinoxes, as well as upon different schools of religious or astronononomical thought. We gather from this an idea of the wonderfully continuous observations which were made by the Egyptians of the risings and settings of stars, because, if the work had not been absolutely continuous, they would certainly never have got the very sharp idea of the facts of the precession which they undoubtedly possessed.[14]

Moreover, Lockyer was of the opinion, once he noted that the earliest Egyptian temples were oriented to the southern stars, that the ancestral, "prehistoric" star-gazers must have come from that direction: "Horus of Edfu, who is represented as leading the victorious hosts who revenge the killing of Osiris by Set, is the ally of the southern-star worshippers whom we have traced from Thebes, possibly to Central Africa."[15] Thus, if we accept these conclusions, continuous star-gazing began in the interior of Africa long before the rise of dynastic Egypt.

All over the world, the remains of ancient observatories abound. Mounds,

---

[14] Lockyer JN, *Dawn of Astronomy*, Cambridge: The M.I.T. Press, 1964, p. 300.
[15] Ibid., p. 376.

monoliths, and medicine wheels testify to the universal enterprise of archaic humanity of monitoring heavenly movements. Wherever one finds these ancient remains—be they megaliths at Stonehenge, medicine wheels in Manitoba, pyramidal mounds in Macon, Georgia, or tilted stone slabs at Namoratunga, Kenya—one is struck by their inevitable location on a level plain. Always, naked-eye astronomy required a clear field as featureless as possible. For those inhabitants of forested or mountainous areas, a large, flat meadow was always sought to aid the work of the star-watchers. Nearly always, since star watching was tied to priestly or shamanistic work, the astro-cult centers or sacred shrines were established on these plains or open meadows.

From the point of view of observing the sky, we always imagine, as Norman Lockyer pointed out, that we are on "top" of the earth. Therefore, "the idea held by all the early peoples was that the earth was an extended plain: they imagined that the land that they knew and just the surrounding lands were really in the centre of the extended plain."[16] However, this "flat" perspective of the earth was a perceptual/conceptual accomodation to sky-watching and, by implication, calendar-making; it does not imply that all early peoples were ignorant of the sphericity of the earth. Evidence has already been cited—and more will be—that the earth's sphericity was known at a very early period, easily 5,000 years before Columbus.

It is important, before proceeding further, to define certain terms necessary to understand the way in which any heavenly body—star, sun, planet, moon—is localized in the sky:

Amplitude: The number of degrees (up to 90) that a heavenly body is north or south of an east or west point.

Azimuth: The angular distance of a heavenly body measured eastward along the horizon from the north point to the intersection of the horizon with the meridian running through the body.

Declination: The distance of a heavenly body from the celestial equator, which is closely connected to the place where this body visibly rises or sets.

Celestial Equator: The equator of earth extended into space.

Ecliptic: The plane of earth's rotation around the sun; also the apparent change of position or path of the sun on the horizon as seen on earth in the period of time (6 months) between the two solstices. This apparent path of the sun between the solstices is caused by the $23^1/_2°$ axial tilt of the earth.

The lives of archaic humans were always regulated by practical requirements. Accurate time keeping, the primary aim of star-gazing, was critical to settled communities where precise foreknowledge of the seasonal conditions that controlled agricultural cycles could mean the difference between feast

[16] Ibid., p. 43.

and famine. Since archaic man was also an inveterate traveller and trader, by land and sea, a star map committed to memory was indispensible for such prescribed movements. It is known, for example, that the native Americans of 1,000 years ago had installed a *continent-wide* trading network. Since no one has ever suggested that the ancient Americans possessed compasses or astrolabes, the stars must have been their navigational guides. The same holds true for ancient sea-farers such as the Melanesian, Micronesian, and Polynesian Pacific Islanders who navigated over hundreds, even thousands, of miles of open ocean and probably made landings on the western coast of what is now South America in the millenia before Columbus.

For a few daring thinkers, it is becoming ever clearer that the entire world was connected anciently by a network of travel, trade, and cultural exchange. No part of the world was inaccessible to wandering humans who maintained contact with far-flung lands at all times. Columbus *did not* "discover" America, nor was he the first to make a sea-borne landfall there. Moreover, as we have already seen and shall explore more fully in the next chapter, the ancients had already measured and mapped the globe. Once upon a time, therefore, there was a world civilization, informed, to be sure, by a different consciousness and different technical perquisites from the modern one. It prefigured the "global village" extant in our own time.

"The Wheel of Heaven": Star-gazing and Astrochronology in the Nile Valley

Like the beginning of Nile Valley history itself, it is difficult to date the beginning of Nile Valley astronomy. Plato accorded Egypt a 10,000-year span of systematic star-gazing; Martianus Capella, 40,000 years. In a sense both are right.

Much has been made of the superiority of Babylonian astronomy to Egyptian, usually by authorities who have considered only the narrowest lines of evidence and already had their minds made up about the question even before addressing it.[17] But careful archaeoastronomy belies this comfortable assertion. Moreover, the learned Greeks of antiquity *unanimously* gave priority to Egypt as the originator of astronomical science. The Nile Valley, indeed all of the northern two-thirds of the Nile borderlands, are admirably suited for extended and continuous naked-eye star-gazing. The weather is mostly dry and clear and has been for many millenia. Vast expanses unbroken by interrupting features are everywhere apparent upon the landscape. Thus, when the stars appear after dusk, they are bright and easily observable for hours at a time, night after night, season after season.

---

[17] See Neugebauer O, *The Exact Sciences in Antiquity*, New York: Dover Publications, 1969, pp. 80-96. Neugebauer is perhaps the most identifiable exponent of the school that has belittled Nilotic science.

For a long time, astronomy was primarily a night-time occupation, and though solar astronomy has been important for at least 6,000 years, astronomy as a whole remains a nocturnal activity. Gerald Massey was of the opinion that, in ancient times, there were three evolutionary phases of observational astronomy, which, in order, were: (1) stellar, (2) lunar, and (3) solar.[18] Norman Lockyer also accepted the primacy of stellar observations and time-keeping based on them because of the comparatively greater positional stability of the stars compared to the sun and moon.[19] Stars could be found in the same seasonal positions for hundreds of years at a time in contradistinction to the sun or moon. However, contrary to Massey, Lockyer took the position that solar observations and time-keeping represented the next important step in the development of ancient astronomy and that precise observations of the sun on either horizon was developed by the Egyptians and taught to the world.[20]

In Africa, as elsewhere, the permutations of the mythos, and therefore of religion, were tied to star movements. The heavenly bodies were, after all, divine insignia, or, accordingly, divinities in and of themselves. The Cosmic Will was thus tracked, monitored, and interpreted through the cyclic movements of the heavenly bodies. Indeed, it may have been the religio-mythic impulse, rather than the empiric (or "scientific") one, that compelled the obsessive need for greater observational precision:

> Why did the [ancient] astronomers strive for precision? Perfection was unnecessary in meteorological or agricultural prediction; the answer lies in the spiritual realm. Religious ceremonies always marked the completion of any natural cycle, and early astronomer-priests believed the deities who represented the forces of nature . . . demanded a strict observance of the rites. When all was in order in the heavens there would be peace on earth, but a disorder necessitated careful corrections in the calendar, lest the timing of good and bad omens occupy the wrong place in human history. Consequently, better observations were in great demand.[21]

Thus, the remote peoples ancestral to the dynastic Nile Valley populations were the first to elaborate a stellar mythos, and the stellar influences never really lost their power there even when, in protohistoric and historic times, the lunar and then the solar mythos became prominent. A clue to this early preeminence of the stellar orientation comes to us from the so-called "Pyramid Texts" of the 5th Dynasty, discovered in the tomb of Unas at Sakkara. These mytho-funerary passages record the earliest formalized religious beliefs that we know of and uncontestably derive from a period far anterior to

---

[18] Massey G, *Ancient Egypt*, vol. 2, New York: Samuel Weiser, 1970, p. 563.
[19] Locker op. cit., pp. 60–1.
[20] Ibid.
[21] Aveni AF, "Old and New World Naked-Eye Astronomy," in *Astronomy of the Ancients*, edited by K. Brecher and M. Feitag, Cambridge: The M.I.T. Press, 1980, p. 63.

the dynastic era. The religious and symbolic thought of the historical era of Nile Valley culture clearly descend from them and the texts are the progenitors of the "Book of the Dead."

The Pyramid Texts represent an amalgam of the rising solar worship of Ra, that emerged as a dynamic religious force in the 5th Dynasty, and the older stellar worship. The Osirian resurrection cult was interwoven into this schema and the beatified royal spirit was destined to become an "imperishable star." Thus, Utterance 373 of the Pyramid Texts says:

> *Khnty-mnwt-f* comes out to you and grasps your hand, he takes you to the sky, to your father Geb. He is joyful at meeting you, he sets his hands on you, he kisses you and caresses you, he sets you at the head of the spirits, the Imperishable Stars. Those whose seats are hidden worship you, the Great Ones care for you, the Watchers wait upon you.[22]

The Imperishable Stars were the northern circumpolar stars that never set below the horizon. They furnished the imagery of a soul that had attained immortality. The Watchers were these same circumpolar constellations that never disappeared from sight, were never extinguished. This stellar imagery pervades the Pyramid Texts, showing that Lockyer and Massey were essentially correct: the oldest astromythic system, at least in Nilotic civilization, was of a stellar orientation.

At a very early time, the ancient star-watchers began to trace out star combinations, or "asterisms," and turn them into constellations. Creating constellations perhaps made it easier to track the movement of the stars and identify them as time keepers. The naturalistic thinking of paleolithic humanity meant that the significant animals in their environment, as symbol-signs and mythotypes, were projected into the heavens and fused with the constellations. The animal mythotypes, or "zootypes," even now so characteristic of the constellations, reflect the hoary antiquity of stellar astronomy.

An example of this truth is the Nilotic Ta-Urt, the oldest identifiable deity in the Nilotic pantheon of *neters*, whose figurines are the earliest known divine representations. She was the pregnant hippopotamous goddess or *neter-t* who was an archaic emblem of the Great Mother.[23] She dominated the celestial planisphere of ancient Egypt. The northern circumpolar constellations revolved around Ta-Urt and she was identical to the constellation we now call Draco, the Celestial Serpent. It is worth noting that the north

---

[22] Faulkner RO, translator, *The Ancient Egyptian Pyramid Texts*, Oak Park: Bolchazy-Carducci Publishers, 1969, pp. 123–24.

[23] See Finch CS, *Echoes of the Old Darkland: Themes From the African Eden*, Decatur, Georgia: Khenti, Inc., 1991, pp. 57–88, for an extended discussion of the "zootypical" old matriarchy.

pole of the ecliptic[24] runs right through Ta-Urt/Draco, around which revolves the celestial north pole, always marked by the star of one of the seven circumpolar constellations closest to the point of the celestial pole.[25] Ta-Urt thus defined the circumpolar movement from which evolved the whole idea of cosmic time. Lockyer has shown that two of the most important and impressive Egyptian temples, the Temple of Mut connected to the Karnak complex and the Temple of Hathor at Denderah, are both ultimately sacred to Ta-Urt in her various guises.[26] It is in fact the two zodiacs at Denderah, the square and the round, that furnish us with some of our best evidence concerning ancient Nilotic astronomical knowledge. Ta-Urt is the central figure of both these zodiacs (see Fig. 45).

As Gerald Massey has indicated, with corroboration from Lockyer, the stellar mythos and stellar time were originally matristic and Sethian. Since it seems to have emerged in the Paleolithic era, not later than 40,000 years ago, it of necessity partook of the ethos of the old matriarchy.[27] This is evident in the figure of Ta-Urt herself or, for that matter, of Draco. Set was the Primordial Son of the old Cosmic Matriarch and took her forms in masculine guise, be it hippopotamous or snake. This identification of Set with the stellar mythos, and especially the northern circumpolar "heaven," was operative even in historical times. In the Egyptian planisphere, the current celestial pole-star, Polaris, was situated in the constellation of *Wep-wat,* the Jackal, our Ursa Minor, beginning around 1,600 B.C. Before that, it was located in Draco as the star α-Draconis.[28] All of these zootypes—hippo, snake, jackal—were Sethian in their masculine guises. So is the bear: the contemporary Polaris that is part of Ursa Minor (Little Bear) is also of Sethian as-

---

[24] The ecliptic pole is defined as the axis 90° vertical to the plane of Earth's solar revolution. The celestial north pole therefore projects at an angle 66°33' from the horizontal plane of Earth's revolution and 23°27' from vertical.

[25] The north pole of the ecliptic is the vertical north pole; the celestial north pole is determined by the 23½° tilt of the Earth's axis so that this tilted pole is marked by any northern star that inhabits a position close to 23½° from vertical. This axial tilt causes a wobble as the Earth rotates, meaning that the tilted axis revolves or "precesses" around the vertical or ecliptic axis. Thus, the celestial north pole precesses around the north pole of the ecliptic in a period of 25,900 years.

[26] Lockyer, op. cit., pp. 182–219. Both of these temples, though 100 miles distant from one another, are oriented to γ-Draconis, one of the most notable stars of the constellation Draco, i.e., Ta-Urt.

[27] See Finch, op. cit., pp. 57–87.

[28] The perceptive reader will ask how can Draco be the site of both the ecliptic *and* the celestial north poles. It is simply that Draco is a long, large, serpentine polar constellation, large enough to inhabit both the ecliptic and, from 5,300 to 1,600 B.C., the celestial north poles. In the Egyptian planisphere, however, the early "Draconian" celestial pole-star (5,300 B.C.) was actually situated in the Thigh of Set so that it was separated from the ecliptic constellation they knew as Ta-Urt.

Fig. 45. The planisphere of Denderah, dedicated to Hathor-Isis-Nut, depicting the equatorial and north polar zodiacs. Though not easy to discern to the casual eye, a knowledge of the 26,000-year precessional cycle is encoded in the planisphere.

pect.[29] Lockyer takes the position that the relative de-emphasis of circumpolar stellar worship in dynastic times was related to the diminishing status of Set in a brave, new solar world dominated by Amen-Ra and Osiris.[30]

We have already suggested that Paleolithic African man began a process of continuous star-gazing as far back as 40,000 years ago. Sometime between 6,000 and 12,000 years ago, star-gazing was transformed into a systematic observational science in the Nilotic lands of Africa. Millenia of observations were codified and one of the results of this was the discovery of the Precession. It is hard now to doubt that in fact these archaic Nilotic Africans had indeed

---

[29] The current Polaris will begin to move away from its polar position, making way for a new polestar, around 2095. It will take about 300 years for the next circumpolar constellation *Cepheus*, to move into the polar position. Cepheus, mythologically, was(is) the Ethiopian king; much earlier he was configured as the *Kaf* ape, emblematic of Thoth, in Nilotic astromythology.

[30] Lockyer, op. cit., pp. 319–20.

discovered this arresting phenomenon, and such a discovery, in the absence of telescopes, applied mathematics, and other technical tools, could only be the fruit of continuous observation over a period of at least 26,000 years. By now, a dozen qualified authorities have carefully investigated the question and have unambiguously concluded that the ancients Egyptians had acquired a precise knowledge of the Precession and had encoded it in their monuments.[31]

There are actually two Precessions, a polar and an equatorial one. They both result from the axial wobble of the rotating earth and are both 25,900 years in duration.[32] The polar Precession is tracked by the movement of the seven circumpolar constellations at or near the position of the celestial pole. Each constellation remains in the vicinity of the pole for 3,700 years (25,900 ÷ 7 = 3,700) before moving on and being replaced. It returns to the polar position every 25,900 years.[33] These seven circumpolar constellations and their polar dates can be listed as follows by their ancient Nilotic designations in italics and modern names in bold:

Shu/**Hercules**,12,700 B.C.: Lion-god and "Kneeler," included the star **Vega** now in the constellation **Lyra**.

Ploughman/**Boötes**, 9,000 B.C.: Also known as the "vinegrower" and included parts of **Ursa Major**.

Ta-Urt/**Draco**, 5,300 B.C.: Hippopotamous goddess was represented by the Serpent in later traditions.

Wep-uat/**Ursa Minor**, 1,600 B.C.: The Jackal in Egypt was turned into the "Little Bear" in Greece.

Kaf/**Cepheus**, 2,100 A.D.: The Ape of Thoth became the Ethiopian king in later planispheres.

Nekhebit-Mut/**Cygnus**, 5,800 A.D.: The vulture-goddess of Egypt was transformed into the swan in late antiquity.

Apesh/**Lyra**, 9,500 A.D.: Apesh the Tortoise furnished the shell on which the first lyres were strung.

A few more words concerning the seven circumpolar constellations. Even while a constellation envelopes the pole, there may be centuries where there is no single star close enough to the polar position to serve as its marker in the current fashion of polaris. Morever, when *Ursa Minor*, and consequently

---

[31] These authorities include: Jean Baptiste Biot, John Herschel, Charles Piazzi Smyth, Norman Lockyer, Gerald Massey, David Davidson, R.A. Schwaller de Lubicz, Livio Stecchini, Giorgio de Santillana, Hertha von Dechend, Jane B. Sellers, and Robert Bauval. Biot, Herschel, Smyth, and Lockyer were four of the most eminent astronomers of the 19th century.

[32] It is important to remember that the starry movements are *apparent* movments caused by the Earth's rotational and revolutionary motion. To an Earth-based observer, the sun and stars seem to "rise" and "set," but actually these are only perceptional phenomena caused (1) by Earth's travel around the sun and (2) its axial rotation.

[33] Because the tilt of the Earth's axis can vary by as much as a degree, the actual lapse of time of the Precession can also vary. Thus, the cycle of the Precession can last anywhere from 25,860 to 25,920 years.

polaris, begin to move away from the polar position, it will take 300 years or so for *Cepheus* to move into the polar position. During that interim also, the world will be, in effect, without a polar marker.

A discrepancy arises in modern circumpolar star charts because usually, only six constellations are drawn on the polar circle whereas there is every reason to believe that in the ancient Nilotic polar planisphere, seven circumpolar constellations were recognized.[34] The Nilotic circumpolar constellation not clearly designated as such on modern circumpolar charts seems to have been comprised of parts of the modern *Boötes* and *Ursa Major* and referred to as the *ploughman* or *vine-grower*. Asterisms change. Certain stars that form part of a particular constellation now often were configured in different asterisms at different historical epochs and in different lands. Consequently, it is often necessary for the modern investigator to reconstruct the ancient asterisms in a manner consistent with the astro-mythic consciousness of those ancient times and places.

The equatorial or equinoctal Precession is tracked by monitoring a circular band of 12 constellations, 47° in width, whose margins are defined by the two tropics. This band of constellations constitutes the conventional zodiac. The configurations of these asterisms were and are so well known as to be commonplace. The equatorial Precession can be followed by observing the constellation rising each year at the spring equinox. Because of the nature of the precessional movement, the circle of constellations moves in a retrograde (counter-clockwise) fashion relative to an observer on earth. Each of the 12 constellations in this band rises at the vernal equinox for about 2,158 years ($25{,}900 \div 12 = 2{,}158^1/_3$) before disappearing below the horizon, to be replaced by the constellation in front of it (because of the retrogression, Aquarius, though "in front of" Pisces, actually moves into the vernal equinox *after* Pisces). Each constellation returns to the vernal equinox every 25,900 years.

As we have already mentioned, each constellation, whether polar or equatorial, was associated with a mythotype and each constellation's mythotype was given a name, story, and symbol. These mythotypes analogically mirrored and embodied the cosmic, natural, social, and psychic rituals of human life. These stellar mythotypes, then, enacted and impelled the unfolding drama of humanity in the heavens, tied to the cyclic rhythms of the universe. They were fused with the symbolic archetypes that emerged out of pre-human instinctual life—becoming signs in the guise of animals—and then assumed final shape as the primordial Forms of the psyche, activating, conditioning, channelling all human thought and action. Thus did astrology evolve, born in the archetypal depths of the human psyche, as the forces of the unconscious moved outward into the universe, projecting onto and merging with

---

[34] The ancient Egyptians even had a name for these which they called the seven *maa-m-gerhu*, i.e., "watchers at night."

the stellar forms that were configured as constellations. Thus did the constellations become the signs of symbolic archetypes written in the book of heaven, to be read as prescriptions for human destiny.

The great antiquity of the circular stellar band known as the zodiac is testified by the very animal symbols that give it its name. The word "zodiac" derives from the Greek "zoion" meaning "animal." Most of what we know concerning the Egyptian zodiac comes from the Temple of Hathor at Denderah which was rebuilt for the last time about 100 B.C. It has been assumed that the zodiacal signs represented in the square and the circular zodiacs at Denderah were transmitted to Egypt from Babylonia and Greece. However, to paraphrase Gerald Massey, the Greeks understood little of mythotypes and their function. They borrowed their mythotypes wholesale from Nilotic mythology, together with their stellar representations, without ever completely comprehending the allegories expressed in them. Moreover, Denderah, home of the famous zodiacs, may have first been built as early as 3,200 B.C. and then rebuilt 1,600 B.C.[35] Indeed, it is thought that the extant circular zodiac of Denderah depicts various dates going back to 1,600 B.C.[36] Thus, it is possible to list the various zodiacal signs by reference to the Nilotic mythotypes they derive from:

Aquarius	Waterman	*Hapi*: The Egyptian name for the Nile depicted as a man (with a feminine breast) pouring water out of two vases. The Waterman was also the Celestial Flood known as *Nu*.
Pisces	Two Fishes	*In* and *Remi*: Two fish neters, one from the upper part of the Nile at Swnw (Syene) and the other from the lower part of the Nile at Qen Mer (Oxyrhynchus).
Aries	Ram	*Amen*: The ram-headed deity whose sacred city was Waset (Thebes) and who, as *Amen-Ra*, was the supreme deity of the Nile Valley for 2,000 years.
Taurus	Bull	*Ptah*: Represented by the bull during the Old Kingdom. *Asar* (Osiris) also took the bull-form.
Gemini	Twins	*Heru* and *Set*: The twin gods, sometimes represented by two heads on one body; male and female twins were *Shu* and *Tefnut*.
Cancer	Crab	*Khepera*: The scarab beetle, transformed into the Crab in the northern Mediterranean.
Leo	Lion	*Atum*: The original lion-headed deity of Egypt.
Virgo	Virgin	*Ast*: The Virgin Isis, mother of Horus who presided over growing of grain. Virgo can also be Neith (of Sais).

[35] Lockyer op. cit., p. 204. Lockyer actually thought that the original temple was older yet, built circa 4,800 B.C.!

[36] Tompkins P, *Secrets of the Great Pyramid*, New York: Harper & Row, 1971, p. 174.

Libra	Scales	*Maat*: *Maat*'s feather determined the outcome of the weighing of the heart on the balance scale in the Judgment scene.
Scorpio	Scorpion	*Serqet*: Scorpion-headed goddess, a form of Isis.
Sagittarius	Archer	*Shu*: The Bowman in Egyptian mythology; *Set* also was a type of Bowman.
Capricorn	Goat	*Menu*: A form of Amen sometimes identified with the goat. *Khnum*, the spiral-horned ram deity of the 1st Cataract also represented Capricorn.

When each of the above constellation-signs inhabited the vernal equinox, it became the ruler of an age that lasted 2,158 years. The symbol and spirit emanating from each sign dominated the age under its rule. Thus, the current Age of Pisces has largely been impacted by the religion of the "Fishermen" or *Pisciculi*, a.k.a. Christianity, under the sign of the Fishes. The previous age of Aries, the Ram, was notable for lamb, ram, and shepherd symbolism, seen to good effect in Judaism, a religion that appeared in world history under this sign.

As seen earlier in this chapter, the polar constellations followed essentially the same regimen. The current polar age is that of the Wep-uat the Jackal, with its co-types of dog, fox, coyote, and wolf. We know this because, according to the Egyptian planisphere at Denderah, the celestial north pole points at the constellation of the Jackal to where it moved around 1,600 B.C. Not coincidentally, the Divine Jackal, Anubis, began, in the same era, to play an important role in the Egyptian funerary ritual and metamorphosed into a son of Osiris by Nepthys. In the Dogon world-system, the Pale Fox plays a central role in cosmic evolution; among the southwest Indians, the Coyote is the central figure in their mythology. In Roman mytho-history (circa 770 B.C.), the legendary founders of the city, Romulus and Remus, were suckled by a she-wolf.

The Fusion of Nilotic Architecture and Astronomic Reckoning

The whole issue of ancient knowledge of the Precession was brought into fresh relief by Giorgio De Santillana of M.I.T., who wrote the introduction to the 1964 edition of Lockyer's *Dawn of Astronomy*:

> When a stellar temple is oriented so accurately that it requires several reconstructions at intervals of a few centuries, which involve each time the rebuilding of its narrow alignment on a star . . . , when Zodiacs, like that of Denderah, are deliberately depicted in the appearance they would have had centuries before, as if to date the changes, *then it is not reasonable to suppose the Egyptians unaware of the Precession of the Equinoxes*, even if their mathematics were unable to predict it numerically. (emphasis added)[37]

[37] De Santillana G, Preface of JN Lockyer's *Dawn of Astronomy*, Cambridge: M.I.T. Press, 1964, p. ix.

As we have seen in the previous chapter, Lockyer was able to show that the temples were constructed in such a way that they functioned as "telescopes" aligned to a particular star. However, they had to be re-oriented by additional construction to follow the star as it moved, by virtue of the precession, to a different position. These considerations are so important in understanding the scope of Nilotic astronomy that it is pertinent to quote Lockyer at length on the subject:

> . . . with a star changing its declination in an average way, the same temple could not be used to observe the same star for more than 200 or 300 years; so that at the end of that time, if they still wished to observe that particular star, they must either change the axis of the old temple, or build a new one . . .
>
> Now this change of direction is one of the most striking things which have been observed for years past in Egyptian temples. . . . As a matter of fact, we find that the axes of the temples have been changed, and have been freely changed; that there has been a great deal of work done on many of the temples . . . in order to give them a twist.
>
> We get cases in which the axis of a temple has had its direction changed, and others in which, where it has been difficult or impossible to make the change in a temple, the change in amplitude has been met by putting up a new temple altogether. We are justified in considering such temples as a series in which, instead of changing the orientation of a pre-existing temple, a new temple has been built to meet the new condition of things.[38]

It is this evidence, as indicated by De Santillana, that led Lockyer to the conclusion that the Egyptians, if they had not known of the Precession when temple-building first began 8,000 years ago, discovered the phenomenon within a few centuries after the initial star-aligned temples were constructed. The practice of constructing two temples aligned to the same star at different periods continued down to the eve of the Christian era. At Karnak, the temple of Seti I was built in 1,300 B.C. and *aligned to Canopus* at 63° S of W; the temple of Khons at Karnak was completed in 200 B.C. and *aligned to Canopus* at 62° S of W:

> Calculation shows that the change of amplitude of Canopus *due to the precessional movement* between 1300 B.C. and 200 B.C. is almost exactly 1°, the difference in the amplitude of the two temples. We seem, then, to have . . . two temples *directed to the same star at different times.* (emphasis added)[39]

In an earlier publication, the author quoted from Peter Tompkins who, in citing R.A. Schwaller de Lubicz, showed how the circular zodiac of

---

[38] Lockyer, op. cit., p. 162–3.
[39] Ibid., p. 185.

Denderah gave a categorical indication that the Precession had been charted in the Nile Valley long before the Greeks.[40] Again, the issues raised are critical to our understanding of ancient Nilotic astronomy and chronology:

> The zodiac [of Denderah] is about 8 feet across, carved in relief on hard stone. The constellations are arranged in a spiral and the symbolic figures are marching counterclockwise in the diurnal direction of the stars as seen from earth. Recognizable mythical figures for the constellations near the pole are a jackal for the Little Bear, an ox-leg for the Great Bear, and a hippopotamous for Draco. Sirius is depicted as a cow in a boat with a star between her horns.
>
> The zodiac is in a circle at the center of which is our north pole. . . . Our north pole is correctly located in the constellation of the jackal or Little Bear, as it was at the time when the zodiac was carved, sometime about the first century B.C. But the zodiac also shows *the pole of the ecliptic*, located in the breast of the hippopotamus, or constellation of Draco.
>
> To Schwaller this explains the spiral formation of the constellations. The mythological figures representing the constellations are entwined in two circles—one around the north pole and one around the pole of the ecliptic. Where these two circles intersect marks the point of the equinox, or due east. The zodiac thus becomes a calendar going back to remote antiquity.
>
> A line due east, which runs between the end of the Ram and the beginning of Pisces, indicates the time when the temple was rebuilt, about 100 B.C. An earlier line runs right through the Ram, indicating a date about 1,600 B.C., at the height of the Amonian domination.
>
> A special hieroglyph on the ring of the zodiac indicates an equinoctal line running through the end of Gemini and the beginning of Taurus—the date of the founding of the empire of Menes and the beginning of the cult of the Bull and the adoption of the new calendar, sometime in the third or fourth millenium B.C. (emphasis added)[41]

That the Precession was known to ancient astronomers at a time far anterior to Hipparchus (187 B.C.), its reputed discoverer, is implied by a much-quoted fragment from Virgil, who lived two centuries *after* Hipparchus, when he referred to "the constellation of the white Bull with the golden horns which opens the year."[42] Now the Roman New Year, like that of other nations, occurred at the vernal equinox. The equinoctal rising of the Bull had ceased more than 2,000 years previous to Virgil's time and had been supplanted by the rising of the Ram at the equinox. Virgil's statement is clearly

---

[40] See Finch CS, *Echoes of the Old Darkland: Themes from the African Eden*, Decatur, Georgia: Khenti, Inc., 1991, p. 120.

[41] Tompkins P, *Secrets of the Great Pyramid*, New York: Harper and Row, Publishers, 1971, pp. 172–4.

[42] Cited by Piazzi Smith C, *The Great Pyramid*, New York: Bell Publishing Co., 1880, 1978, p. 377.

an anachronism but *does* refer to a time, more than two millenia past, when nations marked their New Year's Day by the equinoctal rising of the Bull. It is plain why certain contemporary historians of science, notably Von Dechend and De Santillana, have found it difficult to doubt that the Precessional phenomenon had been known long before Hipparchus's time.[43]

Finally, there is the case of the Dogon people of the Bandiagara Cliffs of southwestern Mali. We find that they refer to the greater and lesser circles of the earth's axis of rotation: "The Earth turns on its axis and simultaneously makes a great circle, *aduno digili*, 'circle of the world,' like a top the rotation of which is accompanied by a circular motion."[44] This passage is an unmistakable reference to the Precession among a Western Sudanese people known to possess astounding knowledge of astronomical matters.[45]

What modern scholarship will be forced to accept sooner or later is that, not only mythology and religion, but also history has been linked to astronomical cycles. It has been said that history has been reduced to "one damn thing after another," the outcome of an exclusive reliance on empirical facts to interpret humanity's Becoming. As noted by one popular novelist, the True has had to yield to the Actual.[46] So many Facts, so little Truth. Still, some are gradually realizing that history wasn't written merely on rolls of papyrus or leaves of pulp-wood but in the very heavens themselves.

There were at least 12 major stars to which the important temples were oriented in the Nile Valley. That a precise alignment to a star or star-group was the prescribed intent of the temple-builders is attested to in surviving Nilotic annals, where the commissioning pharaoh is made to say:

> I have established the surveying rod and I have grasped the point of the peg; I hold the plumb-line with Sesheta. I turn my sight toward the course of the stars; I make my eyes enter the constellation Meshket [Thigh/Great Bear]. The Time Measurer stands next to his hour-clock. I have established the four corners of your temple.[47]

Each star had several temples in different locations oriented to it, meaning that the mythotype associated with that star was sacred in each place. Sirius, for example, had seven different temples aligned to it from Denderah to Naga (Cush). Phact, a pre-Sirian heliacal riser at the solstice, had no fewer than 12 different temples oriented to it, from Thebes to Napata (Cush). The

---

[43] See De Santillana G and Von Dechend H, *Hamlet's Mill*, Boston: David R. Godine, 1977.

[44] Griaule M and Dieterlen G, *The Pale Fox*, translated by Stephen C. Infantino, Chino Valley, Arizona: Continuum Foundation, 1986, p. 508.

[45] See Griaule and Dieterlen, passim; also Temple RKG, *The Sirius Mystery*, New York: St Martin's Press, 1976.

[46] Trevanian, *The Summer of Katya*, New York: Ballantine Books, 1983.

[47] Translated from Obenga, op. cit., p. 267. For citations of additional inscriptions, see Lockyer, op. cit., pp. 173–9.

alignment stars of Nilotic temples thus included: (1) Sirius, (2) Phact, (3) Dubhe, (4) γ-Draconis, (5) Arcturus, (6) Vega, (7) Capella, (8) Regulus, (9) Spica, (10) Antares, (11)α-Centauri, and (12) Canopus.[48] The constellations to which each of these stars belong are: (1) Canis Major ("Great Dog"), (2) Pisces ("Fishes"), (3) Ursa Major ("Great Bear"), (4) Draco ("Dragon"), (5) Boötes ("Herdsman"), (6) Lyra ("Lyre"), (7) Auriga ("Charioteer"), (8) Leo ("Lion), (9) Virgo ("Virgin"), (10) Scorpio ("Scorpion"), (11) Centaurus ("Centaur"), and (12) Argo Navis ("Ship Argo").

What must be appreciated is that, with the exception of γ-Draconis, *each* of these stars is the alpha star of its constellation. A constellation's alpha star is its brightest member. Gamma-Draconis, though not an alpha star, was important because, for a period after 5,000 B.C., it was the star in Draco closest to the polar position. However, as we saw in the previous chapter, the alpha star of Draco (α-Draconis) *did* become an alignment star for the Descending Passage of the Great Pyramid when it began to move toward the polar position about 3,340 B.C.

As has already been noted, the original Nilotic names for many of the constellations were different from the later Greek ones though, in the Zodiac of Denderah of about 100 B.C., there seems to have been a conscious attempt to reconcile the Nilotic constellations—in name *and* configuration—to their Greek counterparts. Doubtless this attempted reconciliation emanated from the Nilo-Hellenic cultural syncretism fostered by the Ptolemies. Nonetheless, there were zodiacal asterisms at Denderah that had come down intact from remote Nilotic antiquity, namely the Bull, the Lion, and the Scorpion. And, as we have already indicated, the mythotypology of the other asterisms was distinctly Nilotic however much they may have been adumbrated with a Hellenic sensibility. Whatever the case, it is worthwhile to connect the known stars and asterisms of the current time to their ancient Nilotic names and forms:[49]

Ancient Nilotic	**Contemporary**
Sahu	Orion
Sep-det	Sirius
Ta-Urt/Rerit (Hippopotamous)	[Parts of] Draco
Meskht (Thigh of Set or the Bull)	Ursa Major
Seshau (Deer)	Cassiopeia
Menat (Great Nurse)	[Parts of] Boötes & Scorpio
Sba en khau [?] (Many Stars)	Coma Berenices
Arit (Portal)	[Part of] Andromeda
Ser[?] (Fleece)	[Parts of] Aries

48 Lockyer, op. cit., pp. 303–14.
49 Allen, op. cit., p. 20.

Geb (Goose)	[Part of] Aries
Iakhu (Lights)	Pleiades
Kaf (Cynocephalic Ape)	Ara; also Cepheus
Repa (Chief or Prince)	[Part of] Virgo
Sab/Wep-uat (Jackal)	Ursa Minor
Stars of the Water	Unidentified
Mena's Herald	Unidentified
Mena's Followers	Unidentified
Necht	Unidentified
Hare	Unidentified

In addition to these, the ceiling of the tomb of Seti I (circa 1,300 B.C.) shows other asterisms whose modern counterparts are not known. These are: (1) An, a form of the martial Horus, (2) a Horus or Ra figure with a disc on his head, (3) another Horus-like figure spearing (4) the crocodile *Serisa*, and (5) a hawk asterism that is either Horus or Ra.[50]

The five planets closest to the sun were, of course, well-known to the ancient Nilotic astronomers. Their names were as follows:

Ancient Nilotic	**Late Nilotic[51]**	**Contemporary**
Sba Shema ("travelling star")	*Heru-Ip-Sheta*	Jupiter
Sba Imenti Dja Pet ("western star that crosses heaven")	*Heru-Ka-Pet*	Saturn
Sba Iabti Dja Pet ("eastern star that crosses heaven")	*Heru-Khuti*	Mars
Sebku ("Sebek, son of Neith")		Mercury
Dja Benu Asar ("Travelling Bennu of Osiris")	*Pi-Neter-Duau*	Venus

Creating Time: The Development of Nilotic Calendars

The most important practical outcome of Nilotic star-gazing was the invention of the calendar of which there were several types. Today we use the calendar to mark the passing of time, to schedule appointments, or to draw up timetables for the minutiae of daily living. We take our calendar so much for granted that its existence is but one of the many banalities of modern life. Outside the circle of astronomers and perhaps historians of science, we have no idea of the immense mental effort that went into the calendar's creation or that it was one of the most important means devised by the an-

[50] Budge EAW, *Gods of the Egyptians*, vol. 2, New York: Dover Publications, 1969, p. 312.

[51] The Late Nilotic planetary names can be translated as follows: *Heru-Ip-Sheta*—"Horus, the Hidden Judge or Messenger"; *Heru-Ka-Pet*—"Horus, Bull of Heaven"; *Heru-Khuti*—"Horus of the Horizons"; *Pi-Neter-Duau*—"The god of the morning." Ibid.

cients to keep in touch with the Movement of Heaven. The ancient calendars "came from the gods" and for that reason were, for most of antiquity, under the exclusive control of the priests. They were probably the most important fruit of the thousands of years of unceasing star gazing conducted in ancient Africa.

We have already discussed in some detail the "Great Year" that the ancient Nile-dwellers, in Egypt and points south, devised from the discovery of the Precession. It is in the Precession, and the zodiac that emanates from it, that one finds the great calendar of "mega-history." There were others and the ancient Nile dwellers went to extraordinary lengths to measure and mark them out with absolute precision.

Like other cultures of antiquity, the ancient Nilotes possessed a lunar calendar, one that is usually considered to have been in use earlier than the solar calendar, although, as we have seen, Norman Lockyer dissented from this opinion. But there are compelling practical reasons to believe that the stellar and lunar calendars preceded the solar. For one thing, the moon and the stars can at night be observed directly, continuously, and indefinitely. Unless the observer wishes to risk permanent blindness, this is not possible for the sun. The sun could only be observed directly on the horizons, that is, at the time of its rising and setting. Indirectly, the sun could be observed by watching the shadows of an upright pole but the moon and stars required no apparatus whatsoever. Indeed, pre-neolithic and neolithic peoples then and now could and did develop an entire system of naked-eye astronomy without observational aids or written records.[52] A certain amount of technical sophistication was necessary for accurate solar observation; this was not a prerequisite for nighttime astronomy.

The neolithic Nilotic villager would have possessed an astronomy and functional—if unrecorded—stellar and lunar calendars prior to the historical period. His innate conservatism preserved the use of lunar time-keeping down to the end of Nile Valley history. Furthermore, survivals of the lunar calendar persist to this very day in every part of the world. The Muslim and Jewish liturgical calendars are still lunar and it is possible to discern, in the pervasive new moon festivals throughout traditional Africa, the continuing presence of this calendar there also.

Monthly lunations are of 29.530588 days duration and the beginning of a lunar month is always reckoned from the start of the new moon phase. The full moon, therefore, always falls in the middle of the lunar month and we find the ancient Hebrews coordinating their equinoctal New Year's Day as closely as possible to the mid-month full-moon, i.e., the 15th of Nisan. A nor-

---

[52] The Dogon of Mali are just one of scores of "neolithic" and "post-neolithic" peoples around the world who possess a well-developed stellar-lunar astronomy without apparent apparatus or written records. Such astronomies are also well-attested among the Australian Aborigines, the Pacific Islanders, and numerous Native American groups.

mal lunar year is consequently 354.36705 days (29.530588 × 12 = 354.36705) and in practice, the length of the months alternated between 29 and 30 days. Indeed, there was a tendency to subdivide the lunar year into six "double-months" of 59 days each.

The lunar year is 11 days shorter than the solar year of 365.25 days. This discrepancy eventually led to the discovery of the Metonic cycle, so-named after its reputed discoverer Meton, in which any particular phase of lunation falls on the exact same day and month every 19 years.

One finds that the Egyptians synchronized all the calendars that they maintained. By the Middle Kingdom, circa 2,000 B.C., the lunar calendar was rationalized to fit the solar one as closely as possible. As Neugebauer informs us, 25 Egyptian civil years of 9,125 days is virtually identical to a cycle of 309 lunar months of 9,124.95 days (29.530588 × 309 = 9124.95). Consequently, the Egyptians devised a 25-year lunar sequence consisting of 16 years of 12 lunar months plus 9 years of 13 lunar months, totalling 309 lunar months.[53] Thus were lunar and solar time-keeping meshed in 25-year sequences, like gears in a watch.

Because of the sacred quality of the number seven, many cultures in in the last millenium B.C. conventionalized the lunar month into 28 days, presumably because this number is evenly divisible by four into weeks of seven days each. Indeed, as Richard Anthony Proctor has shown, this quaternary division of the lunar month follows the easily recognizable "quartering" of the moon and the telltale ancient partiality to the sacred seven probably impelled the adoption of the 28-day month.[54] The month of 29.530588 days of the normal lunar calendar is termed the "synodic" month, determined by the time interval between successive new moons as seen from earth, but there are three *additional* months known, currently of little practical significance save to astronomers: (1) the "sidereal" month of 27.321661 days, (2) the "tropical" month of 27.321582 days, and (3) the "draconic" month of 27.21222 days.[55] The month of 28 days (and a fraction) represents a mean between these months and the synodic month. This factor may have also influenced the conventionalization of the lunar month into 28 days.

This 28-day month had the advantage of ramifying into a lunar year of 364 days composed of 13 months of 28 days each, a figure just one day short of the solar-based civil year of 365 days. The simplest way of equilibrating this type of lunar year with the civil year would be to make the last month a 29-day month. However any of this may be, we find both in the lunar "macro-

---

[53] Neugebauer op. cit., p.90.

[54] Proctor RA, *The Great Pyramid*, London: Chatto & Windus, Piccadilly, 1883, pp. 201-42.

[55] The sidereal month is measured by the time required for the moon to return to the same place against the background of stars. The tropical month is measured by the time required for the moon to return to the same celestial longitude. The draconic month is determined by the time it takes the moon to return to its intersection with the ecliptic.

year" of 309 months and in the lunar year based on 28-day months, the presence of 13-month years in each cycle. This result may be one reason why the number 13 possessed such "magical" significance as seen in the still-surviving idea that 13 is "unlucky." This "unluckiness" of number 13 may have been caused by the official establishment of the solar calendar based on *12* months at the expense of the lunar calendar.

All this might go far to explain the famous fable of Aesop, "The Tortoise and the Hare." Mythically, the tortoise always represented the earth, the hare the moon.[56] The race of the tortoise and hare represents the respective revolutionary movements of these two bodies. The hare, as the moon, is fast because it completes a revolution around the earth in about 30 days. Indeed, in the fable, the hare is depicted as literally running circles around the tortoise. The earth, as the tortoise, is much slower, completing its revolution in 365 days. But what do we find in the dénouement of the fable? The hare, in his over-confidence, decides to take a nap short of the finish line, while the tortoise continues to plod along until he crosses it. Astronomically, the "finish line" is the completion of the year of 365 days; the lunar year, whether of 354 or 364 days, falls short of this complete year. Thus does the hare also fall short of the finish line while the tortoise crosses it. Lending weight to this interpretation is the circular construction of the race-course. Read astromythically, Aesop was preserving a record of the official establishment in antiquity of the solar year—thereby superseding the lunar one—as the basis for the administrative *and* liturgical calendars in the Nile lands and elsewhere.[57]

But there seems to have been a sacred character to the number 13 at variance with its reputation for ill-luck. For example, we find that Jesus and his 12 apostles comprise a group of 13. Moreover, the Mayan-Aztec calendar was based on the number 13 and the United States began with 13 original colonies. Parenthetically, it could be surmised that the number 13 might also have been significant because 13×10^3 is exactly *half* of the cycle of the Precession of 26,000 years (rounded off).

The development of the (solar) calendar has been called one of the half-dozen most significant achievements in human history. That this development occurred in the Nile Valley is so well attested as to be beyond contention. Our own calendar, with minor modifications by Julius Caesar in the 1st century B.C. and Pope Gregory in the 16th century A.D., is the very one elaborated in the Nile Valley six or more millenia ago. And, as we have al-

---

[56] Neugebauer informs us that in certain lunar calendars, the final alternation between 29 and 30 days was skipped in favor of two consecutive 29-day months. This was called *saltus lunae*, i.e., the "jump of the moon," giving one indication, among many, why the moon was mythically identified with the hare. See Neugebauer, op. cit., p. 8.

[57] Adding weight to this explanation is J.A. Roger's assertion that Aesop was but a contraction of "Aethiop," meaning that Aesop was himself a Nilote. Aesop was born around 560 B.C. See Rogers JA, *World's Great Men of Color*, vol. 1, New York: Collier Books, 1972, pp. 73-80.

ready noted, it was in the Nile Valley that observations of the rising and setting of the sun on the horizon were carried out with such perfection that the resulting calendrical precision has never been improved upon.

It is clear that by the end of the 5th millenium B.C. the perfected calendar was in place in the Nile Valley. It possessed three different values: (1) 360 days, (2) 365 days, and (3) $365^1/_4$ days. Though there are those who have maintained that these early Nilotic peoples had first measured the solar year at 360 days, that seems unlikely. A gnomon of some kind must have been erected to mark the solar passage through the year and it would have been just as easy to have marked and counted 365 days as it would 360 days. Even if these early "calendrists" had miscounted the year at 360 days, within a year or two they would have realized their error and made the necessary correction.

The year was figured at 360 days + 5 epagomenal days, otherwise known as *tep renpt* by the ancient Egyptians. It seems evident that the 360-day configuration was maintained for ease of divisibility. Thus, this Nilotic solar year could be divided into 12 months of 30 days each, to be completed by the addition of the 5 epagomenal days. The cyclical nature of the year as a result of the circular orbit of the earth around the sun was evidently well understood by these ancient Nilotes and it is therefore difficult to attribute to coincidence their functional 360-day year and the geometric convention of a 360-degree circle.

However, the year is not exactly 365 days but rather $365^1/_4$ days. Actually, even the length of this year depends on whether one is referring to the "tropical" year or the "Sothic" year. The tropical year, measured from solstice to solstice, is 365.242392 days long; the Sothic year, measured between separate heliacal risings of the star Sirius, is 365.25 days long, revealing an 11-minute difference between the length of the two years. The heliacal rising of any star is one that occurs just prior to the rising of the sun at dawn. The heliacal rising of Sirius occurred for many thousands of years in the Nile Valley at or near the time of the solstice *and* the beginning of the rise of the Nile flood, which accounts for its overwhelming, indeed sacred, importance in Egyptian calendar-keeping.

The Egyptian calendrists elected to use the Sothic calendar as their official solar calendar but in a very ingenious way. Instead of correcting the calendar every four years as we do by adding a day to the year, they let the calendar correct itself. That is to say, the civil calendar was maintained at 365 days which meant that it slipped backward in relation to the Sothic year one-quarter day per year. Within four years, the two calendars would diverge by one day, within eight years two days, and so on. It would therefore take the civil calendar 1,460 years to re-coincide with the Sothic calendar and thus we find that these Nilotic calendrists kept track of both calendars continuously in intervals of 1,460 years.

The author has cited evidence in an earlier publication that this system was put in place by the Egyptians no later than 4,241 B.C.[58] It must have been the result of 1,460 years of anterior observation. It has also been noted that the Egyptians seemed to have maintained a "super" cycle of 36,525 years consisting of 25 Sothic cycles of 1,461 years. This Super Cycle of 36,525 years is attested in the surviving fragments of Manetho.[59] Now a Sothic cycle of 1,460 years of $365^1/_4$ days is exactly the same as a cycle of 1,461 years of 365 days, each giving a product of 533,365 days. It is not without interest that there are also 25 years in the soli-lunar "macro-cycle" of 309 months and there are 1,461 of these "macro-cycles" in Manetho's Super Cycle of 36,525 years. Here we have yet more proof of the astonishing ingenuity of the Nilotic calendrists and chronologists; every one of the cycles they kept track of was meshed with every other. Thus, stellar, lunar, and solar cycles were all interdigitated, one cycle correcting and calibrating the others in a continuous clock-like mechanism. One begins to think that few civilizations, if any, understood the nature and function of time as deeply and completely as Nilotic civilization. In the Nilotic sense, time did not "pass," time revolved; time was created and recreated anew. In such a schema, there could be no sharp distinction among present, future, and past. Time was symbolically configured "uroborically" as a snake with its tail in its mouth, forming a perfect circle.[60] Therefore, what segment of the encircling snake constituted past, future, or present was entirely arbitrary.

Lockyer took the position that the ancient Nilotes had discovered the solar year and the Sothic cycle rather before they started constructing temples to pinpoint the exact length of the solar year. The outstanding examples of solar temples are to be found at Karnak where the axis of the main temple, as we have seen, is oriented to the setting sun at the summer solstice while a smaller temple in the complex is oriented to the rising sun at the winter solstice.[61] As we saw in the previous chapter, the main axis of Karnak temple was used in "telescopic" fashion to convey a thin beam of sunlight to the sanctuary where it flashed for a few seconds as the sun set behind the western hills. An identical flashing phenomenon occurred on the inner portion of the temple aligned to the sun rising at the winter solstice. This arrangement allowed an exact determination of both solstices and an exact measurement of the tropical year.

That Karnak itself was aligned to the setting solstitial sun indicates the manner in which myth and religion were intertwined with solar time-keep-

---

[58] Finch, op.cit., pp. 115–28.

[59] Cited in Finch, op. cit., p. 127.

[60] The term *uroboros* is one used by Erich Neumann in his book *The Origins and History of Consciousness* (p. 10) to describe the state of primordial totality or "allness" which the ancients represented by the figure of the snake swallowing its tail.

[61] Lockyer, op. cit., p. 78.

ing. The main temple at Karnak was sacred to Amen. The word *amen* means "west" and "hidden." In its solar aspect, it therefore refers to the sun setting in the west and commencing its "hidden" journey through the underworld at night. This aspect of the sun is also "Osirian" and shows us why the cult of Amen and that of Osiris were so closely aligned from the Middle Kingdom onward. It shows us too why Set, the Enemy of Osiris, also became anathemized by the Ammonians. And yet the thing was more complex than it might seem, because prior to Amen, the site and the temple had been sacred to Apit, a form of the hippopotamus goddess Ta-Urt. Apit was the mother-wife of Set and one of the smaller adjoining temples at Karnak was indeed oriented to the north polar hippopotamus (Ta-Urt). No matter how antagonistic the Ammonians and Osirians were to Set, it was impossible to displace him altogether. His presence and influence were too pervasive.

We might recall from the previous chapter that the Great Pyramid seemed to have incorporated both solar and stellar elements. We saw how the Descending Passage at the time of its construction was aligned to the north polestar of 3,340 B.C., α-Draconis. Moreover, the Grand Gallery had been aligned, prior to the completion of the pyramid, to the southern stars. At the same time, the pyramid was found to swallow its shadow at the vernal equinox, signalling the first day of spring. Thus did the ancient Egyptians create a means for marking with great exactitude the solstices and the equinoxes. It is no wonder that their calendar became the standard one for most of the civilized Old World west of the Indus.

The Decans and the Invention of "Clock Time"

Otto Neugebauer, who won a significant reputation as a vociferous detractor of Nilotic science, was nevertheless compelled to say: "This [Egyptian] calendar is, indeed, the only intelligent calendar which ever existed in human history."[62] He was also compelled to admit that the division of the day into 24 hours was also an Egyptian invention (see Fig. 46). Down to about the 3rd century B.C., the length of the Egyptian hour was not fixed but varied from season to season, though the total number of 24 daily hours remained constant. Now the hour was the most important unit to measure the day and the Egyptian word for "day" was *hru*. Neugebauer tells us that the "equinoctal" fixed hour of 60 minutes duration was devised in "Hellenistic" times, that is to say, Egypt in the time of the Ptolemies (323–30 B.C.).[63] Thus, the Greek word for "hour," *hora*, may have derived from the Egyptian word *hru* because of the hour's status as the basic subunit of the day. And, indeed, "hour" itself, since it derives from *hora*, ultimately may descend from the Egyptian *hru*.

---

[62] Neugebauer, op. cit., p. 81. He obviously ignored the Mayan calendar.

[63] Ibid.

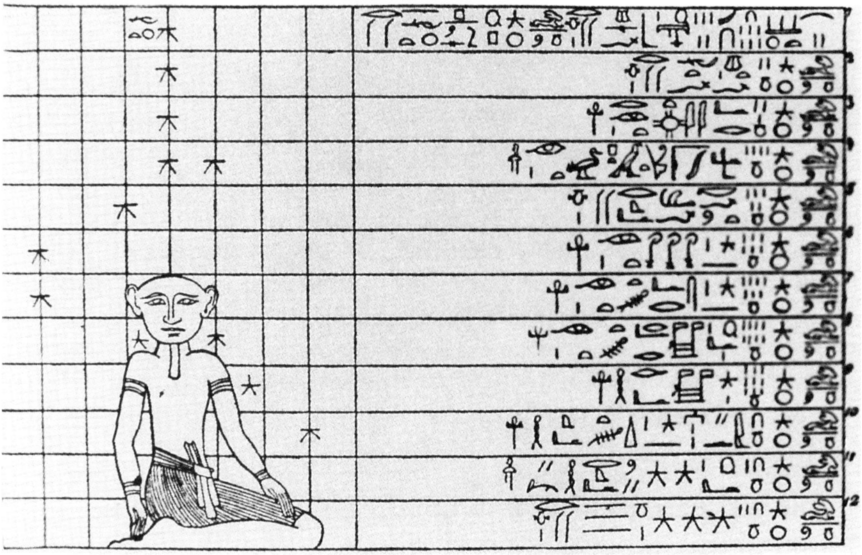

Fig. 46. Star chart showing the positions of important stars at different hours of the night. The chart could serve as a clock and a calendar. It was the Egyptians who first divided the full day into 24 hours.

The solar calendar of the Egyptians provided for 30-day months, each composed of three 10-day "weeks." These 10-day Egyptian weeks were known by the Greeks as the "decans." There were 36 of them to fill the functional year of 360 days and there is evidence of them at least from the time of the Middle Kingdom. But these decans were also astronomical divisions that generated the 12-hour night and, indeed, the 24-hour day. Each decan was defined by a constellation encompassing 10° of the ecliptic, totalling 36 constellations and 360° in all. As Neugebauer informs us, the decans belonged to a sky-belt parallel to and south of the ecliptic. Referring to the decanal asterisms found on Middle Kingdom coffin lids, Neugebauer says:

> These pictures represent the sky with the decanal constellations inscribed on them, arranged in their ten-day intervals throughout the year, forming 36 columns with 12 lines each for the 12 hours of the night. The name of a specific decan moves from column to column, each time one line higher. Thus there originated a diagonal pattern which is the motivation for the name "diagonal calendars" for these texts. . . . In fact, we have here not a calendar but a star clock. The user of this list would know the "hour" of night by the rising of the decan which is listed in the proper decade of the month.[64]

[64] Ibid., p. 83.

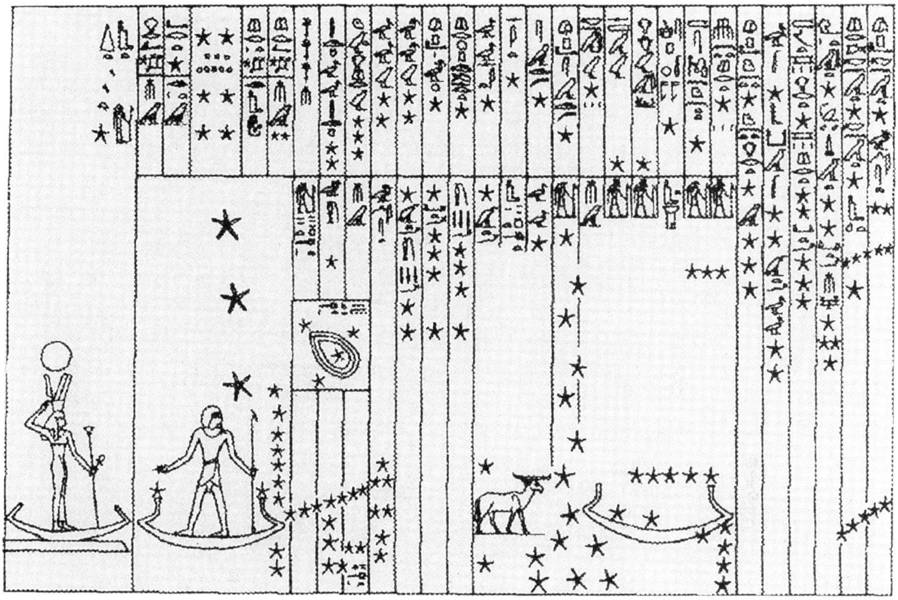

Fig. 47. Star chart on the ceiling of the tomb of Senmut (18th Dynasty). Sahu-Orion is shown beneath the three stars of his Belt. Sirius follows Sahu-Orion. In the top register, the decans are indicated.

It is the phenomenon of the heliacal rising of certain stars, that is, stars that rise just prior to dawn, that helped the ancient Egyptians devise the decans and indeed the 12-hour night (see Fig. 47). As we have seen, Sirius rises heliacally at or near the summer solstice, heralding not only the beginning of the day but also the beginning of the year. Practically, this heliacal rising of Sirius signalled the end of the last hour of night but, due to earth's rotation around the sun, Sirius rises heliacally only for a few days before it begins to appear progressively earlier and another star after it begins to rise heliacally. This star in turn serves as the indicator of the last hour of night for a few days before it too begins begins to rise earlier, leaving another star to begin rising heliacally and serving as the last hour of the night.

In practice, each heliacally rising star served as the indicator of the last hour of night for 10 days before moving "backward" and rising earlier. But in giving up its place after 10 days to a succeeding star rising heliacally, it could then serve as the indicator for the *next to the last* hour of the night, also for 10 days. Ten days later, rising earlier yet, it could serve as the indicator to the *third to the last* hour of night for 10 days. This sequence could be followed every 10 days until the night was complete, permitting the decans to serve as a star clock and helping the ancient Nilotes time the night. What is more, this star clock, moving at intervals of 10 days, was correlated to the 10-day week of the Egyptian calendar. Thus, knowing the week that one was in would allow an observer to know what time of night it was by finding the star governing that hour of the night for that span of 10 days.

We have already noted that the Nilotes had created 36 decans which, consisting of 10 days each, filled a functional year of 360 days. The epagomenal days possessed their own constellations and stars. Neugebauer points out that if we experienced absolute darkness from sunset to sunrise and night and day were equal all year—a situation that is approximated only at the equator—an observer would always see one half the celestial sphere rising each night, giving 18 decans. But at any distance from the equator, the variable length of night and the presence of twilight reduces the number of observable stars.[65] Thus, at the time of the heliacal rising of Sirius at the onset of summer, only 12 decanal constellations can be seen rising during darkness and this is the reason, according to Neugebauer, that the Nilotic night was given 12 subdivisions. Neugebauer says further: "The 12-division is . . . not an arbitrary choice of units, but the consequence of the decimal order of the civil calendar."[66]

Neugebauer points out that while the night was allotted 12 hours, the day was allotted but 10, an hour each given to dusk and dawn. Nonetheless, the schema ramified into a 24-hour diurnal system, forming the basis for the day-measure that still abides all over the world. That 12 decans are the most that could be observed nightly at the latitudes of Egypt may at least be partly responsible for the division of the year into three seasons: (1) the inundation, (2) the sowing, and (3) the harvest. While it is clear that the seasons followed an agricultural sequence, the tripartite division of the year might also have followed three sequences of 12 decans to give 36 decans in all.

The decanal sequence commenced with the heliacal rising of Sirius so the constellations comprising the decans had to be visible in a zone defined by the star Sirius and the constellation Orion, somewhat to the south of the ecliptic. In dynastic times, Sirius was easily the most important star in Nilotic astronomy and was identified with Isis. Orion, called by the Egyptians *Sahu*, was hardly less important and was identified with Osiris.[67] As we have come to expect, myth, astronomy, calendar-keeping, and time were all intertwined in the Nilotic system.

These star-clocks/decanal calendars are attested in several places in ancient Egypt from a period between 2,100 to 1,300 B.C. A number of Middle Kingdom coffin-lids are inscribed with them. The ceiling of the tomb of Senmut (c. 1,460 B.C.), vizier and chief architect of Hatshepsut, also depicts

---

[65] Ibid., p. 85.

[66] Ibid., p. 86.

[67] Richard Allen tells us that in Egypt, Orion was known as *Smati-Osiris*, a Barley God. Elsewhere, the midnight appearance of Orion heralded the beginning of the season of grape gathering which, since Osiris was identified with grape cultivation, is additional evidence of Orion's Osirian aspect. Allen RA, *Star Names: Their Lore and Meanings*, New York: Dover Publications, 1963, pp. 306–8.

the decanal system. Finally, the West Bank temple of Medinet Habu, where there is a cenotaph of Seti I, has preserved a picture of the decans.

We might finally note that the ancient Nilotes had subdivided the hour even further by aid of the *clepsydre* or water clock.[68] It is of interest that the Edwin Smith Surgical Papyrus, its original edition dating back to 4,000 B.C., mentions pulse-taking as a diagnostic aid. Pulse-taking can only be accurately performed when there is a means of measuring the passing of small units of time such as our seconds. Clearly the ancient physicians had such a mechanism and some think that a small clepsydre served this function.

African Astronomy and Time-Keeping Beyond the Nile Valley

Though we have not specifically dealt with Nubio-Cushite astronomy, there is little reason to doubt that it kept pace with that of Egypt. First of all, as Diodorus attests, astronomy descended the Nile from the lands south of Egypt, i.e., from Ethiopia. Lockyer confirmed this assertion of Diodorus by his demonstration of the primacy of the southern star cults in Egypt. Moreover, as we saw above, he found that the Nubian temples in the all-important cult centers of Napata and Naga were as precisely aligned to important stars as any of their Egyptian counterparts. Beyond these facts, little has come to light directly concerning Nubian astronomy for reasons already mentioned: Nubia remains largely untouched by archaeology.

Relatively little is known also about astronomy in sub-Saharan Africa, though in a few places preliminary evidence is revealing that African archaeoastronomy is potentially an immensely fertile field for research. One archaeoastronomical site that has been studied is Namoratunga II on the western shore of Lake Turkana in northern Kenya.[69] Lynch and Robbins, who brought the evidence to light, were spurred to investigate it by their awareness of numerous megalithic/monolithic sites in nearby Ethiopia and their knowledge of a calendar used by contemporary Cushitic-speakers in the region of the stellar-lunar type. This calendar in and of itself is without known parallel in or out of Africa. It is based on the rising of seven stars or constellations synchronized with the lunar phases throughout the 354-day lunar year. The stars and constellations so utilized are, in order:[70] (1) Triangulum[c], (2) Pleiades[c], (3) Aldebaran[s], (4) Bellatrix[s], (5) Central Orion[s], (6) Saiph[s], and (7) Sirius[s].[71] Lynch and Robbins describe the functioning of this unusual calendar as follows:

---

[68] Lamy L, *Egyptian Mysteries*, New York: Crossroad, 1981, p. 75.

[69] The definitive monograph on this site is by Lynch BM and Robbins LH, "Namoratunga: The First Archaeoastronomical Evidence in Sub-Saharan Africa," *Science* (200), May 1978, pp. 766–8.

[70] Individual stars will be indicated by [s] and constellations by [c].

[71] The present-day Borana, a Cushitic people who are a branch of the Oromo living on the Ethiopia-Somali border, are the most prominent keepers of this extraordinary calendar. The Borana names for the pertinent asterisms are: *Lami* (Triangulum), *Busan* (the Pleiades), *Bakkalcha* (Aldebaran), *Algajima* (Bellatrix), *Arb Gaddu* (Central Orion), *Urji Walla* (Saiph), and *Basa* (Sirius). See Obenga, op. cit., p. 280.

> For half the year they identify months by the rising of each of the seven stars or constellations in relation to the new moon. Each star or constellation appears successively in conjunction with the new moon in the order Triangulum, Pleiades, Aldebaran, Bellatrix, Central Orion, Saiph, and Sirius. Only Triangulum is utilized for the second half of the year beginning when Triangulum rises in conjunction with a full moon. Each successive month is identified by Triangulum's relation to progressively declining phases of the waning moon. Such months are approximately 29.5 days long.[72]

Lynch and Robbins were directed to the site known as Namoratunga II by the Turkana people of northern Kenya and there they found a collection of 19 stone monoliths set in rows, tilted at angles to the ground. The stones were set in line with one another at "non-random" distances in a manner that suggested to the two investigators that they were aligned with certain stars (see Figs. 48–49). They knew that the site could be dated to at least 300 B.C.[73] and therefore decided to calculate the azimuths of 300 B.C. of the seven stars/constellations that determine the Cushitic calendar to see if these alignments matched them. The precession would have caused the azimuths to have dramatically shifted since then and indeed they found azimuthal shifts of up to 12° from 300 B.C.

What Lynch and Robbins found astonished them and confirmed their initial theory. There were apparently three siting stones which they numbered as Stone 1, Stone 5, and Stone 18. From each of these stones, it was possible to discern alignments with 15 of the other stones that correlated remarkably well with the calculated azimuths of the rising of the seven designated stars/constellations at 300 B.C. There were 12 such alignments, five of them with three stones, seven with two. According to Lynch and Robbins, the three-stone alignments furnished greater observational accuracy. Whatever the case, 9 of the 12 alignments were exactly accurate; three of them showed an error of just 1°. The results are shown in Table 1.[74]

There was one additional alignment of some significance: Stones 1 and 5 were oriented exactly to true north. Stone 11 had fallen down but the authors think that it must have been part of the Stone 5→10 alignment to Central Orion (see Table 1). The only stone that did not participate in any alignment was Stone 19, the smallest one at the site. It seems that this stone was never intended to serve the purpose of orientation since it rose less than seven inches from the ground.

With all the evidence cited above, one is immediately struck by the precision obtained by these pre-Christian East African astronomers. They must

---

[72] Lynch & Robbins, op. cit., p. 767.

[73] Ibid. Namoratunga II was clearly connected to a site 130 miles south southwest of it known as Namoratunga I. Namoratunga I has radio-carbon dates of 2285 $^{+}$ 165 years B.P.

[74] Ibid.

TABLE 1

Star or Constellation	Present Azimuths	Azimuths in 300 B.C.	Stone Alignment	Error
1. Triangulum (Beta)	35° N of E	23° N of E	#1 → 7	0°
2. Pleiades	24° N of E	14° N of E	#1 → 8 → 12	0°
3. Aldebaran	17° N of E	9° N of E	#1 → 2 → 9	0°
			5 → 13	0°
4. Bellatrix	6° N of E	1° N of E	#1 → 3 → 4	1°+
			17 → 18	0°
5. Central Orion	2° S of E	10° S of E	#5 → 10	1°
			16 → 18	1°
6. Saiph	10° S of E	13° S of E	#5 → 6 → 7	0°
			14 → 18	0°
7. Sirius	17° S of E	17° S of E	#5 → 6 → 9	0°
			15 → 18	0°

have accumulated quite a store of stellar observations over a long period of time to have created an observatory of such accuracy using the azimuths of star-risings. Orion, quite naturally, was the central celestial landmark.[75] This constellation is the most striking one in the heavens, visible in most latitudes north *and* south of the ecliptic. However, what makes this Cushitic calendar and observatory practically unique is the importance of the northern constellation Triangulum whose risings, synchronized with the latter half of the lunar calendar, furnished a very accurate time-keeping mechanism. There may not ever have been another calendar quite like it anywhere.

The calendar to which the Namoratunga observatory is tied is as much lunar as it is stellar. One is therefore compelled to consider the significance of the 19th stone to which the two investigators could not assign any practical function. However, whenever the number 19 is encountered connected to anything lunar, the Metonic Cycle is inevitably called to mind. As we have seen, it takes the full moon (or any phase of the moon) exactly 19 solar years or 235 lunar months to reappear in the exact same day and month of the calendar year and the discovery of this cycle is usually attributed to Meton (5th century B.C.). Lynch and Robbins do not seem to have considered the possible Metonic character of this megalithic arrangement. However, at this stage of knowledge of Namoratunga II, the possible relation of these 19 stones to the Metonic Cycle must remain inferential.

The dates for Namoratunga II, and therefore of the Cushitic calendar related to it, take us back to 300 B.C., but the very nature of this calendar plants a germ of suspicion that it is immeasurably older. Its stellar-lunar nature is in itself powerful testimony to an advanced age. If Namoratunga II's connec-

[75] Bellatrix, Central Orion, and Saiph all belong to Orion.

Fig. 48. Namoratunga in northwestern Kenya. The group of tilted stones, 19 in all, represent part of an ancient observatory.

tion to the Metonic Cycle could be conclusively proven, that would demonstrate an interrelation with the solar calendar; but that inference aside, this calendar exhibits no significant solar elements. It is a calendar of a people who relied almost entirely on the stars and moon for their time-keeping and these facts alone make the whole thing pre-solar. It is well-known that adherence to a calendar, and the astronomical system that goes with it, can persist for many millenia. Though our Namoratunga dates are but 2,300 years old, the site may represent the remains of a system that had been handed down for ages, quite possibly from Late Paleolithic times.

Though the "ethnoastronomy" of other African peoples has not been extensively studied, a few ethnologists have managed to preserve some tradi-

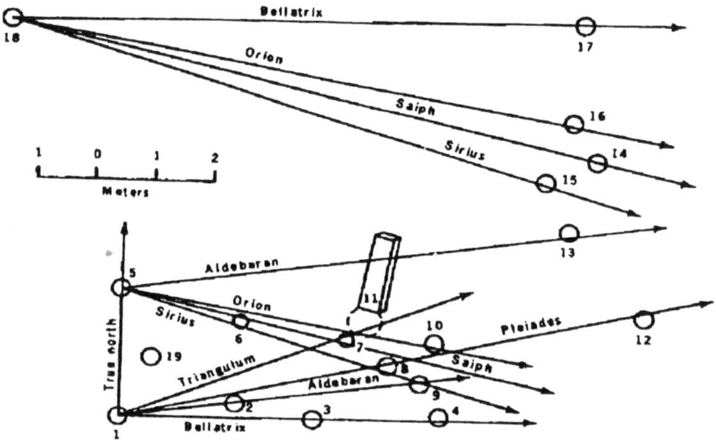

Fig. 49. Diagram of the astronomical alignments of the monoliths at Namoratunga.

TABLE 2

Mbochi Asterisms	Meaning of Astral Names	Contemporary Asterisms
1. Toli		Cepheus
2. Ebvunyukale		Cygnus (the Swan)
3. Kusu	"Tortoise"	Auriga (the Charioteer)
4. Mdzoabombo		Ursa Major (Big Bear)
5. Abvande asare abengi asare	"Three Dogs running with three hunters"	Perseus
6. Angondo tsambo	"Seven young virgins"	Pleiades
7. Kumambolo	"Star of the Shepherd"	Venus
8. Ondongo a dusu tso	"Red-eyed Pepper"	Jupiter[78]
9. Ndye ephoya l'otupha		Milky Way

tional astronomical data. The Fang of West-Central Africa (Cameroon and Gabon) possess a 12-month calendar and have identified several asterisms known to modern astronomy.[76] Their *Tolo* is the constellation Cepheus, *Tolo a Zok* is Ursa Major, *Mone Tolo* is Ursa Minor, *Nkienye* is the Southern Cross, and *Ndzong Bekuma* is the Milky Way.

Theophile Obenga has performed a valuable service for the ethnoastronomy of Africa by reproducing the traditional asterisms of several peoples of Central Africa.[77] The Mbochi of Congo have one of the more well-defined astronomies in the region (see Table 2).

There *are* other megalithic sites in various parts of Africa and Obenga has made the entirely sensible recommendation that they too should be investigated for possible astronomical alignments.[79] The limited material we have so far considered, whether at Namoratunga or in the Congo, suggests there is a vast, almost untouched ethnoastronomical terrain remaining to be explored in Africa. Like Namoratunga, such an exploration is bound to hold some revealing surprises.

Conclusion

The last three chapters, detailing the mathematics, architecture, and astronomy of Africa—with special reference to the Nile Valley—have featured a recurring leitmotif: the ancient scientists managed to weave all these technical disciplines into a common cloth. In the truest sense, each domain of knowledge could be compared to a facet of a diamond, imbedded in a whole that was greater than the sum of its parts. There is a theme of oneness, of

[76] See Obenga, op. cit., pp. 282-4.
[77] Ibid., pp. 282-99.
[78] This is an obvious reference to the large red spot of Jupiter.
[79] Ibid., p. 290.

interrelation coursing through the realized wisdom of old Africa; no "specialty" ever subsisted in compartmentalized isolation, no "expert" ever narrowed his lens on a single branch of learning. As we shall see in Chapter 6, this unified thinking informed the ancient *oikumene* ("ecumene").

The knowledgeable reader will doubtlessly note that no mention was made in this chapter of the Dogon and their astronomy. This was not an oversight; the ramifications of Dogon cosmological knowledge are so staggering that they require a separate treatment. We will meet the Dogon and their cosmology formally in Chapter 7.

Appendix

The Bauval-Gilbert Thesis

Recently a book written by Robert Bauval and Adrian Gilbert, entitled the *Orion Mystery*, has entered into the literature of Nile Valley astro-architecture, revealing for the first time the stellar alignments of the so-called "ventilation shafts" projecting north and south out of the Queen's and King's Chambers of the Great Pyramid.[80] Of no little importance is the authors' confirmation of the ancient Egyptians' knowledge of the Precession. They also affirm—and this makes perfect sense on the face of it—that the three major pyramids of Giza were constructed according to a unified plan. In effect, these ancient Nilotic builders had known before they started that they were going to build three pyramids, all according to a strict plan and orientation. The fanciful idea that the pyramids were merely monuments to pharaonic hubris loses whatever shred of credibility that may still be attached to it.

But central to their thesis is the demonstration by careful measurements, aided by state-of-the-art robotic technology, that the shafts in the King's and Queen's Chambers were astronomically aligned.[81] Bauval and Gilbert had doubted early on that the shafts were simply for ventilation because those projecting from the Queen's Chamber do not communicate with the outside at all. In any case, Rudolph Gantenbrink's angular measurements showed them that the southern shaft of the King's Chamber, at 45°, was oriented perfectly to Orion's Belt in 2,475 B.C. and the northern shaft of the King's Chamber, at 32°28', was oriented precisely to α-Draconis 2,425 B.C. The southern shaft of the Queen's Chamber, at 39°30', was oriented exactly to Sirius in 2,400 B.C. and the northern shaft, at 39°, accurately to the beta star of Ursa Minor.

These investigations have stimulated renewed interest in ancient Nilotic astro-architecture, even among certain segments of academic Egyptology. However, in spite of these impressive findings, a serious problem emerges in the Bauval-Gilbert thesis concerning the dating of the Great Pyramid. On the strength of the shaft alignments in the King's and Queen's Chambers, Bauval and Gilbert advocate a dating of 2,450–2,475 B.C. for the building of the Great Pyramid. Such a dating is far too low to seriously entertain for the Great Pyramid.

In order to appreciate the difficulty surrounding the low Great Pyramid dates advocated by Bauval and Gilbert it is necessary to consider the Descending Passage. Piazzi Smyth 130 years ago showed that the Descending Passage, opening on the

---

80 Bauval R and Gilbert A, *The Orion Mystery*, William Heineman, 1994.

81 Ibid., pp. 168–72; the authors teamed up with German engineer Rudolph Gantenbrink who was the first to explore the shafts with a miniaturized, wheel-mounted robot.

pyramid's north face, is inclined at an angle of 26°17' and at that angle would have been aligned exactly to α-Draconis in 3,340 B.C.[82] As seen in Chapter 3, this is about the time this star first moved toward the polar position. It would have crossed the same spot again around 2,170 B.C. as it was beginning its movement away from the pole in the course of the Precession. As Piazzi Smyth noted, the Great Pyramid therefore would have to have been built at or around one of those dates, 3,340 B.C. or 2,170 B.C. Piazzi Smyth favored the lower date but, given the evidence adduced by Bauval and Gilbert, a date of 2,170 B.C. is impossible unless the Descending Passage was cut *after* the pyramid was built.[83] But other evidence tells us that the Descending Passage *could only* have been cut at the time of the building of the pyramid, meaning that it would have to have been cut, and the Pyramid built, in 3340 B.C. or thereabouts.

How do we know this? It is uniformly agreed that the Descending Passage was hollowed out using the polestar a-Draconis as an orientation marker to true north. It also served to establish the polar meridians. As we saw in Chapter 3, Richard Proctor explained the method by which the angle of the Ascending Passage, 26°17', the *exact* angle of the Descending Passage, was obtained.[84] Using a pool of water sitting atop the plugged Descending Passage below where the Ascending Passage was to be cut, the light of Alpha-Draconis was reflected upward and oppositely to the (northward) Descending Passage, allowing the Ascending Passage to be cut southward, upward and away from, but at the same angle as, the Descending Passage. As Proctor showed, the Ascending Passage, ending as the Grand Gallery, debouched onto the 50th concourse and was used as an observational "telescope" of the southern heaven, *before the pyramid was completed*. In this, Proctor was corroborating the testimony of the Neo-Platonist Proclus who affirmed that the Great Pyramid had functioned as an observational platform for astronomers before its completion.

Unlike the northern polar regions of heaven, the southern polar regions have a dearth of stars that can serve conveniently as polestars. This is particularly true when looking southward from the northern hemisphere and there are long ages when the southern pole has no star to mark it. Thus, the method revealed by Proctor was the best and most efficient ancient method for obtaining the southern polar meridian. If the Descending Passage was oriented precisely to the north pole, which it was, the Ascending Passage, *of necessity*, would be oriented to the southern pole whether a star indicator was there or not. Orion and Sirius, though they inhabit the southern heaven, do not mark the southern pole. We can summarize all of the relevant considerations as follows:

1. The Descending Passage on the north face could only have been built in 3,340 B.C or 2,170 B.C.
2. The angle of construction of the Ascending Passage depended solely on that of the Descending Passage and it can only have been built at the time the Pyramid itself was built because construction was temporarily halted at the 50th concourse, where the Ascending Passage (Grand Gallery) let out, for astronomical measurements.

---

[82] Piazzi Smyth C, *The Great Pyramid*, New York: Bell Publishing Co., 1978, pp. 371-3.

[83] See ibid., pp. 374-9, for Piazzi Smyth's reasoning for preferring the lower date of 2,170 B.C. He links it to the double alignment phenomena that puts α-Draconis at the polar position aligned with the Descending Passage and the Pleiades directly above the apex of the pyramid at 2,170 B.C.

[84] See Proctor RA, *The Great Pyramid: Observatory, Tomb, and Temple*, London: Chattus & Windus, 1883, pp. 119-23, for this discussion.

3. If the Ascending Passage had to have been built at the time of construction so did the Descending Passage.
4. On the strength of Bauval and Gilbert's evidence, radiocarbon analysis of organic remains from the Great Pyramid, and a host of archeological and documentary proofs, 2,170 B.C. is far too late for a construction date for the pyramid, meaning that it must have been constructed in 3,340 B.C. with the Descending Passage serving to align the pyramid with the north polestar.

The point of emphasizing the coterminous construction of the two Passages with the Great Pyramid is that it follows, accepting the Bauval-Gilbert dates for the Chambers' shafts (circa 2,450 B.C.), that these shafts by contrast must have been cut *after* the construction of the pyramid. This conclusion becomes even more convincing when it is recalled, as discussed in Chapter 1, that radiocarbon dating of organic materials extracted from the Great Pyramid indicate an age at or older than 3,000 B.C. Thus, shaft dates of 2,450 B.C. do *not* indicate the construction date of the Great Pyramid itself.

The reason for cutting these shafts in the 25th century B.C. has been plausibly explained by Bauval and Gilbert. The orientation of the southern shaft of the Queen's Chamber toward Sirius shows that there was an "Isiaic" character to the proceeding; likewise, the orientation of the southern shaft of the King's Chamber to Orion reveals an "Osirian" inspiration. Though Osiris was an important *neter* from the beginning of Egyptian dynastic history, it is evident that his status—and that of Isis—began to be paramount, particularly in the funerary ritual, right around this time, i.e., the 25th century B.C.

It is in this way—the post-construction cutting of the Chambers' shafts 500 years afterward—that the disparate facts concerning the orientations of the passages and shafts can be reconciled.

6

African Navigation and Cartography in the Ancient "Oikumene"[1]

We start with the large Mixtec Negro head from Oaxaca. . . . The black color and the features, such as the thick lips, leave no doubt in anyone's mind that the artist has portrayed a Negro. No artist can "invent" authentic races of mankind, such as the types we are examining. The implication is simply that early America was the meeting ground of various races of men from the Old World who were eventually absorbed into the modern Indian populations.
—Cyrus H. Gordon, *Before Columbus*, p. 22

Tremendous strides have been made in amplifying our awareness of the neglected subject of African transoceanic voyages by the seminal studies of Harold Lawrence and Ivan Van Sertima between 1962 and 1976.[2] It had previously been assumed as a matter of course that Africans were incapable of ocean travel. The indispensible labors of Lawrence and Van Sertima show us instead that the subject of African navigation demands deeper investigation.

The subject has ramifications beyond simply the re-interpretation of African history, for there is a pronounced tendency to ignore or minimize the possibility or significance of the transoceanic voyages of *all* non-Europeans. The message clearly is that, prior to the Iberian voyages of the 15th century, *no* other people (with the possible exception of the Vikings) had ventured across the Atlantic *or* the Pacific. Mostly, the argument is a "common sense" one; few pre-modern peoples—in Africa or anywhere else—possessed sufficiently advanced watercraft to make transoceanic voyages. Where shipbuilding was developed enough, there was a dearth of navigational know-how.

---

[1] The word "oikumene" is Greek—rendered in English as "ecumene"—and means the "whole Earth" or, more loosely, "one world."

[2] Lawrence's classic monograph on the subject of African transatlantic voyages appeared in the June-July 1962 issue of the magazine *The Crisis* under the title "African Explorers of the New World." Van Sertima's *They Came Before Columbus*, published in 1976 by Random House, has emerged as the most definitive book on the subject of pre-Columbian African migrations to the New World. More recent articles on the pre-Columbian African cultural influence on Meso-America, including essays by Lawrence and Van Sertima, appear in *The African Presence in Early America* (1987), edited by Van Sertima and published by the Journal of African Civilizations Ltd. of Highland Park, New Jersey.

Where there was sufficient shipbuilding *and* navigational skill—as among the compass-using Medieval Chinese (800–1450 A.D.)—there was lacking sufficient daring or interest to cross the Pacific or the Atlantic.

These are all *a priori* judgments that crumble on closer inspection. They vastly overrate the kind of shipbuilding technique necessary for making craft capable of sustaining transoceanic travel. The peerless mariners of Melanesia, Micronesia, and Polynesia have for centuries been travelling across hundreds and thousands of miles of open ocean in out-rigger canoes, relying on their expert knowledge of wave-forms, currents, and the stars. After all, where *did* the original Hawaiians come from?[3] A molecular biologist, Douglas Wallace of Emory University, has shown that the mitochondrial DNA of certain groups of Amerinds of South America, whose ancestors arrived there 17,000 years ago, cluster *most* closely to Melanesians and other South Pacific Islanders.[4] Clearly, these aboriginal South Pacific Islanders *did not* arrive in South America via the Bering Isthmus; the transpacific route was the *only* way they could have arrived there: "This suggests that there may have been a second migration that did not come overland from Siberia, but by a coastal route from southern Asia, possibly even across the Pacific . . . "[5]

In recent years, demonstrations of the ocean-going capability of these ancient South Seas watercraft have been carried out. In 1976, a replica of a Polynesian sailing craft, named *Hokule'a*, was constructed and—without using charts or instruments—sailed between Hawaii and Tahiti. The feat was repeated in 1980 and, in 1987, a 16,000 mile pan-Polynesian voyage was undertaken and successfully completed by the *Hokule'a* which relied entirely on traditional navigational methods. The name *Hokule'a* represents the old Hawaiian name for Arcturus, one of the main navigational stars of the old South Pacific mariners.[6]

In this century alone, dozens of small, primitive craft, including rafts, dugout canoes, rowboats, and other assorted one- and two-man sailing craft have successfully completed transoceanic voyages. Some of these intrepid navigators had only fishing line, net, harpoon, and fluid containers and with these elementary accoutrements nourished themselves entirely from sealife and rainwater.[7] Sophisticated ships are not required for ocean travel.

Secondly, those who deny the possibility of pre-Columbian oceanic travel also overestimate the degree of navigational skill required. On Thor

---

[3] Hawaii was populated in waves from Tonga and surrounding islands, a distance of 3,000 miles from Hawaii over open ocean.

[4] Toner M, "America's Discovery: How the Indians Arrived," in *The Atlanta Journal and Constitution*, February 13, 1993, p. E1.

[5] Ibid.

[6] See Lewis D, *We, the Navigators: The Ancient Art of Landfinding in the Pacific*, Honolulu: University of Hawaii Press, 1994, for descriptions of the voyage of the *Hokule'a* and ancient South Pacific navigation.

[7] See Kehoe AB, "Small Boats Upon the North Atlantic," in *Man Across the Sea*, edited by Riley, Kelly, Pennington, and Rands, Austin: University of Texas, 1971, p. 276.

Heyerdahl's *Ra* expeditions, his crew on each occasion carried only *one* experienced seaman. This is not to imply that the pre-Columbian shipping lanes were filled with amateur sailors, quite the contary, but the ancient mariners were able to traverse immense distances with the heavens furnishing their only compass. Particularly in the southern Atlantic, powerful currents could have caused—and undoubtedly did cause—accidental drift voyages from the coast of Africa to South America. But we are concerned here with the purposeful transatlantic (and trans-pacific) voyages that occurred recurringly before the time of Columbus.

Finally, there is the assumption that ancient seamen, even those with ocean-worthy ships and superior navigational prowess, lacked the daring or desire to cross the Atlantic or the Pacific. This does not make sense in light of the voyages of the pre-Christian Phoenicians and the medieval Chinese. The Phoenicians were known to have sailed regularly through the Strait of Gibraltar to the British Isles for tin and also to have established colonies on the western coast of Africa as far south as the Senegal River. The distance between Tyre or Sidon (Lebanon) and the west coast of Africa is greater than that between West Africa and America. Moreover, Herodotus tells that, in 600 B.C. the pharaoh Necho commissioned a Phoenician captain to lead a fleet to circumnavigate Africa from east to west, a feat of seamanship lasting three years—accomplished nearly 2,100 years before Vasco da Gama sailed around Africa from west to east (1498). Furthermore, the legendary Chinese admiral, Cheng Ho, otherwise known as the "Three Jewel Eunuch," commanded hundreds of three-masted ships, some with five decks that, in a series of seven voyages, sailed throughout the western Pacific and Indian Ocean. In 1417, Cheng Ho crossed the Indian Ocean and landed in East Africa.[8] Even earlier, in the 11th century, the Chinese had invented the compass and mastered the technique of sailing into the wind with a swiveling mast. Basil Davidson tells us that by that time, "Chinese ships were technically capable of sailing wherever sailing ships could go."[9] Thus, from the 10th to the middle of the 15th century, China was *the* most accomplished maritime nation in the world.

It is beyond the scope of this chapter to rehash in full the arguments favoring pre-Columbian transatlantic and transpacific Old World contacts with America. Scholars of the caliber of Harold Lawrence, Alexander Von Wuthenau, and Ivan Van Sertima have already made the case on the basis of iconographic, ethnobotanical, and documentary evidence. We take the argument as proved or, at the very least, "not falsified." What we have said up to now is designed merely to remove the *a priori* objections to pre-Columbian transoceanic contact. Our main focus in the remainder of this chapter is to ex-

[8] Davidson B, *Lost Cities of Africa*, Boston: Little, Brown & Co., 1971, pp. 188–9.

[9] Ibid., p. 184.

plore African knowledge of the world and its dimensions through navigation and exploration.

Nilotic Shipping and Navigation

Sailing craft of large size are known from as far back as the pre-dynastic period in Nilotic history. The most ubiquitous early ship-building material was papyrus reed (the ocean-going capability of reed boats will be examined in detail when we consider the *Ra* expeditions) though wooden plank ships seem to have been known from the earliest times as well. But, clearly, papyrus reed boats were the earliest type to traverse the Nile, and the characteristic shape of these craft, with upturned bow and stern, became so ingrained in Nilotic boat-building methods that even the wooden plank ships of later ages were constructed in the "papyriform" style (see Figs. 50 and 51).

Be that as it may, a papyriform boat (presumably of wood) is depicted on a pre-dynastic vase with 40 oars and two cabins. Indeed, quite a few predynastic vases depict similar boats which may stretch back in time to 6,500 B.C.[10] These boats were employed for commerce between Egypt and Syria-Palestine. We know this because: "By the late Predynastic Period trade goods from Syria-Palestine are found throughout Egypt, and by the First Dynasty large beams of imported wood are being used in buildings, beams which would have been most easily transported by sea."[11] Similar boats also appear on artifacts from the pre-unification Ta-Seti dynasty and the earliest representation of a sailing boat in the Nile Valley is the one carved on the stone censer from Qustul (Ta-Seti) reported on by Bruce Williams (see Chapter 1).

Papyrus reed was used in every facet of shipbuilding. The body of the boat was constructed of ingeniously woven papyrus; sails, matting, ropes, and cloth were woven from papyrus rind. Papyrus is scarce in Egypt today, but in ancient times it grew in profusion along its river banks. These papyrus-reed boats were the most common of the smaller craft used for travelling up, down, and across the Nile, but, as we shall see, they could be and undoubtedly were used for ocean-going voyages as well. From the 6th Dynasty (c. 2,600 B.C.) there is a depiction of a papyrus reed boat with 32 oarsmen, showing that they often attained significant size.[12]

Wooden plank boats were constructed since before the dynastic period. The Nilotic techniques for construction of wooden plank ships are shown

---

[10] See Petrie WMF, *Prehistoric Egypt*, London: British School of Archaeology/Bernard Quaritch, 1920, pp. 3–4; 16–22.

[11] Vinson S, *Egyptian Boats and Ships*, Buckinghamshire, U.K.: Shire Publications, 1194, p. 20.

[12] Erman A, *Life in Ancient Egypt*, New York: Dover Publications, 1971, p. 480.

Fig. 50. Pre-dynastic ship with papyrus sail, circa 6,500 B.C. (Photo: Petrie)

to best advantage in the ships of Khufu, dating back to before 3,000 B.C., that were buried in four pits on each side of the Great Pyramid.[13] For reasons still not entirely clear, Khufu had full-sized wooden boats disassembled into more than 1,200 fitted components and hermetically sealed in the four pits adjacent to the pyramid that bears his name. The first—and to date only—such boat to be uncovered was disinterred in 1954.[14] When pieced together it measured 142 feet in length, 25 feet longer than Columbus's flagship, the *Santa Maria*. All the pieces of the boat were lashed together transversely instead of longitudinally as is usual with sewn boats; not a single nail was used to construct the ship. Typical of Old Kingdom boat-building, the ship was steered by two oar-rudders lashed to either side of the stern, and motor power would have been provided by oarsmen rowing 10 vertical oars—five to a side—lashed to the fore of the ship. The cabin was placed amidships and the craft was constructed of cedar wood in perfect "architechtonic" lines, evincing a knowledge of ship engineering as sophisticated as anything seen up to Columbian times.[15] The manner of its construction revealed that the ship itself was not fashioned for ocean-going voyages but its style and shape descended from prototypes that had been designed to sail the ocean.

In the words of Steve Vinson: "In the Old Kingdom, Egypt became a fully-fledged nautical power."[16] Fleets of ships were constructed to sail on the

[13] See El-Baz F, "Finding a Pharaoh's Funeral Bark," *National Geographic*, April 1988, pp. 514–33, for an account of the discovery and reconstruction of Khufu's pyramidal ship. Of the four pits around the Great Pyramid designed for the interment of pharaonic ships, only two retained their contents down to recent times.

[14] A second pit with a similarly disassembled wooden boat has been investigated, *without* being exhumed, by a photographic probe. Ibid.

[15] Ibid., p. 550; see also Heyerdahl T, *The Ra Expeditions*, New York: Signet, 1971, p. 20.

[16] Vinson, op. cit., p. 21.

Fig. 51. Late pre-dynastic Nilotic boat showing typical single square-sail.

river and the high seas[17] and numerous canals were excavated to allow these craft access to all parts of the kingdom. The First Cataract was dredged during the Sixth dynasty and it is thought that the capital of the Old Kingdom, Memphis, was connected by a canal at least 10 miles long to the pyramid site at Giza.[18] Later, between 1,200 and 600 B.C., one and possibly two canals to the Red Sea were completed, anticipating the building of the Suez Canal by de Lesseps by 2,500 years.[19]

Old Kingdom Egyptian ships could be rowed, paddled, or sailed, depending on wind and current conditions. From the Old Kingdom onward, Nilotic ships employed but one rectangle-sail, originally of papyrus cloth and later of linen, attached to a boom to preserve the rectangular shape (see Fig. 52). The linen sailcloth of Egypt was so durable and functional that it became an item of export to Phoenicia.[20] The early masts were usually created by binding two slender poles together, with strong ropes anchoring the mast fore and aft on the bow and stern. Cedar from Lebanon was used to build these ships and later (according to Herodotus), thorn wood was used to construct the larger cargo ships. Though, as we have seen, boats could be sewn together with strong papyrus rope, large plank ships built after the Old Kingdom were fastened with metal and wooden nails and bolts.

Ancient Nilotic plank ships were not constructed according to the keel-and-skeleton principle typical of wooden ships after the Medieval period. Dilwyn Jones describes a "shell-first" method in which the hull and sides of the ship were assembled with relatively short planks laid adjacent to each other in a brick-like arrangement and fastened together tightly with mortises-and-tenons and/or dovetails.[21] Jones thinks that this method was probably

---

[17] Vinson reports that the Sun Temple of the Fifth Dynasty pharaoah Sahure shows reliefs of ships "apparently designed for use on the high seas." Ibid., p. 23.

[18] Ibid., p. 8. Recent news reports indicate that at least part of this canal, lined with stone, has been uncovered.

[19] Ibid.

[20] Cited in Wilkinson JG, *The Manners and Customs of the Ancient Egyptians* (Volume II), London: John Murray, 1878, p. 209.

[21] See Jones D, *Egyptian Boats*, Austin: University of Texas Press, 1995, pp. 72-80, for a detailed description of ancient Nilotic boat-building methods.

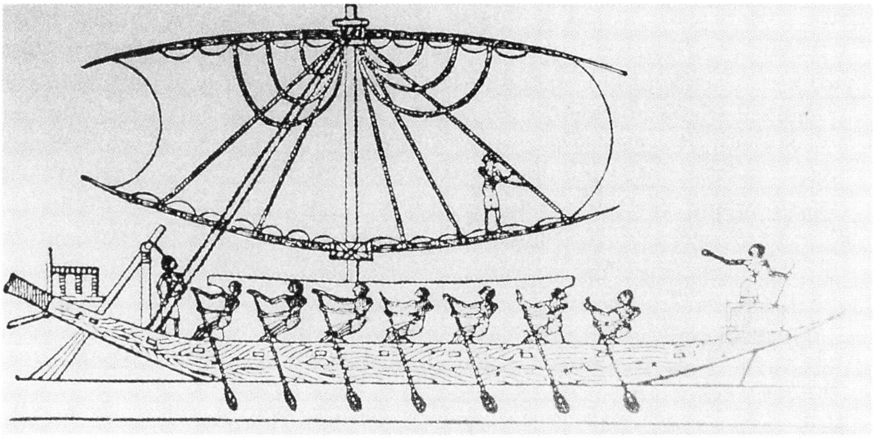

Fig. 52. Typical Egyptian single-sail ship also powered by rowers. A single helm is controlled by one man. This is evidently a river craft with the boatswain standing fore measuring the depth of the river.

invented along the Nile and, in that locale at least, was dictated by the dearth of tall trees. As noted above, Nilotic ship-building was heavily dependant upon the cedar logs (now all but extinct) imported from what is now Lebanon. Lateral stability of the finished craft was achieved by the insertion of a series of cross-beams from fore to aft. Longitudinal stability, in the absence of a keel, was assured primarily by the use of a hogging truss, described by Jones as,

> . . . a heavy rope which was attached to a cross-beam in the bows and then carried aft along the length of the deck on several forked stanchions and made fast to another cross-beam situated below the after-deck. A lever thrust between the strands enabled the crew to twist the rope like a tourniquet and apply the necessary tension to keep the overhanging bow and stern from sagging.[22]

By the Middle Kingdom, significant advances in shipbuilding were realized.[23] The ship's hull assumed a round-bottomed, spoon-shaped contour. Longitudinal stability was enhanced by a midline plank, spanning the ship fore and aft at deck level. The two-rudder system was replaced by one rudder that was easily manipulated by a single helmsman. The mast was erected as a solid single pole rather than two poles lashed together at the top. The sail was stretched between an upper and lower yard (boom) and fastened to the mast by moveable rope rings, allowing it to be raised or lowered at will. Further refinements in the New Kingdom period led to wider and longer

---

[22] Ibid., p. 41.
[23] See Erman, op. cit., pp. 485-7.

sails with more sophisticated rigging. More numerous ropes could be fastened to a round lath box attached to the top of the mast (see Fig. 53). The median cabin asssumed longer and higher dimensions. All of these innovations immensely facilitated Nilotic sailing skill and oceanic voyages in the Mediterranean, the Red Sea, and the Indian Ocean as far as Ceylon are attested.[24]

Nilotic sailing ships of extraordinary size and capacity were reported by ancient commentators. Some Egyptologists think that the barge of Hapshetsut, used to transport her famous obelisks, may have attained a length of 311 feet with a cargo carrying capacity of 7,300 tons.[25] According to Diodorus, the largest ship in the fleet of Senwosret (Sesostris) of 1,900 B.C. measured 480 feet in length (more than four times the size of Columbus's *Santa Maria*). This must have been the flagship in a fleet said to have consisted of 400 ships.[26] In Ptolemaic times (300 B.C.) particularly, substantial ships were a common feature of the pharaonic navy. A single ship of Ptolemy Philopator attained a size capable of holding more than 7,000 men.[27]

Clearly, until the end of the New Kingdom, Egyptian navies mastered both the Nile and the eastern Mediterranean. The earliest recorded naval battle occurred early in the Old Kingdom during the reign of the pharaoh Djer (Zer) who used his ships to war on truculent Nubian principalities to the south. Tomb scenes from the Sixth Dynasty official Weni show amphibious assaults on an enemy.[28] The most famous naval Nilotic naval victory was that won by the sea captains of Rameses III of the 19th Dynasty against the "Peoples of the Sea." These warring seafarers, the terror of the Mediterranean, represented a confederation of northern Mediterranean peoples who had conquered most of the coast of Western Asia before attacking Egypt. They were repulsed with great slaughter and no northern or eastern power attempted to invade Egypt for another 500 years. Diop thinks that remnants of the Peoples of the Sea scattered westward across the Mediterranean littoral of Africa and became the ancestors of the Berbers.[29]

The ships of the navy of Rameses III show innovations that set them apart from all previous Nilotic shipping. The hulls are longer and lower, sterns are more elevated to give the helmsman a better view, hogging trusses are no longer needed because the interior structure of the ships have been considerably strengthened, and sails are no longer attached to the lower yard

---

[24] See Wilkinson, op. cit., p. 208, for citation of a reference from Pliny testifying to Egyptian voyages as far east as Ceylon.

[25] Jones, op. cit., p. 65.

[26] Wilkinson, op. cit., p. 214.

[27] Ibid., p. 212. These ships, however, would have owed their inspiration to the Greek triremes, introduced into Egypt sometime in the 6th century B.C.

[28] Vinson, op. cit., p. 25.

[29] See Diop CA, *Civilization or Barbarism*, trans. from the French by YL Meema Ngemi, Brooklyn: Lawrence Hill Books, 1991, p. 34.

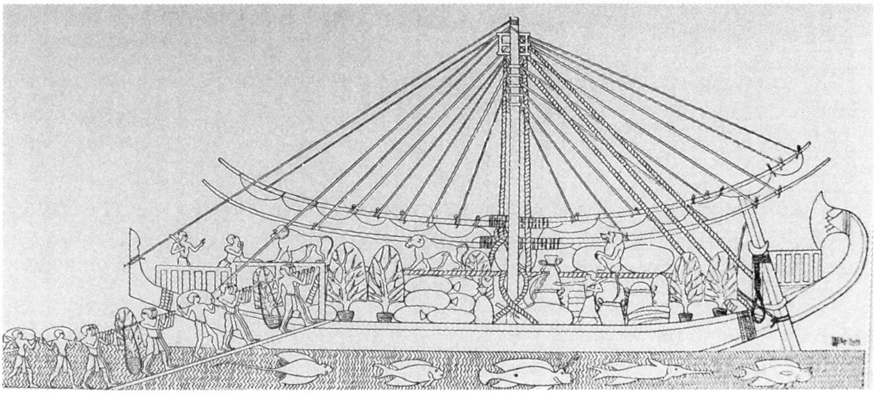

Fig. 53. The lading of a ship of Queen Hatshepsut, displaying the typical features of an 18th Dynasty sailing vessel.

but remain "loose footed." This latter feature allowed for better control of the sail from the deck and selective shortening of the sail for greater maneuverability.[30] Some of these advances had already been achieved prior to the time of Rameses III but it seems evident that national peril served as a catalyst for the consolidation and further refinement of these naval innovations.

Vinson observes that Aegean peoples like the Cretans and early Greeks were profoundly influenced by the shipbuilding methods of ancient Egypt.[31] Indeed, the ships of various seafaring nations of the Aegean and eastern Mediterranean are so close to Egyptian prototypes that it is difficult to avoid the presumption that shipbuilding and seafaring can be added to the list of skills and techniques that flowed northward out of the Nile Valley to the countries of the Mediterranean.

All that we have outlined above shows that the Egyptians were among the ancient leaders in designing ocean-worthy ships and in the development of seaborn navigation and sailing. Much of their early experience in shipping was undoubtedly riverine but they garnered sufficient experience upon the Mediterranean and the Red Sea to have made the necessary modifications of their river-craft to design ships capable of navigating the ocean. Their ships, particularly in the period between 2,000 and 1,200 B.C., plied the Mediterranean Sea and Indian Ocean seeking trade, and their eastward adventures carried them at least as far as India. After 1,200 B.C., their maritime enterprise was eclipsed by the rise of the Phoenicians, who, by that time, had become bolder and more active seafarers than the Egyptians. But, prior to the advent of the Phoenician era of maritime dominance, transoceanic voyages eastward or westward were entirely within the capability of Egypt.

---

30 See Jones, op. cit., pp. 59–60 for a fuller description of the remarkable navy of Rameses III.
31 Vinson, op. cit. pp. 37–41.

The *Ra* Expeditions

Which brings us the to the *Ra* voyages of Thor Heyerdahl and his crew.[32] Though eschewing the vitriol of academic partisanship, Thor Heyerdahl has been a thorn in the side of cultural isolationists ever since his *Kon Tiki* voyages in 1947 demonstrated the capability of ancient Peruvians of completing westward voyages across the Pacific on balsa wood rafts.[33] Heyerdahl in the 60s was struck, in the course of his travels, by the geographic range of reed boats from Africa to the Andes. Even in the modern era, users of reed boats could be found in Ethiopia, Chad, the southern Sudan, Sardinia, Morocco, Mesopotamia, and Bolivia. The earliest reed boats are depicted on the Egyptian monuments and vases and go back to pre-dynastic times. Heyerdahl realized that the continued usage of reed boats—at least in the Old World—could be discerned in those places that still revealed vestiges of an ancient Nilotic or Phoenician influence. New World reed boats duplicated those found along the Nile in both form and function. This would have been but an interesting curiosity were it not for other cultural parallels between Nilotic and Andean peoples.

In 1966, the journal *American Antiquity* published an article by an isolationist anthropologist detailing no fewer than *60* definable cultural parallels between the ancient Peruvians and the ancient Nilotes.[34] The author concluded, however, that since two continents and an ocean separated the two cultural sites, these findings constituted *prima facie* evidence that a whole complex of similar or identical cultural traits could be developed independently by peoples completely isolated from one another in time and space. The underlying presumption was that ancient peoples were incapable of traversing such distances by ship and thus, for practical reasons, contact was impossible. Indeed, as we have seen, it was these presumed practical barriers that formed the underpinnings of the isolationists' whole argument. Heyerdahl, in view of the astounding number of correspondences between cultures separated by 10,000 miles, was motivated, as he had been on the *Kon Tiki* voyages, to prove that "primitive" ancient ships were capable of crossing the Atlantic Ocean and establishing contact between otherwise widely separated peoples. In his mind, the parallel use of reed boats in the Old and New Worlds was one of the most conspicuous of the shared cultural traits.

---

[32] See Heyerdahl, op. cit.

[33] It is, however, of some interest that Heyerdahl never seems to have considered the possibility that a reverse migration to South America eastward across the Pacific could have occurred.

[34] Cited by Heyerdahl, ibid., pp. 39-40. The article in question was entitled "Diffusion and Archaeology," by J.H. Rowe in *American Antiquity*, January, 1966. As Van Sertima observed in *They Came Before Columbus* (p. 154), Rowe was not interested in merely refuting the diffusionist argument but ridiculing it by including trivial correspondences between the two cultures as part of his 60 parallels. But even after winnowing away the silly comparisons, some startling parallels remained, even to his isolationist eye.

A number of ancient authorities reported that the Egyptians had made extended oceanic voyages on reed boats. Egyptologists of the 19th century routinely disparaged the credulity of the ancient writers since it was "self-evident" that no reed boat, no matter how river-worthy, was suitable for sea voyages. Heyerdahl's *Ra* voyages showed with utter clarity the palpable dangers of armchair theorizing, of arriving at conclusions in the absence of demonstrable facts, of indulging in *a priori* reasoning. The experts told him that papyrus (1) could not float and (2) would, floating or not, dissolve in seawater within two weeks. True, in a laboratory tank a papyrus reed will sink; sitting in sea water a reed will dissolve. But Heyerdahl pointed out that a sheet of iron will also sink but that did not stop the *Queen Mary* from sailing on the ocean. What the experts had never done was actually construct a reed boat and test it on the ocean; several tons of reeds organized into a structural complex behaves very differently than does a lone reed, as Heyerdahl was to show.

For his first expedition, Heyerdahl enlisted the services of Buduma boat builders from Lake Chad, who still maneuvered about on reed boats. Heyerdahl's Buduma informants assured him that reed boats could be built to any size. Boats large enough to ferry 40 tons, 80 cattle, or 200 men across Lake Chad had been built within living memory. The actual papyrus used to build *Ra I* had to be airlifted from Lake Tana in the Ethiopian highlands; papyrus no longer grows in sufficient abundance along the banks of the Egyptian Nile.

Though the Buduma boat-builders no longer constructed boats exactly along Egyptian lines—the stern was always left "sawed-off"—they agreed to build *Ra I* to ancient Nilotic specifications under the direction of a specialist in ancient Nile boats. Heyerdahl describes how the boat took shape:

> Four rolls were . . . built into both sides of the longest central roll; the a second, similar layer of nine rolls was attached on top of the first. Over this an extra roll was added along each side of the deck as a bulwark. Three central rolls of the bottom layer were eight inches thicker than the others, and bulged out below, forming a sort of broad keel.[35]

The cabin was woven as a single wickerwork structure 12 feet long and 8$^1/_2$ feet wide, with a height allowing a modern average-size man to stand with his head bent. Both the bow and the stern were curved upright in classic Nilotic fashion, and from the curved stern was attached a cable fastened like a "harpstring" to the deck. The mast was anchored to the mid-section of the bow by an equally stout rope.

It was the "harpstring" of the stern that was eventually to cause problems

[35] Ibid., p. 156.

for the crew of *Ra I*. Try as they might, no one, not Heyerdahl, the boat specialist, nor the Buduma builders, could determine a function for the harpstring except to preserve the curved shape of the stern. There were no written instructions from the temple and tomb scenes to indicate its function and the Buduma builders eventually removed it, never to replace it, because it was getting in the way of their work.

The crew that Heyerdahl recruited for this adventure was also very revealing because, of the seven men, only one qualified as a professional sailor. Heyerdahl, on the strength of his *Kon Tiki* voyage, had some familiarity with sailing on the open sea but he could hardly be considered a seaman. Thus, the experiment was to be conducted with a crew composed largely of neophytes, not experienced mariners wise to the often capricious ways of the sea.

If anything, *Ra I* was overprovisioned. Over 100 ancient-style jars containing everything from salt meat and dried fish to eggs and assorted dried fruits were stored on the craft for the duration of the voyage. Condiments, spices, and seasonings of every description made the trip. Drinking water was also stored in sufficient quantities to sustain a voyage of several months.

The practical difficulties of the voyage became evident the first day out from the launching site at Safi on the coast of Morocco. The heavy wooden rudder-oars cracked from the pounding taken from the waves and swells. It proved to be a lesson not fully absorbed until after the second voyage—that of *Ra II*:

> . . . we had used too thick a rope to secure the rudder-oar top and bottom. We should have tied it with a thinner rope at the bottom so that the rope would snap, like a safety catch, before the oar shaft. Only upon our return did we learn that this was precisely what the ancient Egyptians had done, but this detail had been overlooked on the ancient design, since everybody had thought that the difference in thickness of the upper and lower ropes had been a mere coincidence in the artist's reproduction. Yet, a re-examination of the old designs showed that this difference in thickness was most consistent. *The Egyptians must have sailed in heavy seas.* (emphasis added)[36]

Moreover, after a night or two on the open ocean, Heyerdahl's crew learned to appreciate practically the ancient seafaring method of obtaining latitude without compass, sextant, or astrolabe:

> . . . they could get latitude by measuring the angle from the horizon to the Pole Star. . . . It's always 90° seen from the North Pole and the same star is right down touching the horizon when seen from the Equator. If you are at 60° north, the

[36] Ibid., p. 341.

> Pole Star is 60° above the horizon. . . . If you can see the Pole Star, you can read your latitude directly from it.[37]

But the real revelation for this shakedown cruise of Heyerdahl and his amateur seamen was the seaworthiness of papyrus itself. Strictly from the point of view of buoyancy and ability to float, papyrus revealed itself to be hardly inferior to wood; in some ways it was superior. Papyrus boats, because of their lightness, tended to ride the waves rather than knife through them. Moreover, papyrus swelled when exposed to seawater, obliterating the interstices and spaces between reed and rope so completely that water could no longer drain through them. At the same time, papyrus was incredibly flexible; the papyrus bent and swayed with the tossing and turning of the sea and never once broke apart, unlike the rigid wooden oars. The prediction that a papyrus craft would be dissolved by seawater proved utterly erroneous. The manner in which the reeds were bound, the absorption of seawater, and the consequent swelling kept the papyrus boat afloat for weeks on end. A reed only sank when it floated away from the papyrus bundles forming the ship.

All in all, everything in the experience of *Ra I*'s crew showed that reed ships like the one in which they were sailing had indeed been built to meet the demands of ocean sailing. This was especially appreciated when sailing under the rigging of *Ra I*:

> There was no doubt about it. The creators of the old Egyptian rigging had prepared their flexible reed boats for meeting with ocean swells. After the third day at sea I was already writing in the expedition journal: "This rigging is the result of long experience in navigation on open sea; it was not born on the calm Nile."[38]

Throughout this marvelous adventure, the crew of *Ra I* made many mistakes and had to learn numerous lessons the hard way. But despite their lack of experience, despite storms, despite sailing on a type of craft that had not been seen on the ocean for over 2,000 years, the only fatal mistake was the one made prior to launching the craft: the cutting of the stern's harpstring. *Ra I* hadn't been on the open ocean very long before the afterdeck began to sag. Not only did this sagging slow the craft's progress, but ultimately it caused *Ra I* to founder less than one week away from its destination in Barbados. The harpstring had not been designed to preserve the curve of the stern but to keep the afterdeck from sagging. It proved to the crew, as so many other things did, that every part of the reed boat structure served an essential purpose; nothing was superfluous or ornamental.

---

[37] Ibid. p. 188.
[38] Ibid., p. 219.

When Heyerdahl made the decision to abandon *Ra I*, the craft had been afloat for more than eight weeks and had covered 3,000 miles of open ocean. It had weathered storms, rough seas, and the relative ineptitude of an untried crew. *Ra I* had confounded all the experts and, incidentally, put a major dent in the dogma of isolationism. Ten months later, Heyerdahl set sail once again from Safi with a crew of eight in another, smaller reed boat—this one made by boat builders from Bolivia—and this craft, *Ra II*, with harpstring intact, completed the transatlantic to voyage to Barbadoes in 57 days. There was no foundering on this voyage; clearly reed boats could and very likely did make transatlantic voyages. There was now less and less reason to reject the conclusion that the reed boats of Lake Titicaca were remnants of the cultural baggage brought by ancient Nilotes from across the Atlantic Ocean.

Nilotes and Phoenicians: Commercial Partners in World Trade

As we have indicated above, up until the end of the 19th Dynasty (c. 1,200 B.C.), the ancient Nilotes designed, built, and manned their own ocean-going sailing vessels. Though they never entirely abandoned oceanic trade, for about 900 years after the 19th Dynasty most of the Nilotic carrying trade was in the hands of Phoenicians. The Phoenicians became the de facto merchant marine for Nilotic commerce.

The Phoenicians were themselves a heavily Niloticized Canaanitic peoples living in loosely aligned trading cities on the coast of Lebanon, the most important being Tyre, Sidon, Beirut, Byblos, and Joppa. Though the episode of Phoenician maritime domination of the Mediterranean begins about 1,200 B.C., the history of some of these cities stretched back more than 1,500 years previously. From 1,500 to 1,200 B.C. they were under the direct control of the pharaohs, which accounts for the preponderant Nilotic influence. However, after gaining political autonomy, the Phoenician trading cities quickly reached a *modus vivendi* with their still-powerful neighbor to the south: they handled all the maritime commerce of the Nile in exchange for access to luxury goods from inner Africa plus a measure of political and military protection. From that point on, the aims and destinies of the Phoenician cities and the pharaonic lands were closely intertwined. Apart from their role as the merchant marine of the pharaohs, they also served their patrons from time to time as diplomatic intermediaries and cultural emissaries to other nations around the Mediterranean. Their own daring and enterprise on the sea, coupled with this symbiotic relationship with the Nile kingdoms, brought incredible wealth into Phoenician hands.

There is a relative paucity of surviving information concerning Phoenician exploits; most comes from reports by Greek and Roman annalists. The Phoenicians, though avid record-keepers, were notoriously reticent about discussing trade, jealously guarding their knowledge of trade routes as state secrets against the encroachment of rival nations. In Carthaginian times a Phoenician captain discovered a Roman vessel shadowing him as his ship approached the Straits of Gibraltar with the obvious intention of discovering the sea route to the tin mines of Britain. Rather than risk an involuntary disclosure, the Phoenician captain ordered his crew to wreck the ship on the rocky shore of southern Spain. He and his men made their way overland back to their base in Cades where the reigning merchant-princes showered them with honors, replaced their lost cargo, and outfitted them with a new ship.

In spite of this Phoenician information black-out, several extended Phoenician ocean voyages are attested by classical authors.[39] The first is a report by Herodotus who was informed in Egypt that the pharaoh Necho in 600 B.C. had commissioned and outfitted a small fleet with a Phoenician captain and Nilo-Phoenician crew to circumnavigate Africa from east to west. This account is a curious case where Herodotus was skeptical of its veracity but modern scholarship has accepted it at face value. Mostly this hinges on the report that the sailors came to a point in their voyage where the sun crossed the sky to the right of them, exactly what would happen if a ship, moving in a westerly direction, rounded Africa at the Cape of Good Hope. The entire voyage took three years, showing that a Nilo-Phoenician expedition had circumnavigated Africa more than 2,000 years before Vasco da Gama.

Later, in the 5th century B.C., the Carthaginian mariner Hamilco reached the British Isles and established the highly lucrative tin trade. About the same time, his countryman Hanno passed through the Straits of Gibraltar on a southward voyage, planting colonies along Africa's western coast. Hanno seems to have reached as far south as present-day Gabon. Descriptions of crocodiles *and* gorillas lend credibility to the presumption that this voyage had reached that far.

What is even more fascinating from this perspective is there are at least six references in the writings of ancient Greco-Roman authors to the existence of lands westward across the Atlantic with mountains and navigable rivers.[40] No land mass other than America could possibly fit such a description, and in nearly every one of these ancient citations the Phoenicians are given credit for voyaging to these lands. The writings of Theompompus (4th century B.C.), Aristotle (4th century B.C.), Diodorus (1st century B.C.), Strabo (1st century A.D.), Plutarch (2nd century A.D.), and Aelian (2nd century A.D.) all contain passages indicating the existence of such transatlantic lands, undoubtedly obtained from Phoenician literary sources.[41]

---

[39] See Mathews SW, "The Phoenicians: Sea Lords of Antiquity," in *National Geographic*, August 1974, pp. 149–89, for an account of ancient Phoenician voyages.

[40] Ibid., 184.

[41] See Gordon CH, *Before Columbus*, New York: Crown Publishers, 1971, pp. 36–49, for a compilation of ancient citations testifying to the existence of transatlantic lands.

There is a small but determined body of investigators who take it as axiomatic that transatlantic voyages did occur in the millenia before Christ. Ivan Van Sertima, Alexander von Wuthenau, and R.A. Jairhazboy have brought forth abundant evidence testifying to linkages between Nilo-Phoenician interests and the Americas. Barry Fell has published two books detailing the evidence that shows pre-Christian Celtic, Greek, and Libyan contacts with America as well. There is, however, a slight difference of opinion between Jairazbhoy and Von Wuthenau on the one hand and Van Sertima on the other: the former assert that voyages from the Nile may have begun as early as 1,500 B.C. and no later than 1,200 B.C. (late Ramessid times), whereas Van Sertima, on the strength of the Olmec pyramids and statuary dating to the 8th century B.C., insists that such overseas contacts between the Nile Valley and Meso-America begin only with the 25th or Ethiopian Dynasty.[42] What seems evident in the voyages postulated either for 1,200 or 750 B.C. is that ancient Nilotes, with their cultural baggage, were carried to America on Phoenician ships with Phoenician captains and mixed Nilo-Phoenician crews.

But even within the small coterie of investigators working on these ancient contacts, another kind of controversy arose. Writers like Constance Irwin, author of *Fair Gods and Stone Faces* (1963), and others who echo her reasoning, like to think of the unmistakable Nubian types evident on the busts and statuary of the Olmecs as representing slaves belonging to Phoenician masters who initiated these ancient voyages. In refutation, Van Sertima quite simply says: "If the Phoenicians had made servants of the blacks, why were these so-called 'captives' and 'servants' given such prominence among the native Americans, dwarfing their so-called masters?[43] Rhetorically, Van Sertima asks, "Do people build monuments and altars and oracles to slaves which surpass in significance, size, and number those representing their masters?"[44] It again testifies to the intractable historiographic prejudice that labels all Africans found outside of Africa, no matter what the context, as slaves of some greater people that carried them there.

The famous Olmec busts of unquestionable "Negro" type, dating back to the 8th century B.C., constitute *prima facie* evidence that the Cushite or Ethiopian dynasty of Egypt is one of the ones that commissioned voyages, under the supervision of Phoenician navigators, that eventually landed in America. If Jairazbhoy is correct in his assumption that such voyages had also been launched 400 years earlier, one has to assume that Phoenician ships and mariners carried their Nilotic partners to America then as they would do again during the 25th Dynasty.

---

[42] See Van Sertima I (ed.), *African Presence in Ancient America*, New Brunswick, N.J.: Journal of African Civilizations, Ltd., 1987, for the most recent airing of these diverse points of view.
[43] Van Sertima I, *They Came Before Columbus*, New York: Random House, 1976, p. 152.
[44] Ibid., p. 153.

As Von Wuthenau and Van Sertima have shown, abundant statuary dating back to Olmec times show both Nubian and Semitic—hence Phoenician—phenotypes. Therefore, conclusions about transatlantic Nilo-Phoenician voyages can be predicated on: (1) the intimate partnership existing between Phoenician trading cities and the Nilotic kingdoms to the south of them; (2) the surpassing skill and incomparable daring of Phoenician seamen; (3) the documented record of their long oceanic voyages in the Indian and Atlantic Oceans. All the evidence considered, the Phoenicians, in partnership with their "Egypto-Nubian" neighbors, must have undertaken transatlantic voyages to America in the millenium before Christ. Everything in the Meso-American Olmec civilization points to such contact. It might be added parenthetically that, taking into account Heyerdahl's *Ra* voyages, some of the ships employed on some of the voyages between the Nile and Meso-America were very likely papyrus reed ships, which, as Heyerdahl proved, were as seaworthy as any wooden craft.

Van Sertima takes the position that the initial oceanic contacts between Africa and Meso-America were the result of accidental drift voyages and that even the initial voyage across the Atlantic launched by the Nilo-Phoenician partnership was not necessarily deliberate.[45] There is no reason to doubt that accidental voyages could and did happen but it seems difficult to postulate that they were culturally significant. The scores of well-attested cultural parallels between Old and New World civilizations, especially complex techniques such as pyramid-building and other examples of megalithic stone-working, the use of reed boats, and and the practice of arcane Osiris-like resurrection rituals, all point toward contact involving highly-trained technicians, artisans, scientists, and priests. It is difficult to imagine such personnel sailing on ships not specifically designated to carry out an expedition of trade, diplomacy, and/or settlement. Moreover, the pervasiveness and variety of Nilo-Phoenician influence stretching from Mexico to Chile might well suggest that periodic and recurring voyages set out from the eastern Mediterranean for America over a span of several centuries.

We should briefly mention the well-attested Meso-American traditions that plainly state that there had been visitors to the pre-Columbian New World from across the sea. These visitors were said to be of two kinds; one group was depicted as black-skinned and the other group as black-haired and pale-skinned. Thus, historical traditions on both sides of the Atlantic attest to ancient seafaring between the Old and New Worlds. It should be said, furthermore, that more recent scholarship has cast doubt on the purported descriptions of these visitors as "blond-haired," as these traditions were originally translated.[46] But these black-skinned and pale-skinned voyagers from across

---

[45] Van Sertima, *African Presence . . .* , op. cit., p. 15.

[46] See Van Sertima, *They Came Before Columbus*, op. cit., pp. 71-87.

the Atlantic would conform well to the evidence pointing to early Nilo-Phoenician voyages.

As already mentioned, work done by Barry Fell and collaborators in the 1970s provides strong evidence of pre-Columbian European voyages to North America, particularly by the Celts and Iberians. The ever-enterprising "western Phoenicians," the Carthiginians, left ample traces of themselves there also. But the evidence pertaining to Old World voyages has a certain geo-ethnic delimitation: the evidence for an African presence in the New World is very abundant in Mexico but relatively less so in North America; contrariwise, the traces of the ancient Celtiberians are most pronounced in North America but lacking south of the Rio Grande. The Phoenician presence, however, spans both New World territories.

Swahili Seafaring in the Medieval Period

In the Medieval period, the most active African mariners could be found in the Swahili coastal cities, those "fair cities of stone" along the East African coast that acquired great wealth between 800 and 1,500 A.D. trading goods of inner Africa up, down, and across the Indian Ocean. As one writer has stated:

> The development of navigation is closely bound up with fishing and the gathering of seafood. This development was to take two directions: advances in boat-guilding and improvement of seafaring techniques, particularly navigational astronomy. The latter could only have been developed by sailing the Indian Ocean, and there is every reason to believe that African sea-farers made their contribution to it.[47]

These cities belonged to a great commercial network that linked the Arabian peninsula, East Africa, India, and China in a mutually profitable trade that was sustained for nearly 800 years, until destroyed without scruple by the Portuguese. The commercial products of the East African interior, i.e., gold, steel, ivory, rhinocerous horn, sandalwood, ebony and leopard skins, coupled with coastal luxuries such as shells, pearls, and ambergris were shipped north- and eastward from the Swahili cities. Fine Ming porcelain, silks, and celadon from China, glassware and monochrome pottery from the Persian Gulf, and textiles and precious stones from India made their way west to the Swahili ports. These goods were consumed locally or re-exported into the African interior. This trade brought such affluence and economic development to these coastal cities that in the 14th century they began to mint their own coinage, replacing the traditional cowrie shells as the media of exchange.

---

[47] Matveiv VV, "The Development of Swahili Civilization," in *General History of Africa IV: Africa from the Twelfth to the Sixteenth Century*, edited by D.T. Niani, Berkeley: University of California-Heinemann- UNESCO, 1984, p. 457.

The ships used by the Swahili cities were patterned after the Arab *dhows*. These East African mariners had acquired enough navigational savvy and had designed such seaworthy craft that, in 1415, the Swahili city of Malindi was capable of shipping a giraffe all the way across the Indian Ocean and the South China Sea to the emperor of China. Indeed, Vasco da Gama was able to find his way across the Indian Ocean to India by employing the services of an East African pilot. He was to repay this favor the very next year by returning to East Africa with a fleet that burned, looted, and destroyed up and down the East African coast.

Van Sertima points out that the Swahilis employed a variety of ships and boats for fishing, coastal navigation, and ferrying.[48] Hundreds of these craft plied the coastal waterways of East Africa. The typical ocean-going craft of the Swahilis was called a *mtepe* and, as already mentioned, was patterned after the Arab dhows. Arab dhows could attain enormous size and the Arabs of this period were hardly less skillful mariners than the Chinese. Mtepes themselves could reach 70 tons and were capable of transhipping great cargoes across thousands of miles of open ocean. The mtepe was strictly a sailing vessel, but a smaller version, a *dua la mtepe*, was equipped with both sail and oars.[49] Since the Swahilis were linked to the wider Arabo-Islamic world and since they had established trading contacts with China, they very likely possessed the compass, an instrument invented by the Chinese that had passed into Arab hands. It should come as no surprise that there is convincing evidence that the Arabs too had made transatlantic voyages to the Americas by the 9th century.[50] Over and again it becomes ever clearer that the New World was linked to the shipping and trading web of the Old World for more than 2,000 years prior to Columbus, and East Africa participated in this global network:

> The maritime trade between the East African coast and islands and the countries on the northern shores of the Indian Ocean provided mutual links and contacts between the peoples concerned, and enriched them. It was part of a world-wide process, since it represented one of the branches of the major trade routes linking west and east.[51]

West African Transatlantic Voyages

East Africa was not the only coastal region of Africa involved in oceanic trade. Evidence has come to light pointing to West African voyages across the Atlantic at least by the 14th century, some 180 years before Columbus.

---

48 Van Sertima, *They Came Before Columbus*, op. cit. pp. 61–2.
49 Ibid.
50 Ibid., pp. 232–41.
51 Matveiev, op. cit., p. 465.

The evidence comes in several forms, two of the most important being documentary and metallurgical.[52] Mandinga annals from the 14th century record the launching of two expeditionary fleets westward across the Atlantic by Abubakari II, mansa of Mali. Moreover, Columbus's logs of his third voyage (1498) report that in the Cape Verde Islands *and* in Hispaniola, he was informed that black mariners from West Africa had been making regular voyages across the Atlantic. The presence of Africans in the New World was confirmed by samples of gold-tipped spears that came into his possession which, when tested in Spain, were found to consist of a special gold alloy manufactured *only* in West Africa.[53]

Though four different types of boats can be discerned in West Africa, they were all developed from the basic dugout template. The simplest dugout craft is made from a single tree-trunk that is hollowed out and shaped to conform to the requirements of moving about on the water. The typical dugout craft is manned by paddlers. A second type of dugout-derived craft is created by cutting down two trees, hollowing and shaping them into "half-boats," and then joining the two half-boats together.[54] This technique allows the construction of longer and wider boats. A traditional Western Sudanic boat of nearly 100 feet long, 14 feet across, drawing seven feet of water, was encountered nearly 200 years ago by Caillie on the Niger.[55] Boats such as this had decks, a central cabin, and were capable of carrying 80 tons burden. The ships called *almadias* of Mansa Abubakari II that crossed the Atlantic in 1312 must have been of a size comparable to and perhaps larger than these aforementioned craft.

The boats described by the first Europeans exploring West Africa do not mention sails; the ubiquitous dugout-style craft were all paddled. These boats were adapted for riverine use and on the Niger at least, sails were of scant utility. Michael Bradley brings forward evidence supporting the notion that large, well-manned canoes of 40–60 paddlers *could* have been propelled across the Atlantic and back.[56] If the ability of West African mariners to use sail is not admitted, then West Africans *must* have paddled their craft back and forth across the Atlantic. The natives of Hispaniola, cited by Columbus, conveyed the impression that West African traders were making regular transoceanic trips. Rowing or paddling across the Atlantic is not as far-fetched as it might seem. Hans Lindemann completed a 52-day voyage across the

---

[52] See Van Sertima, *They Came Before Columbus*, op. cit., and Lawrence, op. cit.

[53] Ibid. Van Sertima and Lawrence thoroughly cover this evidence in their respective writings.

[54] See Malloy SC, "Traditional African Watercraft: A New Look," in *Blacks in Science: Ancient and Modern*, edited by Ivan Van Sertima, New Brunswick, N.J.: Journal of African Civilizations, Ltd., 1983, pp. 163–76, for an interesting monograph on the subject of West African boats.

[55] Ibid., p. 169.

[56] See Bradley M, *The Black Discovery of America*, Toronto: Personal Library, 1981, pp. 92–119.

Atlantic in a dugout canoe in 1957, exploiting nothing except his own paddles and the westward currents that skirt the coasts of Western Africa and carry anything that floats inexorably to America.[57] The return voyage could take advantage of the Gulf Stream current that carries anything afloat northeastward toward the Iberian peninsula.

But Bradley and others think that fleets such as those launched by Abubakari II were most likely powered by sail. In fact, Bradley adopts the opinion that West African mariners could sail into the wind:

> I believe that the ancient Egyptians and black Africans were familiar with the principles of windward sailing because of the distinctive shape of their paddles. The blades were shaped like a greatly elongated heart and came to a point at the bottom. . . . The paddles common to Egyptian and black African peoples are too long and too narrow for truly effective paddling, but they do have a high aspect ratio. I believe their shape was a compromise so they could serve different functions: paddling and use as lee-boards.[58]

Probably the most amazing pre-Columbian West African seagoing expeditions were those of Abubakari II, mansa of Mali, ruling in the first decade of the 14th century. The story of this "Mariner Prince," whose consuming interest in overseas exploration anticipated that of Portugal's Prince Henry the Navigator by nearly a century-and-a-half, is recounted and analyzed closely by Van Sertima and Lawrence. He seems to have outfitted two different fleets with a total of 2,400 almadias to sail across the Atlantic. On his second voyage, he accompanied the fleet with much of his royal retinue, a commitment to sea-faring not even evinced by the more famous Prince Henry of Portugal. Abubakari never returned to his empire and his place was assumed by the fabled Mansa Musa, whose pilgrimage to Mecca—with its lavish displays of prodigious wealth—remains the most storied one in history. In fact, it was Mansa Musa who related the story of his predecessor's seafaring exploits to the Egyptian emir, Ibn Hajib, and this exchange was reported in afteryears by the writer Al-'umari.[59]

It is clear from the narrative that the fleet was composed of sailing vessels. These almadias may have exhibited some similarity to the ocean-going dhows of the Arabs but at all events were of sufficient size to carry tons of men, equipment, provisions, specie, and trading goods. The Mandinga presence in the New World left its traces in the language of Meso-America.[60] Also, as seen above, traces of West African gold metallurgy were alluded to by Columbus and, according to Leo Wiener, the myriad uses of pipes and tobacco in the

---

[57] See Kehoe, op. cit., p. 276, and Van Sertima, *They Came Before Columbus*, op. cit., p. 62.
[58] Bradley, op. cit., pp. 105–6.
[59] Cited by Lawrence, op. cit., p. 203.
[60] See Lawrence, ibid.; also Wiener L, *Africa and the Discovery of America* (3 volumes), New York: Kraus Reprint Co., 1971 (originally published 1920).

New World betray a West African influence. Lawrence and Van Sertima show that West African cotton goods were introduced into the New World and became a kind of currency.[61] Moreover, intact groups of blacks were found living in parts of the Caribbean and Central America by Spanish conquistadors invading the New World in the wake of the Columbian voyages.

It is not easy to tell whether Abubakari's expeditions represented the first Mandinga voyages but they were certainly not the last, since West Africans were travelling overseas to America up to Columbus's time. This fact alone lends support to the supposition that at least some of Abubakari's fleet made it back to Africa to tell the tale, probably loaded with enough trade goods to make further voyages attractive to adventuresome traders on Africa's west coast.

Mapping the World

In some sense, all that has been written in this chapter up to now has been preliminary to establishing the presence of an ancient world network. Says Cyrus Gordon, " . . . from the start, all high civilizations formed one ecumene."[62] This global diplomatic and trading network remained intact until the 15th century. Then, inaugurating what would later be termed the "Age of Discovery," the Iberian Peninsula sent out shock troops of conquistadores east and west pillaging and plundering. The Spanish and Portugese did *not* create the transworld network of trade, they merely dismantled the one already in place, traceable at least to a millenium before Christ, for one of their own devising for their exclusive profit. Two centuries later, the Dutch, English, and French were outmaneuvering the Iberians and, in taking over their mercantile routes, conducted "mopping up" operations, completing the European takeover of this transworld network, again exclusively for European profit. The work was done so thoroughly everywhere that, in time, the whole world entered into a kind of collective amnesia regarding the existence of the ancient web of global relations. Thus, until recently, Europeans could—without fear of contradiction—insist that they were the first to link the world by transoceanic travel and also that they were the first to map the globe.

One vital document reopens the entire question to a radical re-examination: the Piri Re'is Map (see Plate 21). Charles Hapgood, who must be the leading authority on the map, had this to say concerning its provenance:

> In 1929, in the old Imperial Palace in Constantinope, a map was found that caused great excitement. It was painted on parchment, and dated in the month of

[61] Lawrence, op. cit., pp. 234-7; Van Sertima, *They Came Before Columbus*, op. cit., pp. 180-91.

[62] Gordon, op. cit., p. 106. For definition of *oikumene* ("ecumene"), see Note 1 of this chapter.

Muharrem in the Moslem year 919, which is 1513 in the Christian calendar. It was signed with the name of Piri Ibn Haji Memmed, an admiral ("Re'is") of the Turkish navy known to us as Piri Re'is.[63]

Hapgood himself is that exquisite rarity: a full-fledged academic (formerly professor of the History of Science at Keene College in New Hampshire) possessed of a mind daring and original enough to entertain and explore hypotheses that took him far beyond the confines of conventional academic thinking. His other gift lay in his ability to take seriously the work of conscientious amateurs. He points out in *Maps of the Ancient Sea Kings* that Copernicus, Newton, and Darwin were *all* amateurs when they made their principal discoveries and only through continued work did they become "specialists" in the fields they themselves created. It is thus noteworthy that his intensive study of the map was done in intimate collaboration with a group of his own undergraduate students.

The Piri Re'is Map that comes down to the present comprises but the western half of a larger map and delineates in astonishing detail the western bulge of Africa and the eastern promontory of South America. What is more astounding is that South America and Africa are represented to scale by the correct longitudinal relationship to one another! European seafarers did *not* have instruments capable of accurately measuring longitude until the middle of the 18th century, nearly 250 years after the map was drawn. Piri Re'is's own commentary tells us that he drew the map from 20 source maps, some dating back to the time of Alexander the Great, and the map was constructed with the aid of mathematics.[64] Moreover, one of the source maps he relied on had been supplied by Columbus.[65] But no map originating from Columbus could have furnished *any* information concerning longitudinal coordinates between the Old and New Worlds.

The painstaking study undertaken by Charles Hapgood and his students to unlock the secrets of the origin of the Piri Re'is Map, as revealed in *Maps of the Ancient Sea Kings*, is a fascinating one to follow. They realized early that the Piri Re'is Map was closely akin to the famous "portolan" maps dating back to the 14th century in Mediterranean Europe. The old portolan maps were remarkable for their accurate detail of the Mediterranean and Black Seas and their surrounding land masses. These maps are criss-crossed by innumerable straight lines that radiate outward from certain centers on the map like

---

[63] Hapgood CH, *Maps of the Ancient Sea Kings: Evidence of Advanced Civilizaion in the Ice Age*, New York: E.P. Dutton, 1979, p. 1.

[64] Ibid., p. 2.

[65] Given the emerging enmity between Spaniard and Turk in the 16th century, it is hard to understand how a map drawn by Columbus would have come into Turkish hands. We do know, however, that one of Columbus's least endearing traits was his venality and he might have made several copies of his New World maps and sold them to the highest bidder, be he Genoese, Venetian, Portugese, or Turk.

spokes from the hub of a wheel. Indeed, these lines emanating from these circular centers evoke the compass. In some cases the centers were decorated in the fashion of compasses and the lines emanating from them were spaced evenly around the circle as on a compass, with each center containing a compass-like 16 or 32 points on its perimeter. It is therefore the prevailing opinion that these compass-like lines aided mariners in the Mediterranean to set their courses by the compass.[66] The Piri Re'is Map displays these same compass-like lines radiating from circular centers situated on the map.

Hapgood, after reviewing the work of ancient map specialist A.E. Nordenskiold, came to accept the latter's conclusion that these portolanos were but copies of maps much more ancient than the medieval period.[67] For one thing, the maps were far too accurate to have been constructed by medieval cartographers *de novo*. Secondly, over the 200 years that the maps were produced (14th to 16th century), they seem to have been suspended in time; that is, the maps in the 16th century are exactly the same as, and no more accurate than, the 14th-century maps. Finally, Nordenskiold could find *no* medieval unit of measure to fit the scale dimensions of the maps. Nordenskiold, though he was at a loss to explain the fact, could see that he was confronted with portolanos that were slavish copies of maps originally drawn before the time of Christ. The Piri Re'is Map was a special portolano and Piri Re'is himself—consistent with Nordenskiold's own conclusions regarding portolanos—had indicated that most of the 20 source maps had come down from pre-Christian Alexandria.

It has already been noted that the Piri Re'is Map is but a fragment of a more complete map that originally had depicted all of Africa, Asia, and the seas surrounding them.[68] Hapgood described the initial procedure undertaken by the ancient cartographer:

> We could see that the mapmaker had selected a center, which he had placed somewhere far to the east of the torn edge of our fragment of the world map, and had drawn a circle around it. He had bisected the circle four times, drawing 16 lines from the center to the perimeter, at angles of 22 1/2°, and he had also drawn in all the four possible squares, perhaps with the intention of using different squares for drawing grids for different parts of the map, where it might be necessary to have different Norths . . . the portolan design had the potentiality of having several different Norths on the same map.[69]

The Piri Re'is Map is not all diagonal compass lines; it does display a series of vertical and horizontal lines that form a grid on its plane surface. One

---

[66] Hapgood, op. cit., pp. 6-7.
[67] Ibid., pp. 7-9.
[68] Ibid., p. 15.
[69] Ibid., p. 15.

of Hapgood's students, Estes, was the first in the investigative group to reason that the vertical lines might correspond to meridians of longitude and the horizontal lines to parallels of latitude. The only way to prove it, however, required finding the precise coordinates of the five projection centers strung in a curved north-south line on the map's Atlantic Ocean. Indeed, the five projection centers were situated exactly on the western perimeter of the original circle drawn by the map's creator (see above quote). Hapgood noticed that lines extending eastward from the five projection points converged at a center off the map and he realized that it would be possible to determine the precise location of the map center by means of trigonometry. It meant, of course, that the original mapmaker had also used trigonometry to construct the map. This point was crucial because map historians had, on the whole, denied that portolanos such as the Piri Re'is Map had incorporated trigonometric projections.

It took Hapgood and his co-investigators nearly three years to solve the map's mysteries. They secured the fortunate collaboration of two sets of experts: (1) the cartographic staff of the Strategic Air Command (SAC) at Westover, Massachusetts, and (2) Richard W. Strachan, professor of mathematics at M.I.T. Trigonometric calculations to narrow down the map's center had to be repeated innumerable times by Strachan. Strachan's calculations made him realize that the Piri Re'is Map was constructed on the basis of plane instead of spherical trigonometry, the basis for the Mercator projection in cartography.

Though the trial and error method that forced itself on Hapgood's group became very tedious, certain features revealed themselves early in the search. It was determined from the outset that the map's center lay somewhere in Egypt, consistent with Piri Re'is's own assertion that a good number of his source maps dated back to the time of Alexander the Great. Alexander had built his imperial capital not in Greece or Macedonia, as might be expected, but in Egypt. With Alexander's accession to power in Egypt in 334 B.C., and for the next 300 years, the scientific and scholarly texts of Egypt were opened to the legions of Greeks eager for learning. Then, and only then, does the Greek scientific "miracle" begin.

Secondly, one of the five projection centers drawn as a wind rose on the map's Atlantic Ocean lay on a parallel that seemed to correspond very closely to where the true Tropic of Cancer lay. A line from this wind rose projecting due eastward would converge on a center in Egypt at or near 23°51' north latitude (for convenience, Hapgood took $23^1/_2$° N, i.e., 23°30' N, as the latitude of this convergence point). This is the latitude of Syene, situated at the First Cataract and representing Egypt's southern boundary in Alexandrian times. As the site of the most important temple of Isis, it was a major cult center in Egypt on the eve of the Christian era. It eventually dawned on Hapgood's group that the selection of the Tropic of Cancer for one coordinate of the map's center made extraordinary practical sense because the Tropic of Cancer is astronomically determined.[70]

After continued trial and error, false starts, and dead ends, the group discovered another astonishing fact: presuming Syene to be the center of the map, the perimeter of the circle drawn by the original cartographer passed very close to the position of True North. Like so many facets of this analysis, this discovery contradicted accepted academic opinion that had never considered the ancients capable of determining True North. This discovery, in hindsight, proved very apt: the North Pole is *also* astronomically determined. Even later, when it became evident that Syene was not the precise center of the map, it is still clear that the ancient cartographer was capable of finding True North.

Hapgood's team came to the point where they were ready to draft a grid onto the Piri Re'is Map to attempt to test its correctness and reliability. This, like the trigonometric calculations, had to be repeated many times over a period of two years in order to narrow down the errors caused by the initial assumptions. In this, the group was rendered invaluable assistance by SAC's cartographic section. Their next discovery revealed much about ancient methods of constructing maps of large parts of the earth's surface:

> We found that some of the positions on the Piri Re'is Map were very accurate, and some were far off. . . . We discovered that the map was a composite, made up by piecing together many maps of local areas . . . , and that errors had been made in combining the original maps. . . . Undoubtedly local maps had been made first, and these were gradually combined, at different times, into larger and larger maps.[71]

As Hapgood says, there was nothing unusual about this procedure since surveying the entire area covered by the Piri Re'is Map at one time would have required immense resources. The Piri Re'is Map is therefore a compilation of compilations: "What Piri Re'is apparently did was to combine this compilation with still other maps—which were probably themselves combinations—to make his world map."[72]

This procedure adopted by Piri Re'is produced some interesting errors and peculiarities. There are 900 miles missing from the east coast of South America, the Amazon River was twice drawn, and there were two equators. In other instances there were two geographic localities twice drawn but in different projections, the cause of the several Norths common to all portolanos.

---

[70] Because of the Earth's tilt when it revolves around the sun, the sun's position at noon appears to move between the Tropic of Cancer on June 21 and the Tropic of Capricorn on December 21. Thus, on June 21, the sun is directly overhead on the Tropic of Cancer, and on December 22 it is directly overhead on the Tropic of Capricorn.

[71] Ibid., p. 25.

[72] Ibid.

All of the anomalies discovered repeatedly convinced the Hapgood team that there was something amiss either with their own assumed center of the map or their designated radius for the original circle. It was only the endless recalculations by Strachan and redrafts of new grids by the SAC experts that enabled Hapgood's team to progressively eliminate the errors and draw closer to a more accurate map center. Finally, after several attempts to localize a center, they were able to find the correct one right at the meridian of 30° on the Tropic of Cancer, lying about 100 miles west of Syene, their earlier proposed center. Thus, it was determined that the center of the map could be localized at 30°E and 23°51'N (Hapgood's 23$^1/_2$°N). Now Alexandria, as we have seen in an earlier chapter, lies on the meridian of 29°54'E which, for all practical purposes, *is* 30° E, so the original mapmaker chose a point in Egypt whose coordinates lay on the meridian of Alexandria where it intersected the Tropic of Cancer. This choice makes perfect sense when we realize that the source maps perused by Piri Re'is go back *at least* to the time of Alexander (c. 340–330 B.C.), when he was building his new imperial capital Alexandria in Egypt. We have already seen that Alexander chose the particular location he did for his capital because it lay on the meridian that defined the western boundary of the Delta, i.e., of Lower Egypt. Clearly, Alexander and his Ptolemaic successors meant for Alexandria to be the new geodetic center of Egypt and so made the meridian upon which it sat the new prime meridian. What better location, according to the thinking of the time, to locate the center of a new imperial map than at the junction of the Alexandrian meridian and the Tropic of Cancer.

Hapgood was convinced that there were *pre-extant* source maps for the Alexandrian ones eventually employed by Pir Re'is.[73] Moreover, particularly after Strachan's analysis, it is difficult to imagine how any of these maps could have been drawn without using trigonometry. Thus, advanced cartography and the trigonometry on which it is based must precede Alexander. Conventional histories of science give Hipparchus (c. 130 B.C.) the credit for discovering trigonometry. But our earlier examination of the Rhind Papyrus and the applied mathematics of the Great Pyramid show instead that trigonometric functions had been worked out in the Nile Valley 2,000 years or more before Hipparchus.

Hapgood's team found that the Piri Re'is Map in its incorporation of the True Pole, the tropics, and the equator revealed knowledge about the dimensions of the entire earth though it displayed an error related to Earth's circumference of about 4$^1/_2$°. Hapgood called this the map's "*Eratosthenian error*" because Eratosthenes was an Alexandrian said to be the first to compute the size of the earth but whose calculations for the circumference of the Earth erred by 4$^1/_2$°. This led Hapgood to the conclusion that the

[73] Ibid., p. 27.

mapmaker drew his map based on the earth dimensions calculated by Eratosthenes. Once this realization set in, it became necessary to redraw the map grid yet one more time to accommodate the Eratosthenian error, producing the startling result that all the previous errors in longitude vanished! To Hapgood this could admit of but one conclusion:

> . . . the Greek geographers of Alexandria, when they prepared their world map using the circumference of Eratosthenes, had in front of them source maps that had been drawn without the Eratosthenian error, that is, apparently without any discernible error at all . . . the people who originated the maps possessed a more advanced science than that of the Greeks.[74]

Who else could the map originators be except the Nilotic predecessors (and teachers) of the Ptolemaic Greek savants, the very ones whose written scrolls and manuscripts now made advanced Greek learning possible?

There's more. Hapgood found that the original mappers of the coasts of Africa and South America depicted on the Piri Re'is Map had utilized a different length for the degree of longitude than for the degree of latitude. This accommodation meant that the square grid of the portolanos had been applied by cartographers upon maps based on another projection. These mapmakers knew that the length of a degree of longitude decreases from the equator to the poles and they accounted for this diminution of the longitudinal degree by spacing the latitudes ever farther apart when moving from the equator to the poles. This is *exactly* the technique used in the Mercator projection, the one upon which all modern cartography is based.[75]

There are many astonishing details about the Piri Re'is Map, not the least of which is the accurate rendering of latitude *and* longitude. Finding latitude is a relatively simple matter, easily mastered by anyone with sustained experience in seafaring. One need only find the angle between the horizon and the polestar. Longitude is a different matter and there was no reliable method for determining it in Europe until 250 years ago. Yet the Piri Re'is Map, compiled from sources in existence more than 2,000 years before the 18th century, displays a ready knowledge of longitude. The map itself shows details of nearly 100 recognizable geographical landmarks. New World islands such as Cuba, Jamaica, Hispaniola, Puerto Rico, the Falklands; rivers such as the Amazon, Rio Negro, and the Atrato are all easily identifiable. In Africa, Cap Vert (Dakar), Cape Juby, the Senegal River, the Canary Islands, and the Cape Verde Islands are clearly rendered. In Europe, Gibraltar, the Azores, and Brest (France) are well marked. What has astonished investiga-

---

[74] Ibid., p. 28.

[75] Actually, other projections have been devised in recent years to counteract the tendency of the Mercator projection to distort the sizes of the continents relative to each other. The most widely used non-Mercator projection is the Peters projection that allows the map reader to appreciate better the true relative sizes of the major land masses of the earth.

tors is that the northern edge of the Antarctic land mass has been delineated. Antarctica was supposedly discovered first in 1818 and, in any case, its northern land coast has been under its ice cap for several thousand years![76]

The Piri Re'is Map—originally, it must be remembered, a world map before being torn in half—is one comparable to the most detailed and accurate of modern maps, at least up to the 19th century. It proves that the ancient cartographers who constructed it possessed precise knowledge of the dimensions and shape of the Earth, its lands, and its seas. But, good as it is, the Piri Re'is Map is, in the opinion of Hapgood and his collaborators, *inferior* to the more ancient source maps from which it is derived. For one thing, the Piri Re'is Map exhibits at least three different projections, a circumstance created by the grafting together of component maps to construct a larger, composite map. Moreover, the original mapmaker(s) was guided by the calculations of Eratosthenes of the Earth's circumference, which produced a small but discernible error. The paradox of this fabulous map can be summed up in Hapgood's own words: "The Piri Re'is Map appears, therefore, to be evidence of a decline of science from remote antiquity to classical times."[77]

Conclusion

It is pertinent here to review material brought to light in previous chapters where we found that the ancient Nilotic mathematicians, architects, astronomers, and geodesists had (1) invented geometry and trigonometry; (2) determined through astronomic observation the poles, the tropics, the equator, the rate of movement of the heavenly vault in ascertaining the heavenly meridian; and (3) precisely measured and mapped the dimensions of all the lands under their polity. All of this work allowed them to establish 5,000-6,000 years ago meridians of longitude and parallels of latitude so accurate that the Great Pyramid, where all this data was encoded, became the geodetic center of the world and marked its prime meridian. Cities, nations, and empires far removed from the Nile established their own geodesy and located their own cult centers in relation to the Great Pyramid. It is clear, therefore, that all the tools and techniques necessary for accurate cartography—global, local, or regional—were in the possession of the ancient Nilotic learned priesthood.

With this chapter's review of ancient African navigation and shipping coupled with its examination of the Piri Re'is Map, a kind of synthesis is achieved. We are able to tie together thematically and empirically the gold and steel metallurgy that formed important trade items for African mercantile enterprises to the trigonometry, astronomy, geodesy, architecture, and

[76] Ibid., passim.
[77] Ibid., p. 31.

transoceanic navigation of Africa that are definably linked to the Piri Re'is Map. It could be said that our investigation of the Piri Re'is Map crystallizes all the work accomplished up to now in this book, as it is a version of more ancient maps created by Nilotic cartographers that reflected the dimensions and shape of the world. These considerations furnish an insight into a possible motivation for the voyage of the Nilo-Phoenician fleet commissioned by the pharaoh Necho in 600 B.C.: mapping the coastal outline of the entire African continent. Indeed, such a voyage of circumnavigation, lasting three years, *only* makes sense if it is assumed that it was sent to survey Africa and collect all manner of information about it.

Time and again throughout this work we have alluded to the existence of a world civilization that antedated the Christian era by hundreds and even thousands of years and was still more or less intact up to the middle of the 15th century. There was indeed an *oikumene* ("one world") encompassing the major cultural centers of antiquity. The nations and peoples of the world had painstakingly elaborated a mutually profitable global network of trade and diplomacy connected by overland *and* overseas routes. Even peoples peripheral to the centers of civilization participated in this oikumene as seen in the existence of the "silent barter" procedure that allowed everyone from Phoenicians to Arabs to trade with unfamiliar peoples in the remoter parts of Africa.[78] Europeans did *not* create transoceanic world trade; they merely superseded the more ancient system and, in the process, eliminated the global economic equilibrium that had abided until they burst upon the scene.

The question might well be asked, how is it we have so little specific knowledge of or data on this ancient system? The beginnings of an answer can be arrived at by considering the fate of the Library of Alexandria. The beginning of the end of the oikumene can be said to have started with the first burning of the Library of Alexandria in 44 B.C. by the legions of Julius Caesar. Though it was rebuilt and restocked, the library never regained its former eminence and it is no coincidence that the scientific achievements of the Greco-Roman world declined noticeably after 44 B.C. There were two additional conflagrations; the final one in the 8th century destroyed the library completely. Only a fraction of the collection once numbering 700,000 manuscripts survived, and much of this only in fragments. The destructive effects of this book-burning on all branches of learning is beyond comprehension. Gradually, the achievements of ancient world science became pro-

[78] In the "silent barter" method, a group of traders would bring their goods to a designated spot and withdraw, signalling their departure by drum rolls. The local people would bring an amount of gold to cover the price of the goods and withdraw, also with a drum signal. When the traders returned to the spot, if they were satisfied with the amount of gold left, they took it and went their way; if not, they would once again signal and withdraw to indicate that they desired more payment. This would go on back and forth until each party was satisfied. Never was there an attempt by either side to take the other's goods until the transaction was complete. Honesty in these dealings was scrupulously observed.

gressively a dimmer memory. The surviving body of Greek literature that managed to be passed down lent itself to the deepening perception that such learning as did exist in earlier times was Greek and Greek alone. In truth, this Greek learning was but a pale after-reflection of a sun that had already set.

Even so, the global trade, diplomatic, and scientific infrastructure had been erected on such solid foundations that much of the ancient system went on as before. The light of learning was kept alive in eastern civilizations such as India and China, and much that had become moribund was revived, particularly in the Arabo-Moorish world. But the eruption of European "manifest destiny," more active and aggressive than the extant older cultures, swept everything before it. Cultural deconstruction and book-burning were carried out on an unheard of scale. With the beginning of the Industrial Revolution, those learned voices in Europe such as Paracelcus, Copernicus, Kepler, and Newton that had expressed admiration and praise for ancient Nilotic thought were stilled. Thenceforth, Europeans began sedulously to cultivate the perception that they were and always had been ensconced at the center of world history and culture.

Picking up where the Greeks, Arabs, and Moors had left off, Europe produced its own prodigious learning, not realizing or wanting to realize that much of the knowledge brought to light, as if for the first time, constituted but a rediscovery of older knowledge and abandoned truths, insidiously covered over by the tides of history. It became easy to assume that before Europe's "Age of Discovery," knowledge of the shape and dimensions of the world had been nonexistent. It became "self-evident" that science, even in its ancient antecedents, was a European creation; that Europe had, in effect, created the "modern mind."

The extraordinary success of this auto-propaganda is measured by the extent to which the rest of the world was persuaded, brow-beaten, or coerced into accepting these reformulated paradigms of history. From 1500 on, African, Arab, Asian, and Amerind alike forgot that they had all once participated in the ancient oikumene, had all traded to their mutual profit, had in some sense shared a common knowledge base, and had been part of a continuous global circulation of people, goods, and ideas. The entire world was blinded by a collective awe of the imposing edifice of pan-European culture. But—paraphrasing the priestly Nilotic mentors of Solon—modern Euroculture is but the grandchild of a world-system immeasurably older.

7

The Myth of Matter: The Unity of Consciousness in the Dogon World System

The problem for us . . . is how the Dogon could have known a host of astronomical facts, all of which are invisible to the unaided eye. In particular, how could they have known about the existence of Sirius B? How could they have known of its incredible density? Its elliptical orbit? Its 50-year period? They have no business knowing any of this.

—Kenneth Brecher, "Sirius Enigmas," in *Astronomy of the Ancients*, p. 109

"They have no business knowing any of this." A strange coda on Dogon astronomical knowledge by an eminent M.I.T. scientist. Yet, given the entrenched Western attitudes toward African scientific knowledge, Brecher's bemusement is entirely consistent. Indeed, it is *not* altogether easy to know how to assess the Dogon, in many ways the most astonishing and enigmatic people in Africa. Their material culture can best be described as "postneolithic"; nor is there any indication that, *qua* Dogon, their culture has ever been anything else. According to their traditions, their migrations began from northeast of Mali about 1,400 years ago, finally leading them to their present habitat—a harsh, forbidding land abutting the Bandiagara Cliffs in southeastern Mali—some 700 years ago. Though clear links to the ancient Nilotic system are everywhere evident in their thought, they seem to have more or less independently elaborated ideas in astrophysics and cosmology that go well beyond what is known of ancient Nile Valley science. Such ideas are advanced enough to compare favorably with certain principles in Einsteinian physics and quantum mechanics. How could a "pre-literate," post-neolithic people arrive at such a high level of thought? How could such a complex system be created without benefit of telescope, microscope, mathematics, or any of the technical paraphenalia that undergird modern science? Brecher isn't the only one racking his brains over that one.

What also to make of the two ethnologists, Marcel Griaule and Germaine Dieterlen, who have brought the Dogon system to light? Griaule, the leader and presiding genius of the team, lived and worked with the Dogon for 25 years, though as he himself tells it, they did not let him in on the deeper aspects of their system for fully 16 years. Clearly, he was being tested, "vetted," as it were. Just as clearly, the Dogon elders had some larger purpose in revealing the profoundest details of their hidden system to outside view for the very first time.

But Griaule and his estimable colleague, Germaine Dieterlen, generally refrained from exploring or discussing the larger ramifications of the system they were uncovering. Perhaps they were merely following the accepted rules of field ethnology: observe, inquire, record, and report—do not interpret too much. Perhaps they knew the trouble they would incur if they attempted to elaborate fully on the epistemological implications of Dogon knowledge. It is impossible to believe, however, that they were entirely oblivious to what they had in hand. They must have known that, sooner or later, their classic work would force a paradigm shift in the history of ideas, not merely for Africa, but for the whole world. It could not have escaped them that they were looking into a pool of African cosmic ideas and seeing some of the most advanced Western scientific and philosophical concepts reflected back at them, concepts that Westerners had always assumed belonged only to themselves.

The Primal Seed: The *Po* and the Origin of Matter and Space-Time

> Their conception of the universe is based . . . on a principle of vibrations of matter, and . . . a general movement of the universe as a whole. The original germ of life is symbolized by the smallest cultivated seed—*Digitaria exilis* . . . This seed, quickened by an internal vibration, bursts the enveloping sheath, and emerges to reach the uttermost confines of the universe.[1]

There is probably nothing in Dogon cosmogony more fundamental than the *po*-seed (scientific name: *Digitaria exilis*). The *po* is many things but in all the many things that it is, it is always the smallest thing. In the Dogon oral records recounted by Griaule, we are told, concerning the World-Egg of Amma, the Creator:

> . . . the oval po was created, which placed itself invisibly at the center [of the Egg]. . . . It is said: "When Amma broke the egg of the world and came out, a whirlwind arose. The po, which is the smallest (thing), was made invisible, at the center; the wind is Amma himself. It is the po which Amma let come out first." Amma's creative will was located in the po, the smallest of things.[2]

Lest there be any doubt concerning the possible "atomic" significance of this facet of the po, we are informed that in the Dogon system, the po is "the

---

[1] Griaule M and Dieterlen G, "The Dogon," in *African Worlds*, edited by Darryll Ford, Oxford: Oxford University Press, 1954.

[2] Griaule M and Dieterlen G, *The Pale Fox*, Chino Valley, Arizona: The Continuum Foundation, 1986, p. 130.

image of the origin of matter."[3] Modern physics has arrived at the same conclusion: "In the structure and behavior of the very smallest things lies the explanation of all matter."[4]

The word *atom* comes from the Greek word meaning "indivisible." Democritus is usually given credit for articulating the concept that matter is composed of agglomerations of incredibly small "grains" or atoms. The concept has survived into modern physics but we also see the idea evident in the po of the Dogon. Furthermore, our modern atom is modelled as a kind of "micro" solar system, with the nucleus of protons and neutrons at the center, in the manner of the sun, around which orbit the electrons in the fashion of planets. Compare this to the Dogon description of the "smallest of elements": "This internal movement of the smallest of all the elements God created [the po] is reflected in space by the spiraling motion of the stars."[5]

The po signifies something more: it is the origin point of space-time itself. This idea is crucial to understand. In 1922, Edwin Hubble discovered that all galaxies are moving away from the earth in all directions at ever increasing speeds; moreover, they are moving away from each other in all directions in exactly the same fashion. This led cosmologists Alexander Friedmann and Georges Lemaître in the 1920s and later George Gamow in the 1940s to propose that the universe is expanding at ever greater speed and that, at one time, all the universe—all space *and* time—was concentrated in a kernel or speck smaller than a proton.[6] This condition is known as a "singularity," a condition of infinite density and curvature of space-time where the laws of physics no longer operate.[7] It was the explosion of this cosmic kernel—popularly known as the "Big Bang"—that inaugurated the creation of the universe.

The po is, in effect, this same singularity, this cosmic seed whose bursting caused the universe to form. We have already seen the Dogon refer to the po as the "image of the origin of matter." They further say: "Due to its [the po's] smallness, it is the image of the beginning of all things. 'All the things that Amma created begin like the little [seed of] po.'"[8] Moreover, the Big Bang—which the Dogon think of as the "Big Burst"—is explicitly attested by them:

---

[3] Ibid.

[4] Gore R, "The Once and Future Universe," in *National Geographic* (163)6, June 1983, p. 710.

[5] Griaule and Dieterlen, *The Pale Fox*, op. cit., p. 445.

[6] See "big bang model" in *The New Encyclopedia Britannica*, Micropaedia Vol. 2, Chicago: Encyclopaedia Brittanica, Inc., 1992, p. 205; also, Gore, ibid.

[7] See Hawking S, *A Brief History of Time*, New York: Bantam Books, 1990, p.49.

[8] Griaule and Dieterlen, *The Pale Fox*, op. cit., p. 131.

> The bursting of the po and the whirling of the spiral in the other direction made [Amma's seat] pivot. . . . Although Amma created the world, he made the po responsible for putting it into motion: by whirling and then acting as a spring, the po took along and then distributed all things in the universe.[9]

Further on we are informed:

> The bursting po is drawn with porridge of emme pilu on the north wall above the platform at the bago during the ceremony of the "tasting of the new millet" . . . . eight lines placed at the four cardinal points show the distribution of the primordial seeds in the realized universe.[10]

Thus, we have before us the Dogon version of the Big Bang singularity from which all creation emerged.

Steven Hawking shows us that, given its expansion, the universe must have begun with a singularity.[11] One might even say the universe must have begun in a "po-state." But Hawking then derives a very interesting conclusion from this primordial singularity:

> At the singularity, general relativity and all other physical laws would break down: one couldn't predict what will come out of the singularity. As explained before, this means that one might as well cut the big bang, and any events before it, out of the theory [of general relativity], because they can have no effect on what we observe. Space-time would have a boundary—beginning at the big bang.[12]

In effect, Hawking is saying that what happened at and before the Big Bang singularity has no practical bearing on what is happening in space-time now; it is "irrelevant" because it cannot be measured or determined by any laws that govern our present universe. The Big Bang singularity is therefore "outside" our existence.

The Dogon do not acknowledge such a limitation. The "existence before existence" is woven into the fabric of their cosmogenic system and the singularity just preceding creation is given tangible reality: "Now, the universe in Amma's womb was still outside of time and space, which were intermingled in a common state of valuation."[13] In the ancient Nilotic system, this "existence before existence" was also posited among the cosmogonists of that time:

---

[9] Ibid., p. 423.
[10] Ibid., p. 438.
[11] Hawking, op. cit., p. 50.
[12] Ibid., p. 122.
[13] Griaule and Dieterlen, *The Pale Fox*, op. cit., p. 201. Note that the Dogon mingle space and time in a way that is entirely "modern."

> Thus spoke the Lord of the Universe: When I became manifest in existence, existence already existed. I came into existence as the Existant, which came into existence at the First Occasion. Having come into existence as the Existant, I existed. Thus it was that existence came into existence, because I was before the first gods that I made, because I had priority over these first gods, because my name was prior to theirs, because I *made the first time* as I made the first gods. (emphasis added)[14]

That which "existed before existence" was *Nun*, the *primum materium*, the abyssal waters, "pre-matter," so to speak. Thus, in African systems, that which pre-existed our space-time is not unrelated to our current existence and is therefore fundamental to their cosmologies.

Signs, Symbols, and the Birth of Space-Time

Typical of African thought, the Dogon concept of space-time, its origin, and its evolution, is of a piece with their mytho-religious ideas. Central to this wholistic precept is the role of signs, the fundamental constituents of symbols. The Dogon say: "Just as Amma began the world by the sign, it is by destroying the signs that he will annihilate it."[15]

In the introduction to the present book, we found that the Dogon called myths *so tanie*, i.e., the "astonishing word." To the Dogon, only by means of the myths, as *so tanie*, can the coherent themes of creation be explained. In this way, we find that the signs are the constituents of symbols and the symbols the tissue of myths. When the Creator wishes to destroy the world, he will first take away the myths by destroying the signs.

In the Dogon explanation of things, Amma created and recreated the world twice over. He created the new universe in this fashion: "Starting from the primordial 'traces,' *bummo*, Amma first drew the 'marks,' *yala*, of a new universe inside his 'womb' or 'egg.'"[16] Analogous language is used in modern scientific writing to describe the formation of the universe: ". . . imagine the big bang. . . . A kernel of the cosmos. . . . researchers led by . . . George Smoot . . . had discovered primordial wrinkles, floating at the very beginning of time. . . . No more than wispy tendrils . . . the ripples . . . are no less than *the handwriting of God*."[17] George Smoot elaborated further in a book describing the monumental discoveries of his team:

---

[14] Passage from the Bremner Rhind Papyrus, transliterated into French by Theophile Obenga, *La Philosophie Africaine de la Periode Pharaonique*, Paris: L'Harmattan, 1990, p. 56. Rendered here from the French by the author (Finch).

[15] Griaule and Dieterlen, *The Pale Fox*, op. cit., p. 110.

[16] Ibid., p. 117.

[17] *Newsweek*, May 4, 1992, p. 76.

> The pattern of wrinkles I saw on the map was primordial. . . . Some of the structures represented by the wrinkles were so large that they could only have been generated at the birth of the universe. . . . I was staring at primordial wrinkles in time, the imprint of creation and the seeds of the modern universe.[18]

As we have already seen, the Dogon conceive of the generation of everything in terms of germinating "seeds":

> The transformation of *yala* into *tonu* (representational diagrams) marks a stage in the formation of the heavenly bodies—visible and invisible—which are the twins of seeds. It prefigures the spiraling worlds of stars that are to fill up the universe of Amma—infinite and yet measurable—when he will "open himself up."[19]

In a modern scientific description of the primal explosion of the universe into existence where matter is created and galaxies start to form, we find the same terminology: "Gravity draws matter to denser regions, which act *like seeds* for galaxies and clusters of galaxies (emphasis added).[20] More and more we find mytho-religious terminology creeping into the theorizing of modern scientists in order to give coherence to concepts that emanate from the investigations into the deepest nature of things. At the outer limits of modern science, myth and matter are merging.[21]

In the Dogon cosmic system, matter comes into existence spinning out of the "primal egg," after the po has burst.[22] This spinning motif seems to underlie the Dogon conception of how space-time comes into existence and evolves. Thus, they say: "Amma's creative will was located in the po, the smallest of things. Like a central air bubble, it spun and scattered the particles of matter in a sonorous and luminous motion, which, however, remained inaudible and invisible. It was less a word than a thought."[23] In the work of creation, the axis of the world emerges spinning: "One sums up this work [of creation] by saying of the egg: 'It spun, opened; the remnant was the . . . axis of the world.'[24]

Again and again, wherever creative forces are at work, there is a creative spinning bringing forth the fibers of matter. It is why the spider is a mytho-

---

[18] Smoot G and Davidson K, *Wrinkles in Time*, New York: Avon Books, 1993, p. 278-9.

[19] Griaule and Dieterlen, *The Pale Fox*, op. cit., p.128.

[20] *Newsweek*, May 4, 1992, p. 76.

[21] As if to acknowledge this truism, George Smoot wrote, "Society hungers for both science and mythology, and the big bang theory is where the two mingle most intimately." Smoot and Davidson, op. cit., p. 18.

[22] Recently the Hubble Space Telescope has sent back an awe-inspiring photograph of a star birth. The caption under the photograph printed in Newsweek Magazine, December 25, 1995/ January 1, 1996, after p. 95, reads: "Stars are born in...immense clouds of hydrogen gas and dust in the Eagle Nebula....Evaporating gas globules—or *EGGs*—may be the incubators of new suns."

[23] Griaule and Dieterlen, *The Pale Fox*, op. cit., p. 130.

[24] Ibid., p. 186.

type that is ubiquitous throughout West Africa. The language that Hawking uses parallels this Dogon idea of the spinning process of creation:

> [After the big bang] the universe as a whole would have continued expanding and cooling, but in regions that were slightly denser than average, the expansion would have been slowed down by the extra gravitational attraction. This would eventually stop expansion in some regions, and cause them to start to recollapse. As they were collapsing, the gravitational pull of matter outside these regions might start them rotating slightly. As the collapsing region got smaller, it would spin faster. . . . Eventually, when the region got small enough, it would be spinning fast enough to balance the attraction of gravity, and in this way disk-like rotating galaxies were born.[25]

These whirling and spinning forces of creation produce a wind, and this wind is crucial to the life of stars. The Dogon say: "When Amma broke the egg of the world and came out, a whirlwind arose. The po, which is the smallest thing, was made invisible at the center; the wind is Amma himself. It is the po which Amma let come out first."[26] And they say: "When life increases, it increases by whirling."[27] Now it has already been said that the po represents the smallest of anything; it is therefore the smallest of stars, i.e., those known as "white dwarves."[28] In this character it is called the *po pilu* ("white seed") which specifically refers to the first- and best-known of the white dwarves, Sirius B, the companion to Sirius. Again we find uncanny parallels to the best knowledge of modern science:

> As the degenerate [white] dwarf and its companion orbit each other, the dwarf's strong gravity pulls gas from the other star. This *swirling gas* forms . . . an accretion disk, which feeds onto the surface of the dwarf, wrapping it in an ever growing mantle.[29]

We find that the birth of any star, including our sun, is heralded by the "winds":

> Our sun, or any star of about the same mass, has a typical history. It begins as a cloud of gas and dust, or nebula. When enough matter falls to the center, the cen-

---

25 Hawking, op. cit., p. 119.

26 Griaule and Dieterlen, *The Pale Fox*, op. cit., p. 130.

27 Ibid., p. 135.

28 Strictly speaking, white dwarves are not the smallest of stars—neutron stars are. In a white dwarf, the mass of a sun is packed into a sphere the size of the Earth; in a neutron star, the mass of a sun is packed in to a sphere the size of a small city. Neutron stars thus possess incredible density and gravity. They spin on their axis at high speeds. There is, however, a suggestion in *The Pale Fox* material that the Dogon recognize the existence of a spinning body smaller than the *po pilu* ("white dwarf").

29 Gore, op. cit., p. 722.

> tral mass ignites and becomes a star, which may go through a "jet phase." *Stellar winds* blow gas and dust away from the star and, in the case of our sun, leave the planets in the clear. (emphasis added)[30]

The creative force of the whirlwind in the Dogon system appears again in their resurrection mythology. The eight great ancestors of all creation are called *Nommo*; one in particular, *Nommo Anagonno*, is represented as the Fish-Nommo that will be sacrificed to counteract disorder in the universe and renew creation. Concerning this sacrificed Nommo's relationship to the po, the Dogon say: "By its spinning the unwound po had poured everything it contained into the *Nommo*'s 'great ark'. . . . This ark was made from the rest of the [sacrificed] victim's placenta."[31] But this ark, according to the Dogon, symbolizes the entire world in time and space and eventually, to fill his creation, Amma causes the ark to descend, moving back and forth like a pendulum:

> While the ark was swinging to and fro, the fact that it was suspended from a chain made it pivot on its axis, back and forth. In this way, the line of its descent made a double helix, reproducing the very movement of life, *of the whirlwind* that quickened the first seed. This movement was maintained by the breath of the ancestors, as it passed through a nozzle. The nozzle is shaped just like this whirling respiration, called "spinning wind," which gave force to the "helix of descent." (emphasis added)[32]

A Universe of Universes

There is a school of quantum theorists that experiment with the "superposition of states." Without going too deeply into technical details, we can note that experimenters have been able to develop a device called a SQUID (*s*uperconducting *qu*antum *i*nterference *d*evice) that can detect electrons moving in more than one state; that is to say, existing in two different forms simultaneously.[33] These results have led to theories that leave an impression that modern physics has moved closer than ever to science fiction. The most intriguing one posits "multiple universes" existing in a "superposed" state relative to one other and that, at the quantum level, particles such as electrons can and do move through more than one universe.[34] Indeed, it is said that at the time of the Big Bang, these universes were all in a superposed state in the cosmic kernel and were created simultaneously at the time of this cosmic explosion. Thus, ours is but one universe among several (or

---

[30] Gore, op. cit., p. 717.
[31] Griaule and Dieterlen, *The Pale Fox*, op. cit., p. 446.
[32] Ibid. p. 463.
[33] See Horgan J, "Quantum Philosophy," in *Scientific American*, July 1992, p. 101.
[34] Ibid., p. 104.

many). These separate universes are self-contained but *not* entirely impermeable to one another. Indeed, there are some avant-garde theorists who attribute all manner of phenomena, ranging from schizophrenia to paranormal phenomena, to the existence of these "superposed" universes.[35] Compare these ideas to the Dogon explanation of Amma's work of creation:

> The number 14, attributed to the spiral stellar worlds produced by Amma, implies the concept of potential reproduction and multiplication: seven is the sum of three, the masculine number, and four, the feminine number. Here it designates the seven Earths and seven Heavens. Amma creates 7 × 2, that is to say, an infinity of worlds. *The superposition of worlds and the concept of the infinite multiplication of stellar universes* are indicated by the fact that the number 28 is attributed to Amma as well; it is called "Amma's number," because one also says: "Amma fourteen above, fourteen below, etc. . . . " The spiral worlds of stars were populated universes; because simultaneously with things, Amma, having given form and movement to the world, created all living beings.[36]

All creation, as we have seen, comes out of Amma's Egg which is the po. Just as the multiple universes postulated by quantum physics are superposed at the point of the Big Bang, so too are the multiple universes that emerge out of the bursting po superposed. What is of surpassing interest here is that the Dogon insist that there is intelligent life elsewhere in our universe and in other universes. Physicist Fred Alan Wolf articulates the same idea:

> What is a parallel universe? Like an everyday universe it is a region of space and time containing matter, galaxies, stars, planets, and living beings. In other words, a parallel universe is similar and possibly even a duplicate of our own universe. Not only in a parallel universe must there be other human beings, but these may be human beings who are exact duplicates of ourselves and who are connected to ourselves through mechanisms explainable only by using quantum physics concepts.[37]

The Cosmic Weaver and the Web of Creation

Among a number of people in West Africa, the spider holds an important place in their mythologies. Among the Akan, for example, there exists a whole constellation of tales around Anansi, the Spider. The spider plays a special role in Dogon cosmogenesis as well. Among the Akan, *Anansi* the Spider is a trickster; among the Dogon, *Dada* the Spider is allied to the "Prince of Tricksters," the source of all Chaos and Disorder in the universe, *Ogo* the Fox. Dada, spins the Fox's thread and assists him in his rebellion

---

[35] See Wolf FA, *Parallel Universes*, New York: Simon & Shuster, 1988.
[36] Griaule and Dieterlen, *The Pale Fox*, op. cit., pp. 193–4.
[37] Wolf, op. cit., p. 21.

against Amma's creation. But this rebellion—this creation of chaos—is, in reality, entropy, the force that pushes the universe ever outward. We know that chaos is intrinsic to the universe for it is the undifferentiated "stuff" from which material space-time arises. In the Dogon scheme, the spider attaches himself to the original Tree, the *acacia*, in order to "weave the words of Ogo":

> Placed at the center of the acacia, the spider wove its threads in a conical spiral for the placement of the warp, and by moving vertically for the coming and going of the passing woof. . . . Through the fabric, it sifted the germs that had been hurled into the universe by spinning. The sound of its work, of the same nature as the word, grabbed them as they passed by and, with them, fastened itself at the crossing of the threads.[38]

The work of the spider has echoes in the "string" theories of physics that have developed in earnest since 1984. In this model, matter is not composed of particles but rather of strands that have no dimension except length. These "strings," as they are called, may have ends (i.e., "open" strings) or may join at the ends (i.e., "closed" strings). In particle theories, the history (or movement) of a particle is represented by a line, but in string theories, matter in the time dimension ramifies differently:

> A string, on the other hand, occupies a line in space at each moment of time. So its history in space-time is a two-dimension surface called the world sheet . . . the world sheet of an open string is a strip. . . . The world sheet of a closed string is a cylinder or tube.[39]

This "world sheet" of string theory is equivalent to the "fabric" woven by Dada the spider (see Fig. 54). But the correspondences are even more specific:

> In string theories, what were previously thought of as particles are now pictured as waves travelling down the string, like waves on a vibrating kite string. . . . The idea was that particles like the proton and the neutron could be regarded as waves on a string. The strong forces between the particles would correspond to pieces of string that went between the other bits of string, *as in a spider's web*.(emphasis added)[40]

This web imagery is constantly echoed by quantum theorists:

---

[38] Ibid., p. 236.
[39] Hawking, op. cit., p. 159.
[40] Ibid., p. 161.

> Quantum theory has shown that particles are not isolated grains of matter, but are probability patterns, interconnections in an inseparable *cosmic web*. Relativity theory . . . has made these patterns come alive by revealing their instrinsically dynamic character. (emphasis added)[41]

Again and again, we find mythic imagery being employed by investigators on the cutting edge of science to describe newly discovered properties of matter. Again and again, we find the "web imagery" appearing in the physical descriptions of the universe:

> A nice image . . . is the way cobwebs, often unseen in ordinary light, become strikingly visible when dew that settles on their strands during the night is lit by the morning sun. The gossamer network of galaxies we see in the night sky is the shimmering dew on a cosmic cobweb, as visible matter outlines the shape of structures of invisible dark matter, to which it has been drawn by gravitational attraction.[42]

Thus, in content and in form, ideas such as these show themselves to be astoundingly close to Dogon thought, a thought ordinarily labeled "mythic."[43]

Dogon Star Systems: The Flow of the "Blood" of Heaven

> . . . in regard to the construction of the universe, the sacrifice of the Nommo determined the creation and the path of the stars, evidence of the blood, of the clavicular seeds and spiritual principles, and of the vital organs when the body was divided up. Summing up this series of events, one says: "Amma was in the po, the seven grains came out of it in spirals. Because of the sacrifice of the Nommo, the stars and seeds were strewn throughout space. . . . In the beginning the stars walked in the sky by following the sacrifice of the nommo.[44]

> Star streaming: In their passage around the Galaxy, stars move in parallel groups called star streams, whose paths cross so that the stars intermingle like marching bandsmen in a military display.[45]

The locution, "As above, so below; as below, so above" constitutes the expression of the Hermetic Principle known as the Planes of Correspondence.[46] In essence, it means that existence mirrors and re-mirrors itself on many different planes; the microcosm is the measure of the macrocosm and

---

41 Capra F, *The Tao of Physics*, Berkeley: Shambhala, 1975, p. 203.

42 Smoot and Davidson, op. cit., p. 286.

43 See Capra, op. cit., for an extended discussion of the closeness of quantum theory to other Eastern systems of thought, i.e., Hinduism, Buddhism, and Taoism.

44 Griaule and Dieterlen, op. cit., pp. 309–10.

45 Moore P, editor, *The International Encyclopedia of Astronomy*, New York: Orion Books, 1987, p. 415.

46 See *The Kybalion* by the Three Initiates, Chicago: Yogi Publication Society, 1930.

Fig. 54. The work of the Spider, spinning the fibers of the universe.

vice-versa. In Chaos theory, the principle of the Planes of Correspondence is expressed as the law of "scalar self-similarity," as we have seen in Chapter 3. The origins of the Hermetic Sciences and Principles can be traced to the dawn of high culture in Africa; indeed their origin may ramify back into Paleolithic times.

In the Dogon system we find the Planes of Correspondence constantly manifested. The seed-stars are exactly equivalent to seed-plants. Amma's Egg, from which space-time emerged, is not inherently different from the female ovum from which springs human life. The ritualistic blood flow of the sacrificial victim that, in death, renews life is but another version of the cosmic current of blood generating and conveying the heavenly bodies in their perpetual streaming through the universe. In sum, depending on which way you are looking, the universe is man writ large or man is the universe writ small.

Recent advances in astrophysics now show that the universe is actually a closed system, meaning that its expansion will eventually stop some billions of years hence and its contraction will begin. Eastern philosophies have already conceptualized our universe as a kind of respiratory system; it expands on inspiration and contracts on expiration. This is consistent with African philosophies such as that of the Dogon whose probing of the universe has shown them a larger version of man there.

It is therefore most intriguing to consider the Dogon conceptualization of the heavenly systems. They recognize an "internal" system of stars—undoubtedly the solar system—that they say corresponds to the flow of blood on the placenta of the sacrificed nommo. As related in *The Pale Fox*, this internal system "is considered to be the driving force of the stellar world, directly influencing the life of man and his development on Earth; it plays a part much like the system of internal organs in the human body."[47] The "external" star system is described as the "path of blood" outside the placenta and is made up of stars much further away, i.e., outside the solar system. This external system of stars also influences life on Earth though not as powerfully as those of the internal system. This external "path of blood," according to the Dogon, includes the Milky Way, "itself the image of the spiral of stars within the 'spiraling star world' in which Earth is found."[48] The Milky

[47] Griaule and Dieterlen, *The Pale Fox*, op. cit., p. 345.
[48] Ibid.

Way, like the preponderant number of galaxies, is indeed a spiral galaxy as the Dogon attest. The "external path of blood" seems to correspond to the orbit of the Milky way around the center of its cluster of galaxies, or, perhaps, the cluster itself around a supercluster center yet farther out in the universe.

The Dogon distinguish between the fixed stars and the planets. The fixed stars are termed *tolo*; the planets are called *tolo tanaze*, i.e., "stars that move across" or "stars that walk separately." Compare this to the Nilotic term for planets, *dja*, meaning "traveller" or "journeyer." Note also that the Greek word *planetoi*, giving us our word "planet," means "wanderer." The Dogon recognize that planets have revolutions and that they in turn have smaller satellites or moons, *tolo gonoze* ("stars that make a circle"), that revolve around them. Indeed, they declare that the planets revolve around the sun, showing categorically that they know the solar system.[49] These planets are the blood of sacrificed things and they circulate in heaven the way blood circulates in living beings. We can list Dogon planets, stars, and constellations as shown in Table 3.[50]

Orion and the Sirian Star System: Nursery of Stars

> The group of stars in which Sirius plays a primordial part includes the constellation of Orion and a certain number of stars located nearby.[51]

As in the Nilotic astral system, there are no asterisms more important to the Dogon than the constellation of Orion and the binary system of Sirius, the Dog Star (see Plate 55). Now Orion's Chariot, otherwise known as the Trapezium, is called "the seat of Amma's foundation" by the Dogon. The Dogon say of it: " . . . in stellar space, the Chariot is the symbol of Amma's seat; it surrounds *atanu*, the Belt. . . . It is related to the basic elements and to the cardinal directions, which presided during the realization of the universe that was 'thought' by the creator."[52] This is interesting considering that the Orion Nebula is considered the birthplace of stars, veritably a "cosmic nursery."[53]

Which brings us to Sirius B. No aspect of Dogon knowledge has caused as much of a stir and created more astonishment than their descriptions of the properties of the star known as Sirius B, *po tolo* to the Dogon. Sirius B,

---

[49] Ibid., p. 347; also p. 514.

[50] These lists are taken from *The Pale Fox*.

[51] Griaule and Dieterlen, op. cit., p. 500.

[52] Ibid., pp. 503–4.

[53] See Mitton J and Mitton S, *Discovering Astronomy*, New York: American Museum of Natural History, 1982, p. 82. The authors report that a completely new star was seen to switch on in the Orion Nebula in 1978. This is one of the rare examples of the birth of a star actually being witnessed.

TABLE 3

Dogon Planetary Names	Meaning of Planetary Names	Contemporary Names
Tolo Tanaze	"Stars that move across"	Planets
1. Yazu Danala Tolo	"Star accompanying Venus"	Mercury (??)
2. Yazu	"Morning"	Venus
3. Yapunu Tolo	"Star of menstruation"	Mars
4. Dana Tolo	"Star of the fontanel"	Jupiter
5. Ani Tolo	"Star of the Hibiscus"	Saturn

Dogon Satellite Names	Meaning of Satellite Names	Contemporary Names
Tolo Gonoze	"Stars that make a circle"	Satellites (moons)
1. Ie Pilu		Moon (of Earth)
2. Gerelle Tolo	"Star of the *gerelle*"	A satellite of Jupiter
3. Bala Tolo	"Star of the *bala* tree"	A satellite of Jupiter
4. Sene Tolo	"Star of the acacia"	A satellite of Jupiter

Dogon Names of Fixed Stars	Meaning of Star Names	Contemporary Names
Tolo		*Stars*
1. Aduno Giri	"Eye of the world"	Polaris (polestar)
2. Aduno Giri Ley	"Second eye of the world"	Southern Cross
3. Amma Bogu Tolo	"Stars of Amma's navel"	4 stars of Orion's Chariot
4. Atanu Tolo	"Three stars"	Orion's Belt
5. Emme Ya Tolo	"Star of the female sorghum"	The 2nd Companion star of Sirius
6. Enegirine Tolo	"Star of the goatherd"	Gamma star of Canis Minor
7. Nay		The Sun
8. Po Tolo	"Star of the *digitaria* seed"	Sirius B: the 1st Companion of Sirius
9. Sigi Tolo	"Star of the Sigui Festival"	Sirius
10. Tara Tolo	"Star of the hyena"	Procyon: the alpha star of Canis Minor
11. Tolo Bani Nenneu	"Sparkling red star"	Aldebaran: the alpha star of Taurus
12. Tolo Dullogu	"Tail stars"	Stars of Orion's sword
13. Tolo Duno	"Grouped stars"	The Pleiades
14. Yalu Ulo	"Bounds of space"	Milky Way
15. Yara Tolo	"Star of the lion"	Sheratan: The beta star of Aries

a white dwarf and virtually invisible to the naked eye, was not discovered until 1862 by modern astronomers and not recognized as a white dwarf until 1914. We have every reason to believe that the Dogon have known of its existence and its characteristics for at least 700 years. Moreover, the Dogon knew, correctly, that Sirius B describes an elliptical orbit around Sirius of 50 years duration. Here again we are confronted with the subtlety of Dogon knowledge. Certain Greeks knew of the orbits of the heavenly bodies around one another, particularly in the solar system, but they visualized them as perfect circles. It was not until Kepler's work in the 17th century that Western astronomers realized that planetary orbits, indeed the orbits of most heavenly bodies, tended to describe ellipses, as the Dogon represent that of Sirius B. But we are still confronted with the question of how, without telescopes, could the Dogon know that, 51 trillion miles away (8.7 light-years), the orbit of an invisible star around its companion was of 50 years duration?

That a nonliterate African people at a postneolithic level of material culture could know so many things about this star is incomprehensible to modern scientists. The late Carl Sagan, celebrated astronomer and writer, advanced the theory that, sometime in the early 20th century, a scientifically literate explorer or anthropologist wandered into Dogon country and, sitting around Dogon hearth fires, exchanged Sirius stories with his hosts. The Dogon regaled him with their Sirius lore and, in return for their hospitality, this unnamed European proceeded to relate all the known scientific data concerning Sirius and its companion star to his hosts. The Dogon are alleged to have remembered every single detail of this European Sirius "myth" and regurgitated it verbatim to Griaule a generation later.[54] Aside from the total absence of a record—oral or written—of such a visit among the Dogon and aside from Sagan's failure to identify by name this fictional European scientist, this preposterous scenario ignores the inconvenient truth that the realization that Sirius B was a white dwarf did not occur until 1914. With the advent of World War I in that year, this datum would not have circulated beyond a narrow band of astronomers for 5–10 years after this discovery, even among the scientifically literate. Moreover, many of the properties of white dwarves that the Dogon describe did not become part of the general body of astronomic knowledge until after Griaule began his researches among them in 1931.[55] Whatever the merits of Carl Sagan as an astronomer, he would have served the cause of objective truth better if he had not fabricated historical mirages to explain Dogon knowledge.

What is it about Dogon knowledge concerning Sirius B that has compelled these mental gymnastics?[56] First, they know that Sirius B is small, the smallest of the (non-neutron) stars, and white. Though they call it *po tolo*, also known as the "star of deep beginnings," they say that it is the heavenly ver-

---

[54] Cited in Adams HH, "African Observers of the Universe: the Sirius Question," in *Blacks in Science: Ancient and Modern*, edited by Ivan Van Sertima, Highland Park, N.J.: Journal of African Civilizations Ltd., 1983, p. 37.

[55] There were two ethnographic expeditions to Dogon country prior to Griaule's, in 1907 and 1921, but in neither instance was the Dogon system revealed beyond its superficial aspects. Nothing of Dogon knowledge of the Sirian star system came to light before 1947.

[56] See Griaule and Dieterlen, *The Pale Fox*, op. cit., pp. 500–8, for a complete review of Dogon knowledge concerning Sirius B.

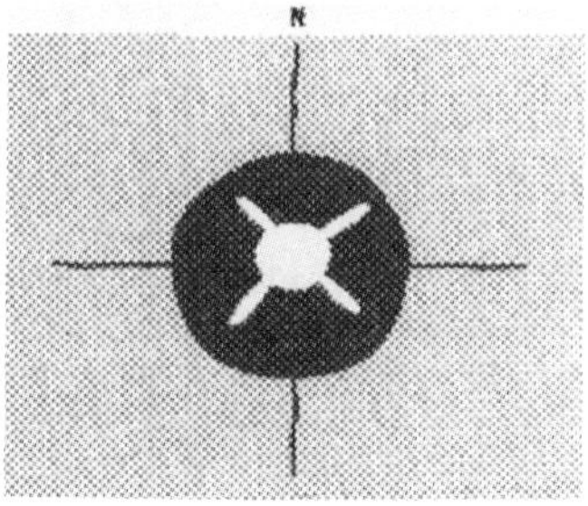

Fig. 55. Dogon representation of Sirius (*Sigi Tolo*).

sion of the *po pilu*, meaning literally "white seed." In fact the term *po tolo* translates literally as "seed star." If we assume that the Dogon tradition from which this lore is derived goes back 700 years, then, properly speaking, *it is the Dogon who deserve credit for having discovered Sirius B and the white dwarf as a category of star*. What is more, they know that Sirius B is incredibly dense and heavy, the densest and heaviest of the non-neutron stars. The substance of a white dwarf gradually converts from a diamond-like carbon material to a metal, similar to iron but different from it; in fact there is no metal like it on Earth. Indeed, the reason a white dwarf burns out is precisely because its core has been converted to metal, therefore depriving the star of the fuel necessary for the fusion reactions that cause stars to burn with such incredible heat. The still-intense heat of a white dwarf is caused not by hydrogen fusion but by the quantum forces affecting the tightly-packed electrons in its dense matrix. But what do the Dogon say concerning the metallic structure of Sirius B:

> The star [Sirius B] contains three basic elements, "air, fire, and water"; the element "earth" is replaced by metal in all its forms, particularly by the metal called *sagala*, somewhat more shiny than iron and of such density that "all the beings on earth together could not lift a small part of it." Hence the star's weight and, since it is so small, its density.[57]

We might be forgiven a little amazement: How *do* the Dogon know this? That white dwarves are composed mostly of an iron-like metallic substance has only been documented in the last generation or so by spectrographic analysis. So much for Sagan's scientifically literate, turn-of-the-century European explorer teaching the simple Dogon people the astronomy of the Sirius star system. We are confronted with a people who, in certain domains of astronomy and cosmology, have no historical peer. There is nothing that comes down to us from the ancient Nile, Chaldea, Greece, China, or the medieval

---

[57] Ibid., pp. 504–5.

Arabic world that remotely compares to this caliber of astronomical knowledge. We are too conditioned by the dominance of the Newtonian-Cartesian paradigm to be prepared to accept that, clearly, there is more than one way to receive or acquire complex empirical knowledge. The word "primitive," therefore, makes no sense in this context for we are faced with a "primitive" people who, in certain respects, seem to operate at a level of cosmic understanding comparable to modern scientists. It is as if Einstein, the quantum theorists, and modern astrophysicists had but re-cleared a road previously opened by an obscure African people tucked away in a half-forgotten corner of Mali. Solomon spoke truly: There is nothing new under the sun.

It goes on. The Dogon seem also to possess a concept that can only be described as gravity. Their description of the interrelationship between *po tolo* (Sirius B) and its companion *sigi tolo* (Sirius) indicates this:

> The movements of po tolo keep all the other stars in their respective places: in fact, one says that without this movement none of them would "stay in place." Po tolo forces them to keep their trajectory: in particular, it regulates the trajectory of Sirius, which is the only one that does not follow a regular curve, and which it separates from the other stars by surrounding it with its own orbit.[58]

The Dogon statement that po tolo regulates the orbit of sigi tolo is born out by scientific research:

> The first white dwarf to be discovered was the companion of Sirius, now called Sirius B. In 1844 Friedrich Bessel discovered irregularities in the motion of Sirius, and concluded that it must have a companion, the pair forming a binary system with a period of about 50 years. *In order for its gravity to swing Sirius around in its orbit*, it was clear that the companion must have a comparable mass. (emphasis added)[59]

We know that gravity is related to the Einsteinian curved universe; the Dogon too say that the universe is "curved."[60]

In a sense, po tolo is the mother-seed of all stars; as the "star of deep beginnings" it is the oldest of heavenly things. We know the universe is at least 12 billion years old and Sirius B (po tolo), because it is now a white dwarf, must have been formed early in the history of the universe. White dwarves represent stars, originally sun-sized, at the termination of their life-span.[61] The exhaustion of the hydrogen and helium that powers stellar fusion for a sun-like star, thereby creating a white dwarf, takes 10 billion years or more. Moreover, the death throes of a white dwarf make possible the birth of other stars

---

[58] Ibid., p. 505.

[59] Moore, *International Encyclopedia of Astronomy*, op. cit., p. 451.

[60] Griaule and Dieterlen, *The Pale Fox*, op. cit., p. 465.

[61] Sirius B is somewhat larger than Earth but has a mass equal to the sun. This means that a teaspoon of Sirius B would weigh one ton on Earth.

and star-systems; all newborn stars are created when older stars die. Our very solar system was created out of the dust of a dying star such as Sirius B.[62] In fact, since Sirius B is the closest white dwarf to our solar system, the Dogon may not be far from the truth when they declare that our solar system—including Earth and everything on it—owes its very existence to Sirius B.

The Dogon explain it this way:

> The movements of po tolo, the heavy embryo of the world and the mark of all creation, are quite active: its contents are ejected by the force of its spinning in forms that are "infinitely small," much like the grains of the po pilu which grow rapidly: in this way, the star, white like po pilu, continues to distribute the germs of of life once contained by the seed it represents. . . . Amma created po tolo the first of all stars. In the (body) substance of po tolo, [all] the things in the [entire] world, their sign exists a bit.[63]

Let us compare the modern scientific explanation:

> As the degenerate [white] dwarf and its companion orbit each other, the dwarf's strong gravity pulls gas from another star. This swirling gas forms . . . an accretion disk, which feeds onto the surface of the dwarf, wrapping it in an ever-growing mantle.
>
> Gas swirling around a degenerate dwarf is accelerated to 5,000 kilometers per second. As this gas strikes the dwarf's thickening mantle, the impact heats the degenerate gas enormously. Temperatures in the mantle rise to the ten million degrees needed to ignite nuclear fusion. . . . But since it is degenerate, the mantle cannot shed the heat by expanding. . . . "Basically a new sun is temporarily created around the ember of a dead one". . . . A nuclear smolder begins to spread. When it hits a flash point, a holocaust engulfs the dwarf.[64]

This "holocaust" is a nova explosion which would appear to an Earth observer as a sudden, brilliant, pinpoint flash of light in the nighttime sky that would persist for several nights. The witnessing of novas—and their "big brothers," supernovas—has been recorded by astronomers for thousands of years. Novas are only possible in binary star systems such as that of Sirius and only involve white dwarves. Most white dwarves go nova many times in the last phases of their existence before vaporizing completely in a Type I supernova explosion.

As intimated above, these nova explosions are critical for the creation of new stars and star systems. Concerning recent observations of Nova Aquilae, infrared telescopy has yielded significant data concerning its activity:

---

[62] The ritual incantation: "dust to dust, ashes to ashes..." intoned at funerals represents a literal statement of how stars die and are reborn.

[63] Ibid., pp. 505–6.

[64] Gore, "The Once and Future Universe," op. cit., p. 722.

> The question about Nova Aquilae that intrigued UKIRT astronomers Longmore and Williams was whether the nova had begun emitting much infrared. That would mean, they believed, that the shell of gas blown off by the nova had cooled enough to form *dust grains*.
>
> "Even the first time we looked at it, we saw excess dust grains," said Longmore. "The gas blown off by the nova couldn't have cooled that fast. That means some of that dust was there already. Hot gas must have been coming off the star for perhaps thousands of years before it went nova". . . . Why is this star dust so interesting? Primarily because it is the *stuff of new stars*. Matter evolves during the lives of stars and emerges largely through their nova, super-nova, and red giant stages.
>
> More complex elements are forged only in the fiery death throes of those stars. Some of the air we breathe . . . may have been released by a nova or red giant. So were many carbon and silicon atoms. Supernovas have created most of the heavier elements. . . . As Sumner Starrfield put it, "The very materials out of which our bodies are made were cooked up in a violent process somewhere in the universe." (emphasis added)[65]

Time and again, the Dogon declare that the po pilu opens and "distributes the elements of the universe."[66] Clearly, the Dogon imagery of grains from the po pilu, i.e., Sirius B, ejected into space and forming the germs of life, finds its echo in the literature of modern astrophysics:

> Condensing solid grains in the gases expelled by novae may be expected to contain . . . nuclides, and so would form a component of interstellar material with an unusual chemical composition. . . . When the Solar System formed from interstellar grains, *some grains derived from novae would be present.* (emphasis added)[67]

What defies the imagination is that the Dogon *describe the nova process affecting Sirius B as a white dwarf*! They say of po tolo:

> On the totemic sanctuaries, po tolo is represented by a circle with six rays marking the six yala of the sex of the po in Amma's womb. The rays are dotted lines indicating that the star—*which will burst, as did the po pilu, and be visible for some time*—can at present not be seen except under special circumstances. (emphasis added)[68]

It is impossible for this author to interpret this material in any other light except the following: the Dogon knew (for at least 700 years) that Sirius B was a mostly invisible white dwarf that periodically underwent nova explo-

---

[65] Ibid., pp. 725–6.
[66] Griaule and Dieterlen, *The Pale Fox*, op. cit., p. 445.
[67] Moore, *International Encyclopedia of Astronomy*, op. cit., p. 293.
[68] Griaule and Dieterlen, *The Pale Fox*, op. cit., p. 507.

sions which spewed matter ("grains") into space that ultimately became the stuff from which other heavenly bodies—including our solar system—were made! From the point of view of astrophysics and cosmology, their intuition that the Solar System and life on Earth ultimately derived from the Sirian star system, at least in part, seems to be a plausible conclusion. Composed of migrating star-dust as we are, a greater or lesser part of it must have come from some white dwarf repeatedly going nova somewhere. Sirius B is the closest white dwarf to us and as such must have gone nova on a number of occasions. Its very proximity would make it the best candidate for the source of our formative star-dust.

Even if is postulated that these dust grains of which we are made emanated from a red giant—another possibility—we know that, once upon a time, Sirius B was a red giant! Red giants usually end their lives as white dwarves. Either way it's sliced, Ariadne's thread still seems to lead to Sirius B.

Our sun is less than five billion years old—young for a star—and po tolo or Sirius B was born more than 10 billion years ago. We can assume, then, that Sirius B began experiencing the decaying effects of advanced age five to seven billion years ago—before the birth of our sun—as it gradually transformed into a red giant and, following that, a white dwarf. When a star turns into a red giant toward the end of its life, it may expand to 100 times its size: when our sun becomes a red giant, it will gobble up the entire solar system as far as Mars. A star in its red giant phase is constantly blowing off its gas mantle out into space but finally reaches a size so massive that it becomes insupportable. The excess mass is exploded off in a nova that turns the red giant into a white dwarf.[69] This white dwarf will itself continue experiencing nova explosions and may even undergo a Type I supernova explosion eventually, causing it to vaporize completely.

Considering the red giant and white dwarf phases together in the life of a star, the repeated nova explosions may continue for billions of years. Each time it happens, gaseous matter containing several of the lighter elements is ejected into space with tremendous force. When this gaseous matter cools, it forms the dust grains already referred to as the constituent elements of new stars. All of the heavier elements, i.e., those above the atomic number of iron, can only be created in the super-heated furnace of a supernova, a million times hotter than a nova. Spectrographic analysis has detected in the sun 70 of the 92 known natural elements on earth. Elements like hydrogen, helium, carbon (the building block of terrestial life), oxygen, and all the elements on the atomic table up to iron, could have been blown in from repeated novas. But somewhere, possibly in the vicinity of the Sirius system, a supernova must have occurred to produce a cloud rushing out at near-light

---

[69] Red giants will often undergo a supernova explosion that turns it into a neutron star, the smallest, densest, heaviest object in the heavens except for a black hole.

speed containing all the heavier elements in gaseous form. Thus, a confluence of gas clouds from Sirius B's repeated novas and another (unidentified) star going supernova may have produced the raw materials necessary for the eventual birth of our solar system when enough cooling had occurred to condense this spiraling cloud into more organized forms. One suspects that when this occurred 4.5–5 billion years ago, this "proto-solar system" was much closer to the Sirius system than presently which the Dogon themselves state was in fact the case. The Sirius system is less than nine light-years away, very much "in the neighborhood" by the reckoning of interstellar distance.

It can be plausibly speculated that Sirius B has contributed an undetermined amount of the formative matter of our solar system. It is clear that the Dogon think so. How do they know? Why should we take their speculations seriously? The answer to the first question is that, at present, there is no answer. The answer to the second is that they have yet to be proved wrong on any of their descriptions of the properties and behavior of the Sirius system. They seem to know as much about it as modern astronomers. The other part of this answer is that the Dogon scenario fits the facts once we understand the interrelation of red giants, white dwarves, novas, and the creation of new star systems.

We cannot leave this subject of the Sirius star system without commenting on the star the Dogon call *emme ya tolo*.[70] The Dogon say that she—for they think of emme ya tolo as the "celestial image of the female 'mother' seed"—is the *second* companion of Sirius! Modern astronomers, as a body, have not acknowledged the existence of a third member of the Sirius system though evidence for the existence of *emme ya tolo* is not lacking. Robert Temple refers to the reputed sightings of a third star in the Sirius system dating back to 1920.[71] Other astronomers think that irregularities in the orbit of Sirius point to evidence of a second companion.[72] In his article "New Light on the Dogon and Sirius," Hunter H. Adams referred to a recent (1979) discovery of a dwarf nova a mere 9 minutes south of Sirius.[73] It is not possible to determine whether this dwarf nova is in reality the second companion of Sirius, emme ya tolo, but Adams points out that one of the Dogon sand drawings show a star, *ie pelu tolo* ("star of the tenth moon"), apparently going nova.[74] Whether or not this is the emme ya tolo referred to by the Dogon, they do seem to have recognized and recorded this dwarf nova near Sirius so recently discovered by modern astronomers.

---

[70] Griaule and Dieterlen, op. cit., p. 506.
[71] Temple, op. cit., pp. 12-3.
[72] Ibid. See also Burnham R, *Burnham's Celestial Handbook*, volume 1, New York: Dover Publications, Inc., 1978, pp. 395-6, for further discussion on the existence of the third member of the Sirius system.
[73] Cited in Adams, HH, "New Light on the Dogon and Sirius," in *Blacks in Science: Ancient and Modern*, New Brunswick, N.J.: Journal of African Civilizations Ltd., 1983, p. 48.
[74] Ibid.

What the Dogon have to recount concerning emme ya tolo just adds to the fascination generated by their knowledge. They say that emme ya tolo is larger by volume but four times lighter than *po tolo* (Sirius B), though both describe a 50-year orbit around Sirius. As mentioned, it possesses an entirely feminine character and is in fact called *yau nay dagi*, "the little sun of women." But its role in the Sirius system is unique, for the Dogon say: " . . . it [emme ya tolo] serves as the intermediary between po tolo and Sirius and that it spins to 'transmit the orders' from the former to the latter, which is watched over by po tolo."[75] The author could not help but think that the way the Dogon conceptualize this system is uncannily similar to the transmission of genetic information. This occurs when messenger RNA transmits the genetic code from DNA to transfer RNA, that will eventually tell the base-pairs of the amino acids in formation how to line up to create proteins, the constituents of living organs.

Dogon Calendar Systems

The Dogon seem to have four time-reckoning mechanisms. Their main calendar is a lunar one but they also maintain a solar calendar. The lunar year is reckoned at 354 days and the solar year at 365 days. But there also seems to be a soli-lunar year of 360 days in which the 12 months are grouped by twos, each pair of months containing 60 days. Their solar calendar, like ours, begins with the winter solstice and the Dogon continue their observances of the solstices and equinoxes throughout the year.

Griaule and Dieterlen do not report directly that the Dogon had devised a method for rectifying their functional solar calendar of 365 days with the actual length of the year of 365.25 days. However, having already explored the ancient Nilotic method of rectification by means of the Sothic cycle of 1,460 years, there seems to be evidence that this system may have survived among the Dogon. We are told that erecting the altar of the resurrected Nommo who, as we have seen, was initially sacrificed to renew the world, is correlated with what the Dogon call the "meeting of Sirius and the sun," i.e., the heliacal rising of Sirius. We are further informed:

> Sirius is the symbol of the ever-living placenta of the resurrected Nommo: the star is also called *albararu*, a contraction of "Amma *albarka*," or "thanks to Amma". . . . On the spatial level, the figure of the Nommo's placenta (Sirius) denotes the collateral directions or "four angles" of celestial space, just as the figure of the Fox's placenta (the sun) denotes the cardinal directions of terrestial space; hence the symbolic value attached to the heliacal rising of Sirius, which figures in the composition of the long calendar of the Dogon.[76]

---

[75] Griaule and Dieterlen, *The Pale Fox*, op. cit., p. 506.
[76] Ibid. p. 350.

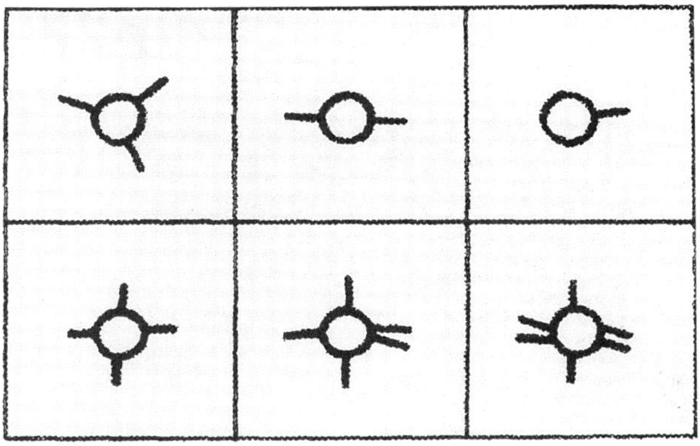

Fig. 56. The six positions and phases of Venus that form the basis for the Dogon Venusian agricultural calendar.

Since the authors do not elaborate on this striking comment, it is hard to deduce anything but that this is the Sothic calendar of the ancient Nilotes. It is one datum, among many, testifying to the evident antecedents of the Dogon in or around the Nile Valley.

Thus far we have identified three main calendars among the Dogon: (1) lunar, (2) solar, and (3) Sirian. But they possess a fourth, one that is probably their most unusual: a calendar based on the phases of Venus. This Venusian calendar is based on what the Dogon call the six positions of Venus (see Fig. 56). The Dogon's agricultural cycle is tied to the appearance of Venus in these positions—that is, the path of the planet through space—at specific times during the year. Thus, it is said of Venus:

> Its positions during the year recall the stages of the germination, growth, harvest, and consumption of the basic cereal grains contained in the first grain, i.e., the annual death and resurrection of the millet. From this perspective, they constitute an agricultural calendar, called *dine bire* . . . [77]

The positions can be enumerated as follows:[78]

Enegirim: Venus as the "star of the goatherds," referring to the time of sowing and germination.

Dige tanu: Venus as the "star of midnight," referring to the time of the millet emerging from the earth.

Donno tolo: Venus as the "star of the west" or Evening Star and refers to the time of the millet that has grown into an ear.

Yazu: The position of Venus just after midnight and refers to the time of the grains formed in the ear ("pregnant millet").

---

[77] Ibid., p. 516.
[78] Ibid., pp. 516, 546-7.

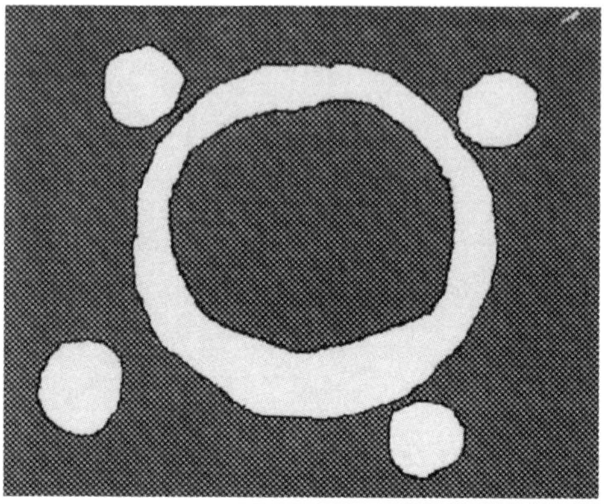

Fig. 57. The four moons of Jupiter as depicted by the Dogon.

Obia: A word meaning "hidden," referring to Venus as the Morning Star, and is the time of the "ripe millet."

Yapunu da: This position of Venus literally translates as "dish of the menstruating women," and refers to the time when the millet is eaten (hence "dead") and thus present in the clavicle of the people it has nourished.

There are some striking features of the Dogon Venusian calendar that seem to resonate with the ancient Sumerio-Chaldean Venusian system. Therefore, some other parallels can be mentioned. The eight Dogon nommo are all fish-beings in the style of Dagon and Oannes of the Fertile Crescent. Indeed, the name "Dagon" is curiously close to "Dogon." Moreover, the importance of the number 60 in their time reckoning, particularly that of the 60-year Sigui festival, recalls the sexagesimal system of the Chaldeans. Finally, the Dogon Venus calendar reminds us that the ancient astronomers of the Fertile Crescent also possessed a calendar based on the Venusian cycle.

Cultural comparisons show an unmistakably Nilotic provenance for many aspects of Dogon culture. Were there additional influences from the riverain civilizations farther east? Among 19th-century antiquarians, George Rawlinson advanced the opinion that the early cultures of the Fertile Crescent were "Cushitic" in origin, tying Sumeria and Chaldea to the Nile Valley in their remote antecedents.[79] Were the ancestral "proto-Dogon" originally part of a geo-cultural matrix that allowed influences from both river valleys to permeate their own? This is not an unviable speculation. Cultural beliefs, practices, and techniques are often very tenacious, surviving the vicissitudes of thousands of years intact. Moreover, when a constellation of cultural practices is exhibited by two distinct peoples, though significantly separated in time and space, the likelihood of shared origins or meaningful past interchange is strong.

---

[79] See Rawlinson G, *The Seven Great Monarchies of the Ancient Eastern World*, New York: John B. Alden, 1885, pp. 33-7.

Fig. 58. Three different representations of the Pleiades by the Dogon.

Astral Miscellany of the Dogon

There are other astronomical data in the possession of the Dogon worth bringing to light before concluding this chapter. The Dogon describe Jupiter and its four moons. Now we know that Jupiter actually has nine satellites but the four major moons are large enough to be seen with a simple telescope (see Fig. 57). Indeed, when Galileo turned his telescope toward Jupiter, he described the four moons. The Dogon also describe Saturn's ring and the turning of the sun, Sirius, and Sirius B on their axes. The Pleiades, i.e. "The Seven Sisters," are another constellation that figures prominently in the Dogon astronomic system (see Fig. 58).

Finally, as a testament to the uncanny sophistication of their knowledge, it is pertinent to reflect on their knowledge of the "entropy" of the universe. Concerning po tolo (Sirius B), they say:

> The star's former position in space is where now the sun is, called "seat of po tolo," po tolo doy. *Like the other stars, it moved away* and is presently at the center of the sky; but it is a center in motion . . . (emphasis added)[80]

In 1922, Hubble noted that on spectrographic analysis, all the remotest galaxies in the heavens seemed to be "red-shifting," meaning that the visible bodies were moving away from him. Wherever he turned to look in the sky, Hubble observed the exact same phenomenon: everything was moving away from his position and everything was moving away from everything else; the remoter the galaxy, the faster it was moving away. This momentous discovery led directly to the Big Bang Theory, permanently changing modern cosmology. The Dogon too seemed to have recognized the phenomenon.

---

[80] Griaule and Dieterlen, *The Pale Fox*, op. cit., p. 505.

Conclusion

Whenever the author presents this Dogon material to listening audiences, the question of "how" invariably arises. We are not dealing here just with symbolic material arising from deep within the Jungian "collective unconscious." The *Pale Fox* shows us that highly complex empirical information has come into the possession of a people lying almost entirely outside modern techo-scientific culture. Since the Dogon do not possess the methodologies or apparatus of modern science, we are driven to conclude that there must be other "pathways" for acquiring such data; veritably, other ways of knowing.

It has always been so. As we have already explained, the rupture between the empirical consciousness of science and the symbolic consciousness of mythology has not yet been 400 years in the making, a dubious legacy of post-Renaissance Europe. And though Newton is said to have inaugurated the process, he himself was not "of" it, for he has been referred to as the "last of the magicians." Relativity and quantum mechanics have perhaps started to re-weave Western science into the fabric of universal consciousness. However, the disunion evident in modern thought never transpired in the Dogon universe; their cosmic system is one of the most viable models extant of the seamless interweaving of empirical science and symbolic myth.

In the next chapter we will explore some aspects of "other-knowing" implicit in this rediscovered unity of science and myth.

8

Quantum Consciousness: The Fifth Dimension?

The exaggerated scepticism in regard to ESP is really without a shred of justification. The main reason for it is simply the ignorance which nowadays, unfortunately, seems to be the inevitable accompaniment of specialism and screens off the necessarily limited horizon of specialist studies from all higher and wider points of view in the most undesirable way.
—Carl Jung, *Synchronicity*, p. 101

The more I examine the universe and the details of its architecture, the more evidence I find that the universe in some sense must have known we were coming.
—Freeman Dyson, from *Wrinkles in Time*

Of his many gifts, one of C.A. Diop's most uncommon was his ability to avoid the traps of paradigmatic thinking (see Plate 22). Perhaps his struggles to penetrate through the web of distortion spun around African history made him realize how self-inhibiting it is to eliminate from consideration avenues of knowing not necessarily consistent with one's own.

If Diop was a religious man, he kept it private. He was raised in a strongly religious atmosphere and in fact was the lineal descendant of the Senegalese Muslim saint Cheikh Ahmadou Bamba, the founder of the Mourides, the most powerful religious brotherhood in contemporary Senegal. It was said that once, when a captive of the French on one of their warships, Cheikh Amadou was refused permission to pray on board. He calmly descended from the ship and went out a little ways to pray on the water! This Jesus-like miracle is recounted with absolute conviction by the Senegalese. Myth or reality? Fact or fable? Once one gets to know some of the adepts of the Western Sudan, the demarcation between reality and illusion tends to become indistinct. In any case, despite Diop's intense religious upbringing, there came a time in his life when he decided to tuck religion, or its outward observances at least, into a hidden corner of himself, away from public view. Few persons knew if Diop espoused any religion or not.

But Diop was steeped in the traditions of Africa, a world where the hidden, occult, and spiritual forces of nature coexist with human beings in an often easy familiarity. The reality of the existence of this paranormal realm in Africa is routinely derided by Westerners. In a de-symbolized, de-mythified, de-spiritualized culture such as contemporary America, it is easy and reassuring to dismiss, belittle, and ridicule any notion of an occult reality. The case is completely different in Africa: to sojourn there any length of time almost inevitably invites contact with such forces. Skepticism in this environment becomes a progressively more difficult frame of mind to sustain. Tellingly, Diop once confided that if even *one* instance of a telepathic or other occult event could be documented, it would be sufficient to refute the entire body of scientific skepticism concerning the reality of paranormal phenomena. In chapter 17 of *Civilization or Barbarism*, Diop devoted a short space to parapsychology in the context of African philosophy.[1]

In sum, Diop showed that theoretical physics from about 1935 had evolved to the point where it could empirically confirm the existence of forces and energies once dismissed as "psychic." Moreover, with mathematical rigor, relativity and quantum mechanics actually predicted the existence of some of these forces. This "Pandora's Box" was first opened by Einstein himself who, in truth, was motivated by the desire to refute the quantum description of matter. Einstein was profoundly uneasy with the implications of the quantum theory that sprung up in the decade after the publication of his General Theory of Relativity in 1915. The work of the original quantum theorists—Niels Bohr, Louis de Broglie, Wolfgang Pauli, Werner Heisenberg, and Paul Dirac—showed that it was impossible to know simultaneously the velocity and the position of an electron orbiting around an atomic nucleus; it was possible to know one or the other with certainty but not both. This meant that it was impossible to characterize an electron—or the atom as a whole—with absolute precision. This limiting factor became known as "quantum indeterminacy" or "uncertainty." The best that could be achieved was a statistical probability that approximately characterized an electron by reference to its velocity and position. Einstein's response to this astounding assertion? "God doesn't roll dice."

Einstein grasped immediately that quantum mechanics was about to dismantle forever the Newtonian model of the universe, where there was order, predictability, and the theoretical possibility of finding the ultimate building block of matter. His own work had already delivered a serious blow to the Newtonian model by showing that space and time are not separate, but rather are interwoven into a seamless fabric referred to as "space-time." Thus, to the ordinary spatial dimensions of length, width, and depth was added time, the fourth dimension. Newtonian gravity was retained in this new model but not in its classical connotation; in Einstein's theory, gravity arises from the curvature of space-time created by any material object. The more massive the object, the greater the curve. Human beings are rooted to the Earth because the Earth is so much more massive than ourselves that the

---

[1] See Diop CA, *Civilization or Barbarism*, translated by Yaa-Lengi Meema Ngemi, Westport: Lawrence Hill, 1991, pp. 366–75.

resulting curvature of space-time pulls us to and "locks" us on the planet. In this Einsteinian universe, there exists an absolute limit, the speed of light, moving through space at 186,000 miles per second.

The Newtonian model is one valid for the objects and events of everyday experience but Einstein's model gives a more complete description of cosmic phenomena. Nonetheless, Einstein saw his model as amplifying and extending Newton's, not replacing it. But quantum mechanics challenged the basic premises of the Newtonian model, in effect saying, things are not what they seem.

One of the consequences of quantum mechanics, hitting at the very essence of the "common-sense" Newtonian model, was the elimination of the dichotomy between subjective and objective. In quantum theory, the properties of an electron or a photon do not exist until they are perceived and measured. Thus, what a photon is going to be—wave or particle—depends entirely on how and when it is measured; it comes into existence as one or the other only by virtue of being measured![2] If the most neutral, unencumbered experiment imaginable could be designed, the results of the reaction would still be altered by the very act of observing it! This means that the perceiver and the thing perceived are indissolubly linked; absolute objectivity is impossible. Moreover, from the quantum point of view, things exist because they are perceived. This realization has created an intellectual crisis that is still being sorted out. What many scientists seem reluctant to face is that, in some sense, consciousness itself is infused into the "stuff" of space-time. The positivistic rationalism, imbued with assumptions of "realistic" objectivity, that has dominated science for the last two centuries is a paradigm requiring radical revision.

All these implications disturbed Einstein profoundly. He then mounted what, at the time (1935), seemed a quirky, almost whimsical, attempt to challenge quantum mechanics by conducting a "thought experiment." With colleagues Boris Podolsky and Nathan Rosen, Einstein concocted what has since become known as the "EPR Paradox" (from *E*instein, *P*odolsky, *R*osen). Simply put, the EPR Paradox poses the question of what would happen if two particles, A and B, collided and, bouncing off one another like billiard balls, moved oppositely to a great distance from one another.

Quantum mechanics rules that the speed *and* location of particle A cannot be simultaneously measured. However, it should be possible, because of the laws governing the momentum of colliding particles, to measure the momentum of particle B to ascertain the momentum of particle A; likewise the position of particle B can be precisely located, reflectively localizing the position of particle A. Since particles A and B may be far apart after the collision, they are not supposed to interact with one another. It should be theoretically possible then to pinpoint the location of B to infer the position of

---

[2] See Horgan J, "Quantum Philosophy," in *Scientific American*, July 1992, pp. 94–104.

A, while, at the same time, measuring the velocity of A, without either particle disturbing or interacting with the other. Thus, by combining direct *and* indirect measurements together, the position and velocity could, in principle, be determined simultaneously. This ingenious reasoning seemed to controvert the central tenet of quantum mechanics, that of "indeterminacy" or "uncertainty," unless some other factor was found to be present:

> It seemed to Einstein that the only way to retain quantum uncertainty across the gap between the particles would be if they were connected by what he called some "spooky action at a distance," operating faster than light and therefore transcending the restraints of his own theory of relativity.[3]

Einstein went on to say that such "spooky action at a distance" was "incompatible" with any valid model of reality. Despite vigorous ripostes from the quantum physicists, most notably Niels Bohr, the EPR Paradox remained a serious challenge to their model for nearly 50 years. Many physicists attempted to grapple with the paradox, but for a long time only J.S. Bell was able to devise an effective means of testing it with a theorem known as Bell's Inequality.[4] Bell's Inequality showed that, if quantum mechanics were correct, then the principle of locality, and therefore "causality" in the classical sense, was no longer valid. That is to say, two particles once connected to one another would continue to interact even at distances of light-years apart. Contrariwise, if locality were to hold up as the fundamental principle of interaction (or causation), then quantum mechanics would be rendered invalid.

The dispute was finally resolved by the French physicist Alain Aspect in 1982. Aspect monitored the behavior of two photons emitted in opposite directions from the same atom and found that, yes, two correlated particles moving away from one another continue to act upon each other, even at great distances. Aspect's experiments have been corroborated many times and in 1990, two English experimenters, Rarity and Tapster, found that even the momentum of correlated particles was influenced quantumly by the EPR effect.[5] Quantum mechanics was "saved." But curiously, it had taken Einstein's famous paradox—implicitly hostile to quantum mechanics—to pose the question that would ultimately establish for certain the place of quantum mechanics in modern physics. It also meant that the principle of "nonlocality," i.e., "spooky action at a distance," was now permanently imbedded in physics. As Diop realized, the principle of nonlocality supplied an empirical foundation for psychic phenomena and, properly speaking, parapsychology could justifiably be considered a branch of modern physics.

What makes all of this so much more interesting is that the validation of

---

[3] Davies P and Gribbin J, *The Matter Myth*, Simon & Schuster/Touchstone, 1992, p. 223.
[4] Cited in Diop, op. cit., p. 369.
[5] Horgan, op. cit., p. 99.

"action at a distance" had already been foreshadowed by the writings of Carl Jung dating back to 1930. On the basis of his own intensive clinical studies on the workings of the psyche, Jung had known for a long time that there were incidents that had occurred in his own life and that of his clients that belonged in the realm of the paranormal. Both relativity and quantum mechanics must have seemed like a godsend to him because here at last were rigorous scientific models that could empirically vindicate these experiences. Jung was extremely sensitive to charges of "mysticism" and constantly sought corroboration for his theories from collaborators in the scientific community. He found one in Wolfgang Pauli, a member of the founding quintet of quantum physicists.

Jung, engaging in his own "thought experiments," invented the term "synchronicity" to describe paranormal occurrences by reference to "the simultaneous occurrence of two meaningfully but not causally connected events."[6] These events were therefore "acausal," that is to say they were mutually interactive in a kind "simultaneity" that did not, however, obey the cause-to-effect relation. Thus, two events connected by synchronicity could, in effect, collapse time and space. Synchronicity, in Jung's view, was at the bottom of divinatory procedures such as astrology as well as precognition and telepathy. Two events related by synchronicity did not, however, *have* to be synchronous. A premonition of an event that eventually comes to pass is a synchronicity but, because there is a time lapse, not synchronous. In any case, Jung emphasized synchronicity as an "acausal connecting principle," as powerful and ubiquitous as cause and effect in objective events.

A word should be said about the relation between synchronicity and superstition. A synchronicity that becomes overgeneralized, overworked, or overextended turns into superstition. This process can be illustrated by reference to the well-known superstition concerning black cats, about which more will be said in the appendix to this chapter. In ancient Nilotic mythosymbolism, the cat, because of its preternatural nocturnal sight, represented the sun by night, i.e., the sun passing through the Underworld, the Land of the Dead. It also was depicted as a form of Isis or Bast in their "necromantic" character. Black cats in particular would be infused with this preternatural aura and we can imagine one or more synchronistic situations arising where the appearance of a black cat coincided in a significant way with someone's death (see appendix). It would be easy then for an attitude of mind to evolve where encounters with *all* blacks cats, even in the most trivial circumstances, would be interpreted as presaging death and therefore as "unlucky."

It is this trivialization of valid synchronicities that gives rise to superstitions and it therefore might be said that a superstition is but an "inconse-

[6] Jung CG, *Synchronicity*, translated by R.F.C. Hull, Princeton, N.J.: Princeton University Press, 1973, p. 25.

quential" or perhaps "inactive" synchroncity. It is important, though, to realize that synchronicities and superstitions are "genetically" related; the demarcating line between them can be so blurred as to be imperceptible. Moreover, there is probably no one entirely free from the influence of superstition and Jung reminds us that it is the over-rationalized personality that is ultimately *the most* susceptible to the suggestive power of superstitions and synchronicities.

Jung's first speculations on synchronicity are remarkable for the manner in which they anticipated and prefigured the EPR Paradox of 1935. In fact, it is not too much to say that the EPR Paradox (unintentionally) constituted *the* quantum statement of synchronicity. We know that Jung himself saw a close connection between synchronicity and the work of theoretical physics. Einstein's "spooky action at a distance" is no more than the "acausal connecting principle."[7] Indeed, it was the shattering of the conventional notion of causality that provoked Einstein's extreme displeasure with quantum mechanics. Later, physicists substituted the term "nonlocality" for "spooky action at a distance." It seems fair enough to say, in the aftermath of Aspect's experimental proofs, that nonlocality and synchronicity are now entrenched in any contemporary model of the cosmos as currently understood.

What continues to surprise is that physicists as a group are slow to come around to accepting this facet of the new reality. Largely, this reticence has to do with accepting the momentous idea that mind or consciousness is as much a component of the fabric of the universe as space and time.[8] The reluctance to embrace this idea wholeheartedly is undoubtedly related to the realization that once consciousness is insinuated into a physical model of the universe, then the "specter" of God raises its unwanted head. Scientists are notoriously hostile to *any* ideas "tainted" by religion entering their cloistered, tightly controlled world. One is reminded of the rejoinder of the French scientist LaPlace who, after presenting Napoleon with his treatise on the nature of the cosmos, was asked where the Creator fit into this scheme, to which LaPlace replied, "Sir, I have no need of that hypothesis."

Most physicists prefer to believe that blind, random processes are the causes of all higher organization in the universe. This belief has caused some of them to embrace the "multi-verse" or "parallel universes" model.[9] This paradigm posits innumerable "superposed" universes in the Primal Seed that all

---

[7] If we can take it as axiomatic that no one is entirely free from the influence of superstition and synchronicity, then it is also axiomatic that no one is entirely immune from "spooky action at a distance."

[8] In the series of 9 books by Carlos Castenada, expounding on the knowledge of the Yaqui sorcerer Don Juan Matus, Castenada is told by his benefactor that the universe consists of "awareness" and what we are or seem to be is determined solely by "perception." If the perception is changed, another reality or "world" appears.

[9] See Wolf FA, *Parallel Universes*, New York: Simon & Schuster/ Touchstone, 1988, for a full-length popular treatment of the subject.

came into existence at the time of the Big Bang explosion. Because an infinitude of possible universes exist, it is possible for there to exist *at least* one universe that contains all the right elements, circumstances, and processes necessary to evolve higher structures, including life. This model is attractive to those physicists wishing to preserve the idea of a random organizing process in a universe free of a supreme directing power. This sort of universe does not require an architect. It allows physicists to say that it is possible for some universe to exist where a couple of tons of brick and mortar can be thrown into the air and when the whole assemblage lands, everything will randomly organize into a house.

Without taking up the huge questions begged by this particular application of the paradigm, the parallel universes model does have some other interesting ramifications. For one thing, it plays havoc with our current linear notion of time. Many universes implies many space-times and therefore many futures. One of the interesting facets of quantum equations—grist for the eager mill of science fiction writers—is that time can flow in one direction as easily as another. From the point of view of real-time events or observations, something called the "collapse of the quantum wave function" occurs that is measured by reference to another similar wave that is a *conjugate wave*. The two waves must "merge" in order for anything "real" to happen. The thing is, these waves come at each other from different directions of time. Referring to the work of quantum physicist John G. Cramer, Fred Alan Wolf explains it this way: "Cramer calls the original wave an *offer* wave, and the conjugate wave an *echo* wave. He also calls the multiplication of the two waves a *transaction*. Thus a transaction occurs—involving an *offer* and an *echo* . . . "[10] The implications of this thinking almost defy comprehension:

> Every observation is both the start of a wave propagating toward the future in search of a receiver event and is itself the receiver of a wave which had propagated toward it from some past observation event. In other words, every observation—every act of conscious awareness—*sends out both a wave toward the future and a wave toward the past.*[11]

Since there is more than one future, all possible futures "receive" the offer wave signal and send back echo waves. But only one future, the one that actually transpires into an active present, completes the transaction via its echo wave. It is the future with the strongest wave "match" for its past that combines with the offer wave from that past to create the present event. It is a situation in which, potentially, a sensitive mind could "remember" the future and "predict" the past. Indeed, according to Wolf, since past and future co-mingle in the present, "quantum physics implies that devices must

[10] Ibid., p. 220.
[11] Ibid., p. 221.

exist that enable us to tune in on the future and resonate with the past."[12] It also implies that the present can influence, perhaps even modify, the past and that it is being pulled along and molded into shape by the future.

One can step outside one's home or office, move to the street, and look in either direction up and down the street to perceive events removed from one in space. What can be inferred from quantum mechanics is that it is possible to do exactly the same in the path of time, i.e., look in either direction and perceive events on either end of the time "street." What is clear from quantum mechanics is that the collapse of the quantum wave function effected by conscious perception is tantamount to the collapse of space-time. This can only mean that not only does mind or consciousness suffuse space-time, it encompasses it. It may well be the 5th dimension. Jungian synchronicity and Einsteinian nonlocality ("spooky action at a distance"), both correlated functions of consciousness, do, as Einstein rightly intuited, transcend the Einsteinian light-speed limit.[13] Therefore, we are right back at the door of the Hermeticists, who said all along that "the Universe is Mind."

It is interesting that 17th and 18th century rationalists like Descartes and Berkeley could not entirely extricate themselves from this arcane manner of thinking. Descartes' way of cutting through the conundrum of "real" existence was to say, "I think, therefore I am." Bishop Berkeley, a century later, echoed this dictum by saying, "To be is to be perceived." Only with the industrial revolution in full swing could the Western mind find the means of completely discarding the unitarian thinking of the ancients.

The African Universe: The All-Mind

In common with virtually all non-Western peoples, Africa has, from the beginning, conceived of the universe as a mental expression of the creator. In Nilotic metaphysics, Thoth or Djehuti is considered the mind and will of the creative Demiurge, and from this Personified Mind emerges the Word that brings all creation into being. The Dogon state simply that the universe is a thought in the mind of Amma, the Creator. Indeed, the one thing smaller than the smallest of all things, the po, is thought and it is apparently this thought that activates the process causing the po to open and the formative seeds of matter to scatter. It is possible to wonder if the fruitless attempts of modern research physicists to find the elusive building block of matter—no matter how small the subatomic particle they detect, there is always another more fundamental one—may be related to the basically mental nature of the universe. Mind is an entity that does not exist in a "particle state" and therefore cannot be resolved down to its fundamental constituents.

---

[12] Ibid., p. 222.

[13] This collapse of spacetime and transcending of light-speed means that, through the medium of the mind, it *is* possible to be two places at once.

Heretofore, the insistence of Africans of the ability of many of their adepts to achieve "magical" effects has been routinely dismissed as credulity and superstition. Yet, everywhere in Africa, there exist men and women who claim to be able to see into the future, divine invisible causes and effects, affect events at a distance, and communicate with other beings—human and non-human—through psychic means.

The author has met a number of such persons on his yearly visits to Senegal. Divination in Senegal is so commonplace that it lacks the element of the fantastic that attaches to it in the United States. The author's own divinatory encounters have convinced him of the extraordinary gifts possessed by the diviners.[14] He has also met traditional healers whose primary renown was their ability to heal at a distance. One such person was said to be able to heal clients 90 miles away from Fatick where he lived. The *saltiges* among the Serer of Senegal are known for their precision forecasting of the future. One was said to have predicted the Persian Gulf War and the border skirmish between Senegal and Mauritania three years prior to each event. The wry comment was also made that, had he been consulted, he could have shown the warring principals a way to avoid both conflicts. The saltiges are not infallible: an unexpected, unforetold drought in the early 1970s caused several of them to commit suicide to atone for the failure of their predictive powers.

The author is not offering these examples as "proofs" in the strict empirical sense. These are anecdotal episodes whose purpose is to illuminate the "quantum reality" that has always pervaded traditional African life. The literature concerning the "synchronistic" ability of traditional adepts all over the world could fill a couple of dozen encyclopedic volumes. However, the diviners and healers are not out for mass conversion to their system of thought per se; they know the measure of their own abilities and it is enough. In Africa, one is bathed in "quantum reality"; the mind-bending discoveries of contemporary theoretical physics would raise a bemused eyebrow among the Serer traditional healers. They would wonder what is all the fuss about? Doesn't everybody know this?

Conclusion

Quantum physics is redefining Western thought, knowledge, and reality. Science is no longer what it once was: an empirical, clockwork method capable of exposing all the secrets of matter. Another "shock to the system" has occurred, one that in some ways is terrifying to contemplate: matter *may be*

[14] These adepts seem to be heirs to arcane methods of training their minds to "tune in" to echo waves from the future, giving clarity and coherence to the faculty of precognition latent in everyone.

mythical.[15] According to the avant garde of modern physics, the true universal "stuff" may be composed of "informational perception," in a word, "mind."

One implication of these developments is that myth and matter are reconjoining. The Dogon say, when Amma created the universe He first installed myths (*so tanie*); when He is ready to destroy it, he will first take them away. As the author has said in another context, it is impossible *not* to think mythically, not to *be* mythic. The rupture that divorced myth and science simply could not last and this means that Western science, realizing it nor not, has simply been chasing its tail for the last 350 years.

Appendix

Synchronistic Anecdotes

Synchronistic events occur or have occurred in the life of everyone. Most of us living in the West ignore them, forget them, or explain them away as "coincidence."[16] Few have allowed themselves to derive any meaning or significance from them. In this appendix the author will relate a few synchronistic events that impacted his life, the full ramifications of which are still not clear. The reader is free to accept them as meaningful or not according to his or her disposition.

Jung spoke of the synchronistic "compulsion of names" first noted by Stekel.[17] For example—referring to his former colleagues—Jung noted that Freud, whose name in German means "joy," was the exponent of the pleasure principle; Adler, whose name means "eagle," formulated the theory of the "will to power"; Jung, whose name means "young," was preoccupied with the symbolic idea of re-birth.[18] The author's own name "Finch" at first did not seem to yield any significant content in this way until he started reading about finches as a genus of birds. There are a species of finches in Africa who are known as weaver birds because of the way they build their nests; four generations ago, the Finches were a family of tailors on Auburn Avenue in Atlanta. The finches as a group of birds are noted as seed-eaters and as I read this I couldn't help but think of my own emerging fascination with the Dogon po, the "seed," and its astronomical correlative po tolo, the "seed star."

This aspect of synchronicity really began to reveal its possibilities, however, when I began to analyze my wife's surnames. Ellen, through her father, is actually a Byrd from east-central Virginia, so it seemed a meaningful coincidence that there had occurred a kind of "avian" marriage of a Finch to a Byrd. But there is more: Ellen's father assumed his maternal grandmother's surname Nixon. The story goes that Aunt Jane Nixon, as she was called by everyone, went down to the spring near her home

---

[15] The author does not means here "fictional" in reference to myth but myth as an expression of reality through analogy. This "translogical" manner of thinking informs all non-Western, indigenous thought.

[16] It should come as no surprise that neither Jung nor the quantum physicists recognized the validity of "coincidence."

[17] Jung CG, *Synchronicity*, translated by R.F.C. Hull, Princeton, N.J.: Princeton University Press, 1973, p. 11^{n}.

[18] Einstein, whose name in German means "one stone," began with Relativity to unlock the secrets of matter, thereby treading the same path as the ancient alchemists searching for the Philosopher's Stone. Napoleon, whose name means "lion of Naples," was born under the sign of Leo.

one day and experienced a profound spiritual awakening that changed her life. She became a highly sought-after "reader" and spiritualist, consulted by Black and White alike. It just so happens that, in German folklore, a *nix* or *nixie* is a water spirit, capable of passing great knowledge on to humans, that is said to live in a splendid palace at the bottom of pools and springs.[19] Moreover, Aunt Jane was originally born into the Wright family; a "wright," by definition, works in wood and the Wright family owned a sawmill.

Under this "compulsion of names," we recall that a man named Martin Luther King became the undisputed leader of the civil rights movement, ascending the mountaintop prior to his death; a man named Malcolm X—the "X" symbolizes "crossing over"—underwent a profound spiritual conversion ("crossing over") that ultimately led to world renown and martyrdom. Marcus Garvey, forever linked in the historical imagination to the ship known as the *Black Star Line*, is thus related to the word "garvey" which is defined in the dictionary as "a small boat off the New Jersey coast."

Other significant synchronistic events have left an indelible imprint. I remember a dream in 1985 in which a fox was staring at me with a quizzical expression, holding out his severed tongue to me. Two years later, a copy of *Pale Fox* came into my hands and to my amazement I read on page 202 that Amma had cut off a part of the tongue of Ogo the Fox, Master Diviner and Master of Chaos. This episode seemed to be an example of a non-synchronous synchronicity.

When I was writing Chapter 3 of this book, I was poring through diverse materials, including Dogon cosmogony and Chaos theory, and was studying a passage concerning Dada the Spider who wove the words of Ogo (Master of Chaos) in the acacia tree as the universe was being formed. I went to bed that evening, my mind spinning with fresh revelations. The next morning my third daughter Rabiah, opened the front door on her way out and there, stretched from the top of the door to its bottom was a large spider in the middle of a sizable web. Rabiah summoned her mother and myself and we all stared at this "apparition" for some minutes until I closed the door with an admonishment not to molest or disturb the spider. It was a windy day and that afternoon when we opened the front door to check on our "visitor," spider and web had disappeared.

About the time of this episode, I was studying the lives and careers of Charles Darwin and Alfred Russel Wallace, the two earliest and most important exponents of the biological theory of evolution. It is remarkable that *both* men began their careers as avid collectors of beetles, a predilection they carried with them to the end of their lives. I didn't attach any particular significance to this fact until I realized that the ancient Nilotic scarab or "dung" beetle constituted the hieroglyphic sign of "transformation" or "evolution." The two naturalists were, unknowingly, enacting through their hobby a revealing synchronicity tying them to the theory that launched evolutionary biology and transformed the modern life sciences.

On March 10, 1994, I finished Chapter 5 of the manuscript to this book, in which I discussed boats and navigation and also briefly talked about South Pacific migrations to Hawaii. Almost exactly three hours later, I received a call from a colleague, Dr. Virginia Floyd, who was vacationing in Hawaii with her family. She had just that day met one of the last traditional boat builders in Hawaii named Keola Sequeira and was so excited by the encounter that, knowing my interest in ancient transoceanic voyages, she called me to tell me of the meeting. In fact, tracing it back, the encoun-

---

[19] *Nixes*, according to the lore, are not usually friendly to human beings but can be prevailed upon from time to time to part with their knowledge.

ter seemed to have been taking place right at the time I was putting the finishing touches on the chapter.

Synchronicity or coincidence? One of Jung's definitions of a synchronicity is a "meaningful coincidence." Thus, these interesting "co-events" in my life—if indeed coincidences they were—were, to me at least, meaningful. Additionally, they were "acausal" and "nonlocal," so fitting more securely Jung's definition of a synchronicity. That these experiences displayed an uncanny content is illustrated by two further episodes that I shall relate.

In the latter part of June 1992, a small group of friends and colleagues had accompanied me to Senegal to meet some of the healers I had been working with. On the first Wednesday of our sojourn in Senegal, the group travelled in two vehicles to the Malango Traditional Healers Center in Fatick, 90 miles southeast of Dakar. I had served as the principal investigator of a just completed knowledge, attitudes, and practices survey among a group of 376 mostly Serer traditional healers and a companion utilization study of 493 of their village clientele in the Fatick region. The project had been funded by USAID-Senegal and was conducted in collaboration with my co-investigator, Dr. Erick Gbodossou of the Experimental Center for Traditional Medicine in Fatick, Senegal and co-workers from the Tulane School of Public Health and Tropical Medicine (see Plate 23). Several members of the group accompanying me to Malango arranged for consultations with selected diviners, one of these was Pamela Foster, a counseling therapist at the Progressive Life Center in Washington, D.C. Toward the end of the consultation, the diviner told her that she and the group were going to have trouble returning to Dakar but that the trouble could be avoided with a small sacrifice of some of her blood and two kola nuts. My friend sought me out, and asked me to make a small incision on her wrist to extract blood to put on the kola nuts.

In Africa, I had learned not to question the word of a known adept, so with a dull knife, we managed to extract a few drops of blood to smear on the kola nuts. The diviner explained that the kola nuts had to be given to the first blind person encountered. By this time I was nervous; on each of my previous three trips to Fatick, I had experienced automobile breakdowns or other mishaps. I wasn't too worried about finding a blind person, however; endemic blindness makes such persons a common sight almost anywhere in the country. As the two vehicles pulled out of Fatick on the highway to Dakar, we all scanned the sides of the road for a blind person, but to no avail. We drove for 45 minutes without catching sight of a single blind person and my own anxiety grew apace. We had relaxed our vigilance slightly while driving through a village called Nguegokh when, suddenly, from the back of the vehicle, came an urgent shout to stop. Pamela had seen her blind man. We stopped and she climbed out of the van, but when we turned around to catch up with her she had disappeared into the village. Two other members of the group got out to follow her trail into the village while I remained with the vehicle. The lead vehicle, chauffered by a Serer mechanic, had turned around to come back when he lost sight of us. Finally, Pamela and her "escort" returned, and the story that was pieced together possessed some strange elements that have not been explained to this day.

When she jumped out of the van to deliver the kola nuts to the blind man, he was nowhere to be found at the place where she had espied him. She is not altogether clear what happened next but she began to ask—*in English* to people who didn't know a word of it—where the man was that she had seen. Somehow comprehension dawned, because she was led to a house in the village where she was taken in and made to sit. In a few minutes the blind man was led into the room and she was able to make him understand that the gift of kola nuts was for him. He evinced no surprise at this and graciously accepted the nuts, as if he had been waiting for them all along. There is no question of the man leaving the scene where she saw him and returning to his house in that short period of time; it was clear that, at the time she saw him, he had not *physically* been at the spot where he was sighted.

As it turned out, Nguegokh was the natal village of our lead driver and the blind man was one of his father's closest friends. The whole group then took the opportunity to pay our respects to the driver's father and his family. The rest of the trip back to Dakar passed without incident.

There were processes at work in this sequence of events that are not entirely accounted for in Jung's rather spare definition of synchronicity. Precognition, (astral) projection, and "sympathetic magic" all seem to have been operative forces in this episode, testifying to the reality of realms beyond realms and realms within realms; in a word, parallel universes. It was if we were witnesses to a transaction between an offer and an echo wave. There was nothing, by the current laws of physics, that makes any of these incidents "supernatural." The EPR effect is sufficient to explain quantumly most of what occurred. These episodes, then, are no longer paranormal, they are rigorously normal by the rules of quantum physics.

The local people did not find anything extraordinary in these happenings. The gift of traditional peoples not yet fully acculturated into modern civilization is their ability to interact with and understand the invisible, quantum world of which true "reality" is composed; our visible, empirical world—one that deifies matter—is an epiphenomenon, a secondary process. It is why the Brahmans and Buddhists insisted that the world we live in is an illusion—*Maya*—and only the facility of seeing the invisible, listening to the unheard, and touching the intangible allows any possibility of glimpsing Reality, i.e., the Unity underlying all Creation.[20]

The second episode occurred on October 31, 1992. I was in Leicester, England, on a lecture tour and was being driven to my place of lodging that night by my hostess, Jennifer Jeffers. The program had finished late; it was about 2 a.m. and we had just finished discussing the book *Pale Fox* when suddenly, across the path of the vehicle, dashed a pale fox. For a brief second the encounter didn't register, but when it did, we both realized we had truly seen a pale-colored fox pass across our vision. Jennifer admitted that it was more than a little odd to see an English fox within the city limits. I was sensitive enough to the possible significance of this event that I called home later that morning to find out if everything was alright. My wife Ellen hesitated a bit and then recounted a "close encounter" in the garage. The previous evening at 9:30 (October 30, because of the time difference between Georgia and England) she raised the garage door only to find a black cat sitting just outside, gazing fixedly at her. Indeed, the cat and my wife stared at each other for some seconds before Ellen said, for no particular reason, "What do you want, Bast?"[21] At that moment, the cat disappeared into the night, never to be seen again. The two occurrences were eerie but neither of us could make a thing out of them, until 10 days later when my wife's mother suffered a fatal heart attack.

Both the (pale) fox and black cat are harbingers of death as we have alluded to elsewhere in this book. The fox is a co-type of the jackal, making it a type of *Anubis*, the guide of deceased souls in Amenta. Moreover, the pale fox among the Dogon, as Ogo, is the author of disorganization and death in the universe. We have already discussed the deathly connotations of the black cat in Chapter 7.

This complex of episodes illustrates the manner in which synchronicities can collapse space. If we entertain the possibility that the appearance of the pale fox in Leicester, England, was synchronistically linked to the appearance of the black cat in

---

[20] See Capra F, *The Tao of Physics*, Berkeley: Shambala, 1975, for a fuller exposition of the relation of ancient Eastern mystic traditions and the "new physics."

[21] *Bast* was the cat-goddess of ancient Egypt. She was a form of *Ast* or Isis.

Decatur, Georgia, USA, and that both of these "epiphanies" were tied to my mother-in-law's eventual demise, it shows that this rather complex synchronicity wove together events separated by thousands of miles. Moreover, taking into account the time differential between England and Georgia, the two "epiphanies" of fox and cat were occurring at about the same time! Thus, there was not only a collapse of space but also a collapse of time! As if this weren't enough, the epiphany of the pale fox occurred on All Souls Day, i.e., Halloween.[22] Further, there is the fascinating juxtaposition of cat and fox which, mythically, represent Isis and Anubis. In Nilotic mythology, Isis and Anubis work in tandem in the resurrection ritual of "The Coming Forth by Day."

What to make of this astonishing sequence of correlated events. Quite possibly, my wife and I are over-interpreting these co-events and inappropriately reading significance and meaning into what rationalists would call "mere coincidence" (or "superstition"). Sometimes the only thing that distinguishes a true synchronicity from "mere coincidence" (or "superstition") is context. Every encounter with a black cat or a pale fox is not synchronistic. But in the events just related, it was the unusual contexts (in time and space) of the encounters that lent them a "numinous" character and made them synchronistic. Whatever the case, everything that happened in that 11-day period impressed itself indelibly upon us.

I cannot bring this subject to a close without touching on an anecdote concerning the remarkable figure of Adji Fatou Seck, of Rufisque, Senegal (see Plate 23). Maam Seck, to use her honorific title, is Senegal's leading *ndeppkat* and quite possibly that country's most powerful spiritual personality. *Ndepp* is the Lebu-Wolof spiritual healing tradition whose antecedents are lost in antiquity. It is directly allied to the Serer *Lup* and its affinities to other such systems among the Akan, Dogon, Yoruba, Fon, and Ibo, to name but a handful, are so close that ultimately, all must derive from a single system. Principally, the activities of ndepp rituals are designed to invoke the power and benign intervention of those cosmic-ancestral spiritual entities called *rabs.*[23]

I first met Maam Seck, in the company of her youngest daughter Néné, when she became a guest in my home in September, 1986. The woman simply emanated power; it was the first time in my life I had ever palpably felt someone's aura. Without prompting, Maam Seck offered to perform cowrie divinations for all present, through the interpreter, and while the experience was electrifying for everyone, one particular episode stands out.

My friend and colleague, Mwalimu Imara, came to meet my guests and, like the rest of us, had a divination performed. As we were only to find out much later, Maam Seck, during the course of the divination, told him that there was another child in his family. Mwalimu, an only child, corrected her by saying that he was his parents' only offspring. Maam Seck consulted her cowries again and repeated that there was another child in the family. Mwalimu again demurred, thinking, "I should know whether I'm an only child or not!" Once again, this time with some firmness, Maam Seck informed him of another child in the family. Not wishing to be impolite, Mwalimu let the matter drop. That night, however, he called his mother to describe this rather peculiar encounter at which point his mother told him, yes, you had a brother who was born before you. However, he died before you were born. A dumbfounded and chastened Mwalimu was to say later on, "*That* got my attention!"

---

[22] "Halloween was thought to be the most favorable time for divinations concerning marriage, luck, health, and death." Quoted from the article "Halloween" in Volume 5 of the *Micropaedia* of the *New Encyclopaedia Brittanica*, Chicago: Encyclopaedia Brittanica, 1992, p. 646

[23] The *rabs* are the same as the *pangool* of the Serer, the *nommo* of the Dogon, the *orishas* of the Yoruba, and the *vodun* of the Fon.

This is a typical ploy by Maam Seck to get the attention of a skeptical client; tell him (or her), early in the divination, some fact or event in his personal past that the client, and *only* the client, knows. Such unexpected revelations by a complete stranger, a somewhat exotic-looking foreigner speaking no English, is not merely disconcerting but sometimes terrifying. One can easily come away from an encounter with Maam Seck with the uneasy feeling that, with her, it is impossible to dissemble. With those 16 cowry shells, each client's life is an open book whose most intimate secrets are readable. Yet Maam Seck exudes more discretion than the confessional priest; she never tells all she sees, only what the client needs for the purpose at hand. Still, as benign a person as she is, those divinations, giving Maam Seck a penetrating "other eye," have been known to inspire fear.

But clearly, in the "quantum" world that Maam Seck enters during her divinations, the "arrow of time,"[24] as we understand it, has little meaning. Through the medium of the cowry shells, she can go in any temporal direction. Or it might be said that she enters into a state of timelessness, where every fact and event of the past, present, and future co-exist simultaneously; where everything is synchronistically woven into the common cloth of existence. Up to now, this "other world"[25] that Maam Seck can occupy at will, was closed to modern science. However, the sages of Africa always say that this "other world" is an expression of a reality more fundamental than any we know.

---

[24] The "arrow of time" is a concept in physics that says that time moves in only one direction, from past to future. However, neither relativity nor quantum mechanics require an arrow of time, an uncomfortable circumstance that has called forth much rationalizing designed to keep the arrow of time in place in modern physics despite its irrelevance at the quantum level.

[25] African cosmogonies actually postulate the existence of multiple worlds of up to 28 ("fourteen above and fourteen below"). These other worlds seem to be the African equivalents of parallel universes. (See Chapter 7 of this book)

Afterword

It is a quantum leap we have made, from the Ishango Bone 25,000 years ago to synchronistic foxes and cats in the 1990's. But if we have learned anything it is that *nothing* is unrelated. This is the emerging lesson of modern science. It is also exactly the message of Zen, the Great Pyramid, and Dada the Spider. The world has endured a profound psychospiritual rent in the last few centuries, one that has become insupportable. It couldn't last; one way or the other the forces that pulsate through the universe were bound to close the rupture. We would have come to this place eventually, no matter how vigorously we attempted to twist and turn, like a squalling child, out of the clutches of whatever destiny animates Earth and the life born from her. It is my profound conviction that the adepts of old foresaw, at least in its main outlines, all that was to happen. They are talking to us now. The future too is trying to pull us to our feet, slapping us in the face as if to awaken a semiconscious drunkard. It is we that now hold together the separated strands of time, each of us working in his or her métier to spin, to weave his or her piece of the web.

Special Section

The author is taking the liberty of printing a letter (with permission) he received while drafting this manuscript that, in his estimation, does much to concisely amplify the aim and content of this book. The letter was written by Mr. Earnie Ferreira of Faifax, Virginia, in March 1996, inspired by a study tour Mr. Ferreira took with the author to Egypt in July 1995. Mr. Ferreira is a professional engineer and was able to examine much of what he saw of the material structures of ancient Kemit with an engineer's eye. His observations were so unusual and insightful that they seemed worthy of wider circulation.

Mr. Ferreira is following a well-travelled path in Egyptological historiography: from the early 19th century onward, scientists, engineers, and technicians, most of them without formal Egyptological training, such as Richard Proctor, Charles Piazzi Smyth, Norman Lockyer, R.A. Scwaller de Lubicz, Cheikh Anta Diop, Giorgio de Santillana, Livio Stecchini, and others, have revealed startling facets of ancient Nilotic civilization by examining its methodologies strictly from the point of view of their techno-scientific content. Egyptologists have never been pleased with such "non-specialist" intrustions by even qualified scientists and though unable to refute the rigorous findings of these investigators, have largely ignored their ramifications.

In a small, but significant way, Earnie Ferreira, like the above-mentioned men, is opening up unheralded vistas for a greater understanding of African science and technology. It is hoped that others like him will turn an eye to the technical and scientific problems met and mastered by ancient Africans, free from the epistemological prejudices of many Egyptologists.

March 10, 1996

Earnie Ferreira
Fairfax, Virginia

Dear Dr. Finch:

I hope this letter finds you in good health. I am amazed at the lightning speed at which time has passed since the *Gift of the Nile III* tour of 1995. The tour left me in an incredible state. When I returned, my direction was so clear that by September of 1995, I was enrolled in an advanced program in information systems at the American University, and I am now preparing for comprehensive exams. Most of my career has been in the field of electronics and telecommunications systems engineering, with special attention to information extraction. The decision to continue is due to a personal desire to further myself in information systems and devote more time to the development of telecommunications systems in Africa. I became very excited by the signs I saw of the origins of technology in Kemet. I hope there will be an opportunity to discuss this further in the near future.

I did not forget about agreeing to write my impressions of the tour, from an engineer's point of view. I started to write immediately upon settling at home, and this is where my troubles started. I have started this task a dozen times already and the more I thought about what I saw in Kemet, the more difficult the task becomes. Given our formal training, we always seem to place so much value in abstraction, only to return to some very fundamental understanding, in solving problems. Yet, what I observed in Kemet *was so obviously the work of masters of fundamentals* (emphasis added).

What really happened to me out there is that I saw everything that I learned in eighteen years of engineering practice and formal training, in which I had to study a lot of mathematics, simply melt before me. As an example, while observing the workmanship of the sacred pool [Karnak Temple] and underground waterways, I remembered the hydrostatic paradox I studied in physics. Physicists for a long time were unable to understand why a liquid poured into a number of vessels of different shapes, and connected in a sequential fashion, will stand at the same level. Today we can explain the hydrostatic paradox. However, this could not have been a paradox for the people of Kemet because such an understanding had to have been reached in order to build the waterways which we can still see today. In addition, in mathematics, I remember lectures on the significance of the "0" and how its discovery revolutionized mathematics, and how it was not possible to count above certain numbers without it. Yet, as I looked at the numbering systems on the walls of Kemet and the precision of their workmanship, it was clear to me that there could not have been a limit on their ability to count to any level. How could they have cut and constructed stones with such accuracy if there were any limitations in their mathematics? In the United States, around the turn of the century, three out of every four bridges collapsed shortly after construction. Yet I saw in Kemet structures which have been standing for thousands of years, with a symmetry and balance unmatched today. While sailing towards the Nubian village, I noticed that the inscriptions on the stones in the water were balanced with the horizon, the lettering was all the same size and the water lines were straight. The stones were not lopsided, I was able to follow the inscriptions without seeing any of them sink into the water at one end. The structures I observed were so meticulously formed that one could only conclude that they were formed with the intention of having them survive as long as the stones would survive.

I went to Kemet with idea that I would observe the works of a primitive people, but there was nothing primitive about what I saw. I could not help but think that NASA would do well to study the way stones were shaped and interlocked in forming the walls of the temples in Kemet. This knowledge would enable them to see a way of securing the tiles in the nose of the space shuttle to enable a safe reentry into the earth's atmosphere. To speculate even further, the passageways in the pyramids and temples reminded me of the passageways of wave-guides for microwave systems. In microwave systems wave-guide design, the straightness of the guides and the seals of the angular joints are critical to the prevention of signal losses or attenuation. The passageways in the pyramids were so straight that I could not help but think of the possibility of the development of a form of communication such as semaphore.

The Cairo Museum was a treasure house of amazement for me. The display of boomerangs were among the many artifacts which surprised me. I had always attributed the development of boomerangs to the aborigines of Australia, though I could never figure just how they developed the tool. However, there were so many different varieties of boomerangs, the size and shape of which indicates an early understanding of different flight patterns. The varieties of boomerangs made me realize that there was a fundamental understanding of aerodynamics.

All of the above made it very difficult for me to respond as I would have liked. It all stands before you with such clarity that one comes to realize that it may well be that the development of civilization may have been brought to a full stop, and started all over again with someone else claiming the original works and developments of the past. This sort of realization is not easy to take. I believe I have a lot to do and must make another visit to Kemet.

Sincerely,
Earnie Ferreira

(Mr. Ferreira's letter has been slightly edited to better showcase the material relevant to the content of this book.—Author)

Also from Charles S. Finch III

Echoes of the Old Darkland

THEMES FROM THE AFRICAN EDEN

Echoes of the Old Darkland examines the paleontological record, cultural anthropology, religious symbolism, and astronomical mythology to trace the origin and evolution of humanity and culture in Africa from prehistory to the Christian era.

Dr. Finch analyzes Biblical names and compares ancient Egyptian and Judaeo-Christian texts to show that Judaism and Christianity are lineal descendants of the old Egyptian religion, itself a product of the incredibly ancient symbolic myths of prehistoric Africa. Thus, Yahweh and Christ both derive from early African divinities—Moses was an Egyptian priest, and the historical Jesus was initiated by the Egyptian Therapeutae (Essenes) in the Western Desert.

Also in this book, the author brings forward evidence that suggests world history may be tied to 26,000-year astronomical cycles that were discovered in Africa 40,000 years ago.

The underlying message of *Echoes of the Old Darkland* is that, in recovering the early story of humanity, all roads lead to Africa. Thus, *Echoes* will not only amaze, anger, and astound, it will also provoke new thought on subjects that are increasingly important as the world enters a new millenium.

$16.95 (paper) • 232 pages • Illus. • ISBN 9629444–0–8

Book Orders
Khenti, Inc.
P.O. Box 361003
Decatur, GA 30036–1003

KHENTI PUBLICATIONS

Order Form

Please send me _____ copies of ***Echoes of the Old Darkland*** at $16.95 each: $_________

Send _____ copies of ***The Star of Deep Beginnings*** at $18.95 each: $_________

Sales tax enclosed (7%, GA residents only): $_________

Shipping & handling enclosed (see below): $_________

Total payment enclosed: $_________

☐ Check or money order enclosed (please make payable to Khenti, Inc.)

Charge my: ☐ Mastercard, ☐ Visa.

Card# ________________________________ Expire date __________

Signature ____

Shipping & handling: $3 for 1 book; $4.50 for 2-5 books; $7 for 6-9 books. (Please write for quantity discounts on orders of 10 or more books.)

NAME ____

ADDRESS ____

CITY ___________________________ STATE _________ ZIP _________

Send this order form (photocopies of form are acceptable) to:

Book Orders
Khenti, Inc.
P.O. Box 361003
Decatur, GA 30036-1003

Early America Revisited

REPLY TO MY CRITICS

By Ivan Van Sertima

Many claims had been made over the years by notable black and white scholars—Woodson, Rogers, Lawrence, Jackson, Bailey, Gordon, Irwin, Jeffreys, Cauvet, Wiener. They had all been dismissed. Some of these claims had failed to convince anyone but the converted because they needed to be backed up by corroboration from many disciplines. To claim, for example, that Columbus saw "blacks" in Haiti is one thing. To prove that those "blacks" were Africans was another, since there are dark-skinned American Indians in tropical zones of America. Botany (cotton), linguistics (the origin of the word *gua-nin*), metallurgy (the metal alloys in the spears the blacks gave the Americans in trade), navigation (African boats tested on the Atlantic), oceanography (the currents that provided an easy circular route for the pre-Columbian West African trade), archaeology (new skeletal finds in the Caribbean of Africans in a grave dated 1250 A.D.)—all these disciplines provided the corroboration that was needed to establish that single claim on a scientific basis. This, Van Sertima felt, would be his original contribution to the subject at the end of a century of speculation—THE DEFINITIVE PROOF.

The disciplines Van Sertima explored in order to provide this proof are highlighted in a letter published in the *New York Times*. In this letter, one of the oldest and most important archaeologists in America, Dr. Clarence Weiant, who was on the site in Mexico when the first Olmec stone heads (some with African features) were discovered, defended Van Sertima against attempts by the British archaeological establishment to discredit *They Came Before Columbus*.

"Van Sertima's work," Dr. Weiant wrote in the *New York Times* (May 1, 1977), "is a summary of six or seven years of meticulous research based upon archaeology, Egyptology, African history, oceanography, geology, astronomy, botany, rare Arabic and Chinese manuscripts, the letters and journals of early American explorers and the observations of physical anthropologists. . . . As one who has been immersed in Mexican archaeology for some 40 years and who participated in the excavation of the first of the giant heads, I must confess I am thoroughly convinced of the soundness of Van Sertima's conclusions."

Ivan Van Sertima (Editor)
Journal of African Civilizations
Africana Studies Department
Beck Hall
Rutgers University
New Brunswick, New Jersey 08903

Approx. 110 pages Book $15.00 ☐
(with illustrations) Tape $10.00 ☐
Available January 1998